Climate Change and Geodynamics in Polar Regions

Climate Change and Geodynamics in Polar Regions covers most of the scientific aspects of geoscientific investigation undertaken by Indian researchers in the polar regions: the Antarctic, Arctic, and Himalayan regions. A firm understanding of the cryosphere region's geological perspectives helps students and geoscientists evaluate important scientific queries in the field.

This book will help readers understand how the cryosphere's geoscientific evolution took place in the geological past, as well as how the climate changed throughout history, and how polar regions were affected by global warming. It also discusses how we might expect polar climate to change in the future. A firm understanding of the cryosphere region's geological perspectives helps students and geoscientists answer some of the most puzzling scientific queries and generate new ideas for future research in this field.

The book is edited by **Dr Neloy Khare**, presently Adviser to the Government of India with a very distinctive acumen in quality science and research in his areas of expertise covering a large spectrum of geographically distinct locations like the Antarctic, Arctic, Southern Ocean, Bay of Bengal, Arabian Sea, Indian Ocean etc. He obtained a doctorate (PhD) in tropical marine region and Doctor of Science (DSc) in Southern High latitude marine regions towards environmental/climatic implications.

Maritime Climate Change: Physical Drivers and Impact

Series Editor: Neloy Khare

As global climate change continues to unfold, the two-way links between the tropical oceans and the poles will play key determining factors in these sensitive regions' climatic evolution. Now is the time to take a detailed look at how the tropical oceans and the poles are coupled climatically. The signatures of environmental and climatic conditions are well preserved in many natural archives available over land and ocean. Many efforts have been made to unravel such mysteries of climate through many natural geological archives from tropics to the polar region. This series makes an effort to cover in pertinent time various depositional regimes, different proxies- Planktic, benthic, pollens and spores, invertebrates, geochemistry, sedimentology etc. and emerged teleconnections between the poles and tropics at regional and global scale, besides sea-level changes and neo tectonism. This book series will review theories and methods, analyze case studies, and identify and describe the evolving spatial-temporal variations in climate and providing a better process-level understanding of these patterns. It will discuss significantly, generalizable insights that improve our understanding of climatic evolution across time—including the future. It aims to serve all professionals, students and researchers, scientists alike in academia, industry, government, and beyond.

Climate Change in the Arctic
An Indian Perspective
Neloy Khare

Climate Change and Geodynamics in Polar Regions
Edited by Neloy Khare

Climate Change and Geodynamics in Polar Regions

Edited by Neloy Khare

CRC Press is an imprint of the
Taylor & Francis Group, an **informa** business

First edition published 2023
by CRC Press
6000 Broken Sound Parkway NW, Suite 300, Boca Raton, FL 33487–2742

and by CRC Press
4 Park Square, Milton Park, Abingdon, Oxon, OX14 4RN

CRC Press is an imprint of Taylor & Francis Group, LLC

ISBN: 978-1-032-25531-6 (hbk)
ISBN: 978-1-032-25657-3 (pbk)
ISBN: 978-1-003-28441-3 (ebk)

DOI: 10.1201/9781003284413

Typeset in Times
by Apex CoVantage, LLC

Contents

Foreword

The Paris Accords in 2015 is an outcome of global concern about climate change, which plays a pivotal role to determine whether the UN mandated Sustainable Development Goals will be achievable by 2030. The polar regions of the Earth are best suited for studying the cause-and-effect relationships of climate change. Geodynamics is one of the many natural factors of climate change, and a broad understanding of the inter-relationships between deep Earth processes and their surface manifestations is a significant advancement in our knowledge. Viewed in these contexts, this book presents a collection of studies related to climate change and geodynamics with a focus on the polar regions. The Arctic polar region situated in the northernmost part of the Earth with the North Pole is characterized by distinctively polar conditions of climate, plant and animal life, and other physical features. Likewise, the spectacular "icy continent" of Antarctica is the southernmost continent with the South Pole being a virtually uninhabited and largely ice-covered landmass. The main distinction between these two polar regions is that the Arctic is an ocean surrounded by land whereas, the Antarctic is the land surrounded by an ocean. The Himalayas, considered to be the Third Pole, aroused general interest to generate field and laboratory data for a better understanding of the geodynamic evolution of these spectacular features and to assess the impacts of global warming on the Himalayan glaciers along with other two polar regions.

Over the past 30 years, the Arctic has warmed at roughly twice the global rate. It is attributed to a phenomenon known as Arctic amplification, enhanced by anthropogenic factors. Similarly, new data indicates that climate change is negatively impacting Antarctica and the Himalayas (e.g. melting of glaciers).

Evidence suggests that the West Antarctic Peninsula is one of the fastest-warming areas on Earth, since Antarctica is large, climate change is not having a uniform impact. It has been observed that some areas experience increases in sea ice extent, on the contrary, in other regions, sea ice is decreasing. The unprecedented warming in the Arctic, Antarctic and the Himalayan region is severely causing changes not only to the physical environment but to the entire ecosystem. It is believed that understanding climate change impacts on all these three poles (the Arctic, the Antarctic, and the Himalayas) is a matter of critical importance not only for the region alone but globally.

The Arctic region is an element of a geodynamic system that includes the ancient Eurasian continent and intensely developing younger Arctic Ocean. The Circum-Arctic terrains comprise a series of complex geological structures, with fragments of Archaean to Paleoproterozoic shields and platforms, remains of orogenic belts of Neo-Proterozoic to Cenozoic ages. Recent seismicity indicates ongoing geodynamic processes. New surveys and data provide insights into Arctic continental terrains, basins and tectonic structures. Antarctica has geometric significance for global plate kinematic studies owing to its linkage to the seafloor spreading systems of the Indian, Atlantic Oceans and the Pacific. On the contrary, the Himalayan region, a site of continent-continent collision and disappearance of the Tethys is yet another site to understand various climate change impacts and geodynamical events apart from the Arctic and the Antarctic. The great mountain ranges of the Himalayas are the Earth's unique features. This region came into existence mainly due to the successive accretions of various isolated continental and arc terranes with the converging Indian and Eurasian land masses during Late Mesozoic-Early Tertiary.

Therefore, the Arctic, Antarctic and the Himalayan region are the hotspots of climate change assessment and are sites for understanding geodynamical processes. **The** present book, *Climate Change and Geodynamics in Polar Regions*, aptly provides a comprehensive account of Indian efforts to help understand the impact of climate change and the geodynamics of the Arctic, Antarctic, and the Himalayas.

The book begins with the Arctic through an assessment of the Geomorphology around the Kongsfjorden and Krossfjorden system, Svalbard, by **Pattanaik et al**. Interestingly the significance of the Antarctic Climates and glacial dynamics using mass balance over the Antarctic ice sheet has been highlighted by **Kumar et al**. On the other hand, **Singh et al**. have detailed the Antarctic Climate history and its relationship with global climate changes gleaning clues from ice core records. Significant attempts have been made to study the glacial geomorphology around Schirmacher oasis, Antarctica by **Pattanaik and Baba**. Similarly, **Chaturvedi and Khare** highlighted the significance of foraminiferal studies from the southern high latitudes in assessing global climate change coupled with **Sunil et al.'s** ionospheric response over Antarctica during the Total Solar Eclipse on December 4, 2021. On the contrary, **Reshmi et al**. dealt with the isotope hydrochemistry of Antarctic lakes and ponds while a limnological assessment of water bodies of extreme Antarctic climatic conditions has been made by **Khare et al**.

In a significant attempt, **Singh et al**. focused on the glacial morpho-sedimentology and processes of landscape evolution in the Himalayan region with special emphasis on the Gangotri Glacier area, Garhwal Himalaya.

Altogether, this book provides a comprehensive, up-to-date account of how climate changes and various geodynamical processes occur in the poles (the Arctic, the Antarctic, and the Himalayas). The book provides a holistic picture of the impacts of rising temperatures over the cryosphere region with a highly cross-disciplinary approach to reflect the importance of the Arctic, Antarctic, and the Himalayas in addressing the global issues of climate change.

It will be of immense value to all researchers keen to understand the science of climate change in these sensitive regions. It will also help the Decision Support System and the development of climate models.

Somnath Dasgupta
INSA Senior Scientist at Indian Statistical Institute, Kolkata and Honorary Professor at Indian Institute of Science Education & Research, Kolkata

Date: May 2022
Place: Kolkata

Preface

The polar regions are the center stage of the evolution of all surrounding continental bodies separated millions of years ago. The entire world is experiencing the looming danger of global warming where the Arctic, Antarctic, and Himalayas have again emerged as the keystone in a changing world. It reinforces the importance of continual changes in these cryosphere regions, their history, and the impact of these changes on global climates.

Geodynamics and environmental geodesy cover the entire gamut of research activities on the deformation of the solid Earth including its fluid envelope. It also encompasses the modelling of the past ice history of the Earth, climate change and its impact on polar ice sheets, and sea-level variations. Similarly, the studies on elastic tidal deformation of the Earth and seismogenic tectonic deformation are also significant components of the research activities which can easily be addressed through geoscientific instrumentation such as Interferometric Synthetic Aperture Radar, GPS etc. Recently introduced space-geodetic techniques such as the Gravity Recovery and Climate Experiment (GRACE) and satellite altimetry provide new observations of the changing nature of our mother planet

Such tools play a significant role to investigate how climate change is affecting the environment, of these cryosphere regions (the Arctic, the Antarctic, and the Himalayas)

It is a fact that many glaciers have retreated and ice shelves have either collapsed severely or are being retreated that formerly fringed the peninsula. Such visible signs of the climate changes over the Arctic, Antarctic and Himalayas are responsible for causing physical changes and the living environment by bringing notably changes in sea level, rates of melting of polar ice caps and even ground water storage. Consequently, the melting of ice sheets has the reverse effect, causing uplift of continents and increases in ocean volumes.

Undoubtedly the study of climate change in the Arctic, Antarctica, and Himalayas is important to enabling researchers to prepare the predictive models accurately and put forth future climate change scenarios and help contribute to the decision support systems for the policy makers.

We need to understand and recognize the warming pattern. If global warming continues, it may not be uniform. In the context of global warming, we must address the more significant issue of climate change and geodynamics in polar regions on which the present book places emphasis.

The book begins with the Arctic regions which are important in geomorphological studies as the region is characterised by the lowest sediment fluxes. While assessing the Geomorphology around Kongsfjorden and Krossfjorden systems, Svalbard, **Pattanaik et al**. pointed out that the Arctic region is characterised by a large sediment store and a favorable topography for sediment transport. The Svalbard archipelago in the Arctic has been widely studied in terms of its landscape, glacial dynamics, sediment transports and paleoclimatic studies. The fjord system is the transition zone between the terrestrial and marine environment, acting as an important archive that will have geomorphological features that have been formed as a result of both terrestrial and marine processes. It is a potential site to study paleoenvironmental changes as the region provides a high-resolution sedimentary archive. Diverse geomorphological features developed in Svalbard are due to the interplay of climate and surficial processes active in that region. A study on these features will provide an opportunity to understand the sequence and mechanism of the surficial processes.

The polar regions act as a sensitive barometer to the climate changes occurring on the global scale. **Kumar et al**. present a snapshot of glaciological studies carried out in Schirmacher Oasis and the Nivlisen ice shelf in Central Dronning Maud Land (cDML), East Antarctica during Indian Antarctic Expeditions. Systematic observations have been meticulously carried out since 1996 to document the changes in the ~9 km long Polar Ice Front along the southern margin of the Schirmacher Oasis. Ice stakes have been installed to monitor the accumulation/ablation and estimation of surface mass balance over a 14000-sq.-km area. Ice dynamics and estimation of velocity using differential GPS were done to calculate the surface mass balance. Ice thickness data using ground penetrating radar is used to calculate the net mass balance in the area. According to them, the Dakshin Gangotri Polar Ice Front (DGPIF) shows an annual retreat of 1.32 m during 2018–19 and a cumulative retreat of 14.9 m since 1996. The polar ice front at Schirmacher Oasis shows an annual retreat of 2.79 m during 2018–19 and a cumulative retreat of 42.08 m since 2001. The average annual accumulation of 0.98 m is observed in the high Polar ice sheet area which is equivalent to 394.01 kg/m² Snow Water Equivalent (SWE). Impeded Polar Ice sheet area shows an average ablation that ranges from 133.12 kg/m² to 183.25 kg/m² SWE. The ice velocity during 2018–19 varies from 1.98 m/y to 51.65 m/y with an average velocity of 21.16 m/y in parts of the High Polar Ice sheet area, whereas the ice velocity was 4 m/y to 97 m/y for the same time during 2017–18. The ice sheet movement with an average of 6.78 m/y in parts Impeded Polar Ice sheet area has been recorded, whereas the ice sheet movement over DGPIF is found to be of the order of 5.23 m/y which reveals a total 6.55 m annual degeneration of DGPIF wall during 2018–19, which is equivalent to 6006.35 kg/m² loss of ice mass at DGPIF area. Out flux of ice mass for estimation of mass balance in the region shows 1.7504 gigatonnes of ice mass loss from the out-flux gate at Nivlisen ice shelf during 2017–2018, as compared to 1.66 gigatonnes during 2016–17. An increase in the ice loss is attributed mainly to a faster rate of movement of the ice sheets during that period. The cumulative ice loss from the area would have contributed to a rise in mean sea level by about 0.0045 mm and 0.0048 mm during 2017–18 and 2018–19, respectively.

On the other hand, **Singh et al**. have provided an overview of the Antarctic Climate history through ice core records. They highlighted that the Ice core studies in the last two decades have made important conceptual advancements in our understanding of the climate forcing factors and climate response at millennial time scales. Antarctic climate and its variability in the last 800 kyr have been revealed through the study of long ice core records obtained from a few important strategic localities in Antarctica. According to them, Ice cores have many advantages over the marine and continental sedimentary records in terms of their completeness; ability to calibrate actual and measured temperature data through proxies. Ice core records have truly preserved the signatures of variations in the rate of snowfall, and temperature, of the region where they are located. In addition, the wind-blown dust and sea salt record, and the trapped air bubbles containing major and trace gases of the prevailing atmosphere, all provide us clues to the climate changes through ages and over wide regions. The ice cores have improved our understanding to our understanding of the various forcing factors and their mutual impact on the climate system. The nonlinearity of the climate system involves an important threshold of some of the forcing factors like greenhouse gas concentration and Meridional Ocean Circulation. The long ice core records have thrown light on such threshold which gave rise to abrupt climate events. Also, the comparison of the Antarctic and Greenland ice core records led to important concepts like bipolar seesaw and lag and lead of the north and south They also opined that while climate change in the low latitude region is governed by a multitude of factors including land use, anthropogenic activities, ocean circulation, greenhouse gas emission and the external forcing like sun's radiation, the climate change at the poles, particularly Antarctica are mainly controlled by greenhouse concentration, orbital forcings, and feedback processes. That probably is a key factor for Antarctic amplification which is a manifestation of a climate signal two to three times the global average. For a clear understanding of the Antarctic amplification, the ice core records provide an excellent opportunity. One of the most significant findings from the long Antarctic ice core has been to understand the nature of the glacial and interglacial intervals including their amplitude and profile. Their comparison of the paleoclimate record of the Antarctic ice cores with mid-latitude marine sedimentary records reveals teleconnections in the climate through the ocean and atmospheric circulation. Also, the abruptness of the climate change in the north pole is not so evident in Antarctica where the changes are rather gradual. The Antarctic Ice core record compared with the Greenland Ice Core record enables us to understand the interhemispheric coupling at the millennial scale. The ice core studies have revealed that during the last glacial period, the Dasngaard-Oeschger events in Greenland and millennial-scale warm events in Antarctica are strongly coupled with a time lag through the Atlantic meridional overturning circulation showing teleconnections between north and south. The bipolar seesaw is to a great extent proved by comparing Byrd (Antarctica) and Greenland ice core records. In addition, the Antarctic climate is connected to the climate of the tropics. They highlighted that the Antarctic sea ice extent is related to the Indian and African monsoon. The Antarctic ice expansion and sea ice extension are related to enhanced summer monsoon. Well, dated stalagmite δ^{18}Orecord between 88 and 22 kyr BP from Yongxing Cave in central China characterizes changes in Asian monsoon (AM) strength and the studies have shown that the record is strongly anti-phased with Antarctic temperature variability on sub-orbital timescales during the Marine Isotope Stage (MIS) 3, thus establishing teleconnections between monsoon and Antarctic climate. Another major contribution to the ice core record has been the Antarctic Isotope Maxima Events (AIM). During MIS 3, the Antarctic climate is marked by some warming events when the temperature has gone up to 1 to 3 degrees Celsius and a high oxygen isotope value. These warming events are known as Antarctic isotopic Maxima (AIM) events and are characterized by gradual warming and cooling. This is in contrast to the Dansgaard—Oeschger events in the North where rapid warming (8–16 degrees Celsius) is followed by gradual cooling to Greenland Stadials. This shows a distinct difference in the pattern of the warming events in the north and south. Several theories have emerged to explain the bipolar seesaw including the development of a southern heat reservoir during Greenland stadials. Thus we see that the Antarctic Ice core records provide us with a great understanding of the climate system, feedback, forcing factors, response system, and global Tele connectivity from north to south and from polar to tropical latitudes. Though the oldest record goes back only to 800 Kyrs, it provides conceptual advancements in our understanding of the cause and effect relationship and physical processes of linkages, that can be applied for time beyond 800 Kyrs.

On the other hand, **Pattanaik and Baba** provided an exhaustive account of the glacial geomorphology around Schirmacher oasis which is an ice-free area of Dorning Maud Land, East Antarctica. This area is sandwiched between a continental ice sheet in the south and an ice shelf in the north. The ice-free region provides opportunities to study lithology, paleoclimate, glacial process and glacial geomorphology of Antarctica. The landscape and the glacial-geomorphological features found in the Schirmacher oasis help to understand the glacial process active in the region. Striations on bedrocks and erratics indicate the glaciers' movements and deglaciation phases of the Oasis. A considerable amount of sediments found in the oasis as till, pattern ground, moraines, block fields, and outwash plains witnessed the past glacial processes. Retreat and advancement of polar ice have resulted in the formation of various landforms such as valleys, glacial troughs, cliffs, moraines, and *roche moutonnées*. Sediment archives from the lakes provide paleoclimatic information about the oasis. Permanent Indian research station Maitri and Russian research station Novolazarevskaya provide a platform for the researchers for conducting various research works in the Schirmacher oasis.

In view of the significant influence of the southern high latitudes on global climatic change, it is important to understand the role of the Southern Ocean and Antarctica (SOA) in climatic change both at present as well as during the geologic past. The role of the southern high latitudes in climatic change during the geologic past can best be evaluated by using foraminiferal characteristics in sediments from beneath the ocean, from lakes, and uplifted on land. Numerous studies have been carried out in which foraminiferal characteristics have been used to assess paleoclimatic/paleoceanographic changes at southern high latitudes. **Chaturvedi and Khare** aim at reviewing the findings of major foraminiferal studies carried out on sediments from southern high latitudes. The changes in foraminiferal characteristics can help to understand the present and past Physicochemical aspects of the high-latitude Southern Ocean. The application of advanced and recently developed techniques on foraminifera recovered from the SOA enabled the reconstruction of the seawater temperature and the extent of the continental ice sheets during the geologic past. The foraminiferal studies from the Antarctic Circum-polar Current and the regions north of it vastly helped to understand the past variations in productivity as well as changes in the positions of the various polar fronts and the production of deep and intermediate waters. Although the surface distribution of foraminifera has been studied from many regions of the Southern Ocean, there still are several gaps in coverage. In addition, the potential of foraminifera recovered from the Antarctic lake sediments has not been fully explored.

Interestingly the total solar eclipse that occurred on December 4, 2021, generated significant electron density variations over Antarctica. **Sunil et al**. tracked and observed the ionospheric responses over Antarctica during this important period. They observed clear electron density depletions are observed along the eclipse totality path followed by gradual recoveries. Total Electron Content (TEC) derived from 35 Global Positioning System (GPS) stations located over West Antarctica was used to study the ionospheric electron density variations during the eclipse. They found that the absence of solar radiations following the eclipse onset resulted in the drop of charged particle density at various ionospheric layers, which in turn resulted in the decrease of TEC values.

Among the major features of the Antarctic landscape, water bodies stand as a potential source of information as most of these water bodies are of glacial origin, relatively small and date from the Pleistocene epoch of the Quaternary Period. During warmer austral spring and summer periods when the ice melts, the Antarctic water bodies receive the majority of their sediment supply. It is, therefore important to understand that the climatic changes influence the sediments and fauna/flora of the Antarctic water bodies. Hence these water bodies could emerge as an important source of paleoclimatic and global change information. To understand the inter-relationship between various aquatic communities, the study of the physical, chemical and biological conditions of the ambient water is significant. Similarly, the detailed study of biotic communities of the Antarctic water bodies is essential because the biota inhabiting the region today is of relatively recent re-colonization, which undoubtedly accounts for some of the distinctive species distribution. Furthermore, the Oasis and Hills in East Antarctica host many lakes and transient ponds in the ice-free regions formed by the advection heat and differential albedo promoting increased melting of Polar ice. **Reshmi et al**. studied one such Oasis, the Schirmacher and Larsemann Hills areas of the East Antarctica region to probe the chemical and isotopic evolution of the lake waters. They noticed that lake water in the Schirmacher Oasis is fed by the glacial melt water with a relatively low ionic load, whereas the lakes from the Larsemann Hills had a high concentration of dissolved ions. Inter ionic variability showed that weathering of silicate rocks is the prime source of ions in these lakes, followed by ion exchange and evaporation. According to them, Isotopic ratios were also distinctly different in the lake water in both regions. Diffusion controlled kinetic effect at the liquid-ice interface for different water isotopologues is the prime determinant of the isotopic composition in Schirmacher Oasis lakes, whereas, in addition to it, evaporative enrichment of heavier isotopes from open water bodies affects the lakes in Larsemann Hills.

Indubitably, any modification of the catchment region gets reflected in the lake system, which is more easily studied from the lake sediments, than the catchment itself because water bodies are natural sumps of large catchment areas. It is important to mention that the limnological studies can also reveal the structural features and geomorphology of the region in which they are situated. Being the largest and one of the deepest water bodies of the Schirmacher Oasis, central Dronning Maud Land region, of East Antarctica, the sediments of Priyadarshini water body provide a continuous record of high-resolution paleoclimatic information. To retrieve the long sediment core the need to get the information about lake bathymetry, bottom topography and an estimation of the distribution, thickness and stratigraphy of the sediments underlying the lake floor utilizing acoustic techniques such as echo-sounding and sub-bottom seismic reflection profiling of the Antarctic lakes is emphasized in the present study. The preparation of a detailed map of the Priyadarshini water body will help to understand the potential pathways for the sediment and water influx to the waterbody. The results of the present study coupled with the results of the sub-bottom profiler will help in marking potential coring locations. Having realised such importance of the Antarctic's water bodies, **Khare et al**. undertook a detailed morphometric assessment of Priyadarshini water body and also attempt to assess the limnological parameters of the water body (Priyadarshinin Lake) under the extreme Antarctic climatic conditions.

While addressing the danger of ongoing global warming on the Himalayan glaciers, **Singh et al**. focused on the paleo-depositional and paleo-climatic conditions. Glacier dynamics and resulting geomorphic features are the characteristics of various stages of glacial fluctuations. Various geomorphic features/landforms such as Lateral Moraines (LM), Terminal/Recessional Moraines (RM), Outwash Plain Deposits (OWP) and Kame Terraces were identified in the field. The field observations and lithological analysis reflect that these morphological features are evolved by the glacial (Gl), glacio-fluvial (Gf), and mass movement activities. The sediment size decreases whereas sorting, roundness and percentage of matrix increase from the Gl-Gf process. These geomorphic features are modified by catastrophic events such as Landslide Lake Outburst Flooding (LLOF) and Glacier Lake Outburst Flooding (GLOF). Therefore, the geomorphic features/landforms are evolved by glacial processes under the direct control of tectonics and climate and further reshaped by LLOF and/or GLOF.

Thus, the present book emphasizes deciphering the climate records in ice cores, geologic cores, and those inferred from rock outcrops. Its chapters on scientifically significant and addressing a specific issue of climate change and geodynamics over the polar region will be of interest to policy makers, researchers, and scientific institutions

Neloy Khare

Date: April 2022
Place: New Delhi

Acknowledgments

It is my great pleasure to express my gratitude and deep appreciation to all contributing authors. Without their valuable inputs on various facets of climate variability over the Antarctic and surrounding Southern Ocean region, the book would not have been possible. Various learned experts who have reviewed different chapters are graciously acknowledged for their timely, constructive, and critical reviews.

I sincerely thank Dr M. Ravichandran Secretary Ministry of Earth Sciences, Government of India, New Delhi (India), for their suppor and encouragement. Prof. Govardhan Mehta, FRS has always been a source of inspiration and is acknowledged for his kind support.

Prof. Anil Kumar Gupta, Indian Institute of Technology, Kharagpur (India) and Dr Rajiv Nigam former Adviser at the National Institute of Oceanography, Goa (India) are deeply acknowledged for providing many valuable suggestions to this book. This book has received significant support from Akshat Khare and Ashmit Khare, who have helped me during the book preparation. Dr Rajni Khare has unconditionally supported enormously during various stages of this book. Shri Hari Dass Sharma from the Ministry of Earth Sciences, New Delhi (India), has helped immensely in formatting the text and figures of this book and bringing it to its present form. The publisher (Taylor & Francis) have done a commendable job and are sincerely acknowledged.

Neloy Khare

Date: April 2022
Place: New Delhi

Contributors

K. K. Ajith
National Atmospheric Research Laboratory
Gadanki, India

Waseem Ahmad Baba
Department of Geology
School of Environment and Earth Sciences
Central University of Punjab
Bathinda, Punjab, India

Vikash Chandra
Polar Studies Division
Geological Survey of India (GSI)
Faridabad, Haryana, India

Subodh Kumar Chaturvedi
Institute of Hydrocarbon
Energy and Georesources
ONGC Centre for Advanced Studies
University of Lucknow
Lucknow, India

A. Dharwadkar
Polar Studies Division
Geological Survey of India (GSI)
Faridabad, Haryana

Prabhu Prasad Dash
ACOAST, Amity University Haryana (AUH)
Gurugram, India
and
Department of Geology
Maharaja Sriram Chandra Banja Deo University, Keonjhar Campus
Keonjhar, Odisha, India

Chetan Anand Dubey
Department of Geology
University of Lucknow
Lucknow, India

Deepak Y. Gajbhiye
Polar Studies Division
GSI
Faridabad, Haryana

Pawan Kumar Gautam
Department of Geology
University of Lucknow
Lucknow, India

Girish Gopinath
Kerala University of Fisheries and Ocean Studies
Kochi, India

Amrutha K.
Department of Geology
School of Environment and Earth Sciences
Central University of Punjab
Bathinda, Punjab, India

Tushar Kaushik
Biodiversity and Paleobiology
Agharkar Research Institute
Pune, India

Rajni Khare
Department of Environmental Sciences & Limnology, Bakatullah University
Bhopal, India (Formerly)

Dhirendra Kumar
Department of Geology
University of Lucknow
Lucknow, India
and
Department of Geology
Central University of South Bihar
Gaya, Bihar, India

Pradeep Kumar
Polar Studies Division
Geological Survey of India (GSI)
Faridabad, Haryana, India

Kirtiranjan Mallick
Department of Geology
Utkal University, Vani Vihar
Bhubaneswar, Odisha 751004

Jitendra Kumar Pattanaik
Department of Geology
School of Environment and Earth Sciences
Central University of Punjab
Bathinda, Punjab, India

M. Praveenbabu
Centre for Water Resources Development and Management
Kozhikode, India

Rahul Rawat
Indian Institute of Geomagnetism
Mumbai, India

T. R. Resmi
Centre for Water Resources Development and Management
Kozhikode, India

Ankush Shrivastava
Department of Geology
Mohanlal Sukhadia University
Rajasthan, India

S. P. Shukla
Polar Studies Division
Geological Survey of India (GSI)
Faridabad, Haryana

Anoop Kumar Singh
Department of Geology
University of Lucknow
Lucknow, India

Ashutosh K. Singh
Department of Geology
Centre of Advanced Studies
University of Delhi
Delhi, India

Dhruv Sen Singh
Department of Geology
University of Lucknow
Lucknow, India

Vikram Pratap Singh
Department of Geology
Indira Gandhi National Tribal University
Amarkantak, India

Devesh K. Sinha
Department of Geology
Centre of Advanced Studies
University of Delhi
Delhi, India

A. S. Sunil
CUSAT-NCPOR Centre for Polar Sciences
Department of Marine Geology and Geophysics
School of Marine Sciences
Cochin University of Science and Technology
Kochi, India

P. S. Sunil
CUSAT-NCPOR Centre for Polar Sciences
Department of Marine Geology and Geophysics
School of Marine Sciences
Cochin University of Science and Technology
Kochi, India

A. K. Swain
State Unit: Odisha, Geological Survey of India (GSI)
Bhubaneswar, Odisha, India

Dhanya Thomas
CSIR Fourth Paradigm Institute
Bangalore, India

Balkrishan Vishawakarma
Department of Geology
University of Lucknow
Lucknow, India

Ashwani Wanganeo
Department of Environmental Sciences & Limnology
Bakatullah University
Bhopal, India (Formerly)

Rajni Wanganeo
Department of Zoology
Government Benzeer College of Commerce & Science
Bhopal, India (Formerly)

Editor

Dr Neloy Khare, presently Adviser/Scientist "G" to the Government of India at MoES has a very distinctive acumen not only in administration but also in quality science and research in his areas of expertise covering a large spectrum of geographically distinct locations like the Antarctic, Arctic, Southern Ocean, Bay of Bengal, Arabian Sea, Indian Ocean etc. Dr Khare has almost 30 years of experience in the field of paleoclimate research using paleobiology Paleontology)/teaching/science management/administration/coordination for scientific programs (including Indian Polar Programme) etc. Having completed his doctorate (PhD) in tropical marine region and Doctor of Science (DSc) in Southern High latitude marine regions towards environmental/climatic implications using various proxies including foraminifera (micro-fossil), have made significant contributions in the field of Paleoclimatology of Southern high latitude regions (the Antarctic and the Southern Ocean) using Micropaleontology as a tool. These studies coupled with his paleoclimatic reconstructions from tropical regions helped understand causal linkages and teleconnections between the processes taking place in Southern high latitudes with that of climate variability occurring in tropical regions. Dr Khare has been conferred Honorary Professor and Adjunct Professor by many Indian Universities.

He has a very impressive list of publications to his credit (125 research articles in national and international scientific journals: three special issues of national scientific journals as guest editor; edited special issues of *Polar Science* (Elsevier), *Journal of Asian Earth Science* (Elsevier), *Quaternary International* (Elsevier), and *Frontiers in Marine Science* as its managing/guest editor. He has authored/edited many books, and has contributed 130 abstracts to various seminars, as well as writing 23 popular science articles and five technical reports). The government of India and many professional bodies have bestowed him with many prestigious awards for his humble scientific contributions to past climate changes/oceanography/polar science and Southern Oceanography. The most coveted award is the Rajiv Gandhi National Award of 2013, conferred by the Honourable President of India. Others include the ISCA Young Scientist Award, Boyscast Fellowship, CIES French Fellowship, Krishnan Gold Medal, Best Scientist Award, Eminent Scientist Award, ISCA Platinum Jubilee Lecture, IGU Fellowship, and many more. Dr Khare has made tremendous efforts to popularize ocean science and polar science across the country by way of delivering many Invited lectures, radio talks, and publishing popular science articles.

Dr Khare sailed in the Arctic Ocean as a part of "Science PUB" in 2008 during the International Polar Year campaign for scientific exploration and became the first Indian to sail in the Arctic Ocean.

1 Geomorphology around Kongsfjorden-Krossfjorden System, Svalbard

Jitendra Kumar Pattanaik,[1] *Amrutha K.,*[1] *and Prabhu Prasad Dash*[2]

[1] Department of Geology, School of Environment and Earth Sciences, Central University of Punjab, Bathinda, Punjab, India

[2] ACOAST, Amity University Haryana (AUH), Gurugram, India; Department of Geology, Maharaja Sriram Chandra Banja Deo University, Keonjhar Campus, Keonjhar, Odisha

CONTENTS

1.1 INTRODUCTION

Polar environments are distinctly sensitive to climatic changes as it belongs to the lower end of the energy spectrum of our planet. Hence, it is susceptible to ecosystem variation, landscape evolution and anthropoid influence. The Arctic region has its characteristic geomorphic features which indirectly indicate the geology, climatic condition and different geomorphic forces that are/were active in the area. The abundance of standing waters is a characteristic feature of the Arctic region (Pienitz et al., 2008). Presence of certain geomorphological features can be used to understand the climatic conditions and geomorphological

DOI: 10.1201/9781003284413-1

agents prevalent in the area. Periglacial land forms such as ice-wedges, rock glaciers, pingos, solifluction, avalanches, debris flows, rockfalls and nivation; glacial land forms like surge glaciers and different aeolian land forms are few example to this. Compared to other coastlines, Arctic coasts have experienced substantial modifications (Overduin et al., 2014). Since the LGM (Last Glacial Maximum), there has been a large degree of glacial retreat from the coastal tract resulting in the predominance of paraglacial processes (Bourriquen et al., 2018; Gjermundsen et al., 2013). Considering the geomorphological studies from the Arctic region, it is important to note that despite having the 30% of the global coastline in the Arctic, only 1% of Arctic coasts have been studied (Lantuit et al., 2010). Arctic mountains are always been important in geomorphological studies as the region is characterized by the lowest sediment fluxes, even though it is characterized by a large sediment store and a favorable topography for sediment transportation (Mithan et al., 2019). The Svalbard archipelago in the Arctic circle has been widely studied in terms of its glacial dynamics and its implications for climate change. However, a wide variety of geomorphological features formed as a result of a single major (ice) geomorphological agent, make it difficult to interpret the underlying geomorphological processes. Svalbard is located in the Arctic sea between latitudes of 74° to 81°N and displays a vast expanse of ice caps (≈ 60%) and valley/fjord glaciers (Hagen et al., 1993). Spitsbergen is the largest island of Svalbard archipelago (Miccadei et al., 2016). Kongsfjorden-Krossfjord systems are located between 78° 40′ and 77° 30′ N latitudes and 11° 3′ and 13° 6′ E longitudes (Svendsen et al., 2002). Kongsfjorden is aligned from southeast to northwest and the orientation of Krossfjorden is from North to South. The coast of Kongsfjorden is defined by the Blomstrandbreen area in the North, Broggerhalvoya toward the South, Ossian Sarsfjellet and Colletthogda in the East (Miccadei et al., 2016). Fjord systems and the lowland areas in the Arctic are important as they experience both terrestrial and coastal processes. Fjords are semi-enclosed marine basins that are deepened by glacier activity and generally represent a transition region between terrestrial and marine environments (Howe et al., 2010).

Numerous fjords of different dimensions are present in the Svalbard archipelago and the longest fjords, Storfjorden, separate Spitsbergen from Barentsøya and Edgeøya. The major fjords of Svalbard are listed in Table 1.1. Krossfjorden-Kongsfjorden system together constitute a basin area of 3074 km^2 and 74% of the area is covered by ice (i.e. 1651 km^2) and 2257 km^2 area island. The system has a glacier volume of 308 km^3 (Svendsen et al., 2002). Kongsfjorden-Krossfjorden systems are located near Arctic and Atlantic water mass and it is influenced by the west Spitsbergen current and freshwater from glacier runoff (Kumar et al., 2018). Seasonal sea ice-rafted, across gravity-driven and the glacier generated sediments control the sedimentation process in the Kongsfjorden-Krossfjorden fjord system. The sediments deposited in the fjords and the landforms formed by glaciers in the region are less modified and hence this is an ideal location for climatic research (Trusel et al., 2010). Geomorphology of the Kongsfjorden-Krossfjorden system is studied by the systematic appreciation of geology and geomorphic processes responsible for landscape evolution.

TABLE 1.1
Major Fjords of Svalbard Based on Length

Sl. no	Name of the fjords	Region in Svalbard	Length (km)	S. N.	Name of the fjords	Region in Svalbard	Length (km)
1	Storfjorden	Spitsbergen	132	19	Lady Franklinfjorden	Nordaustlandet	25
2	Wijdefjorden	Spitsbergen	108	20	St. Jonsfjorden	Spitsbergen	21
3	Isfjorden	Spitsbergen	107	21	Bellsund	Spitsbergen	20
4	Van Mijenfjorden	Spitsbergen	83	22	Brennevinsfjorden	Nordaustlandet	20
5	Woodfjorden	Spitsbergen	64	23	Raudfjorden	Spitsbergen	20
6	Wahlenbergfjorden	Spitsbergen	46	24	Smeerenburgfjorden	Spitsbergen	20
7	Tjuvfjorden	Edgeøya	45	25	Ekmanfjorden	Spitsbergen	18
8	Rijpfjorden	Nordaustlandet	40	26	Grønfjorden	Spitsbergen	16
9	Duvefjorden	Nordaustlandet	35	27	Sassenfjorden	Spitsbergen	15
10	Lomfjorden	Spitsbergen	35	28	Tempelfjorden	Spitsbergen	15
11	Austfjorden	Spitsbergen	32	29	Lilliehöökfjorden	Spitsbergen	14
12	Billefjorden	Spitsbergen	30	30	Vestfjorden	Spitsbergen	12
13	Hornsund	Spitsbergen	30	31	Adlersparrefjorden	Nordaustlandet	10
14	Krossfjorden	Spitsbergen	30	32	Möllerfjorden	Spitsbergen	9
15	Liefdefjorden	Spitsbergen	30	33	Magdalenefjorden	Spitsbergen	8
16	Van Keulenfjorden	Spitsbergen	30	34	Recherche Fjord	Spitsbergen	7
17	Dicksonfjorden	Spitsbergen	30	35	Fuglefjorden	Spitsbergen	6
18	Kongsfjorden	Spitsbergen	26	36	Kobbefjorden	Spitsbergen	3.5

Note: The locations of the fjords are indicated. Some of the branches/tributaries/distributaries of the major fjord system are also included here.

1.2 GEOLOGY AROUND KONGSFJORDEN-KROSSFJORDEN SYSTEM

The Kongsfjorden-Krossfjorden fjord system is located near a major tectonic boundary and hence the region has diverse petrology (Streuff, 2013). This tectonic boundary separates the Northwestern Basement Province to the northeast and the Cenozoic fold- and thrust belt of western Spitsbergen to its southwest (Bergh et al., 2000). Igneous, metamorphic, and sedimentary rocks are reported from the Kongsfjorden-Krossfjorden system. According to the Norwegian Polar Institute, the eastern boundary of the island is dominated by sandstone, siltstone and shale with small patches of gabbro or metagabbro rock type. These layered rocks are of the Devonian age. The southern boundary of the island has the maximum diversification and exhibits sedimentary rocks such as chert, silicified limestone and sandstone and metamorphic rocks like marble alternating with other metasediments. Sandstone, siltstone and shale with gabbro or metagabbro patches are also present here along with bituminous shale, phyllite and metapelites schists. Phyllites and quartzites are often found locally with layers of other rocks. These rock types are of Paleogene to Neogene, Middle Jurassic–early Cretaceous, Triassic–Middle Jurassic, and Carboniferous and Permian sequences. The southwest boundary of the island is characterized by limestone and/

or dolostone, conglomerate, tilloid rocks, garnet-mica schist, calc-pelitic schist and marble, quartzite, and other high-pressure metamorphic rock types. The northwestern boundary is majorly occupied by gneisses and schists. Patches of marble with phyllite and metapelitic schist are often observed along this boundary with granite or granodiorite patches. The bedrocks in this region belong to Mesoproterozoic. The Kongsfjorden-Krossfjorden fjord system is surrounded by phyllite and meta-pelitic schist with patches of marble. Quaternary continental and coastal deposits overlie the bedrock sequences. Surficial continental deposits include till and diamicton, talus, rockfalls, fluvial and beach deposits.

1.3 LANDSCAPE TYPES

Landscape in the Svalbard region can be divided into several categories depending upon the geomorphological processes and the localities where it is found. As per the scheme adopted by the Norwegian Polar Institute, landforms are classified into various categories as listed in Table 1.2, Sl. No. 1, under terrestrial landforms. Other landscape types are listed under glacial landforms and surface deposits (Table 1.2). Depending on the landforms that found in the different localities, these features can be grouped under coastal landforms, landforms developed in the slope, mountain ranges and valleys.

1.3.1 Coastal Landforms

Sediment distribution by retreating glaciers and the efficiency of the fluvial system to transport sediment towards the shoreline controls the overall evolution of the coastal region (Bourriquen et al., 2018). Coastal zones with continuous sediment supply have undergone coastal progradation and the areas that lost the sediment supply experienced coastal recession (Bourriquen et al., 2018). Most of the coastal part of the system is

TABLE 1.2
Different Landscape Types and Geomorphological Features Found in Svalbard

Sl. No	Landscape types	Geomorphological feature
1	Terrestrial landforms	Coastal low land, sandur, or river flat within coastal lowland or U-shaped valley, moraine field, ice field and ice cap, valley glacier and glacier cirque, glacial denudation flat, open, and hilly landscape, mountainous landscape with rounded shapes, plateau mountainous landscape, edge dominated and alpine mountainous landscape
2	Glacial landforms	Fjord, Pingo, pattern ground, glaciers, U-shaped valley, moraine field, Drumlins, and glacial flutes
3	Surface deposits	Ice, recent moraines, moraine and till, glacial-fluvial and fluvial deposits, marine deposits, gelifluction deposits, slope deposits, weathering materials
4	Other features	Thermal spring, karst or thermo-karst landforms, exposed bedrocks

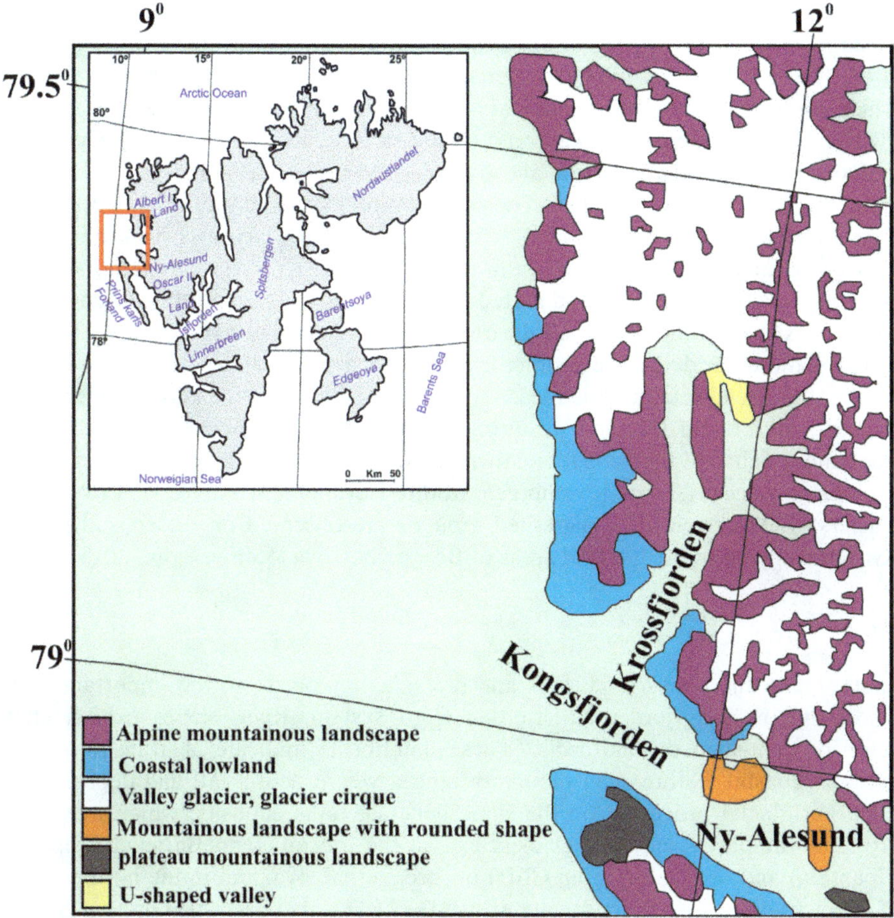

FIGURE 1.1 Map showing important landscape types and geomorphological features in and around Kongsfjorden-Krossfjorden system.

covered with Quaternary deposits of moraines, marine shore, and fluvial deposits (Ingvaldsen et al., 2001; Kumar et al., 2018).

The coastal geomorphology of Svalbard is dominated by Coastal cliffs, spit-barrier lagoons, deltas, flats and raised beaches. Wind, rain or/and wave action in coastal areas erode the soft rocks, resulting in the formation of cliffs that belongs to hard rock remnants of the shore lithology. Longshore drift in the coastal areas results in spit-barrier lagoons. These common features associated with the coast are formed in accordance with the variation in sediment supply, longshore drift and exposure to fetch (Brückner and Schellmann, 2003). River incisions of bedrocks in the coastal area form steep scarps and channels perpendicular to the coast (Berthling et al., 2020).

Coastal lowlands with well-developed raised beaches are a characteristic feature of Svalbard. However, its distribution is not uniform, and they are on wide display along the western boundary. The genesis of raised beaches is related to post-glacial isostatic rebound (Berthling et al., 2020). The western coastline of Spitsbergen is characterized by a series of marine terraces. Strandflats are common in coastal areas, which are mostly bounded by mountains and low coastal cliffs with beaches towards the sea. Raised beaches are preserved on the strand flats (Streuff, 2013). Currently, the coast is covered with linear and pocket-sized rocky cliffs and gravel beaches, including Lagoons and deltas. The western rocky shoreline of Kongsfjorden-Krossfjorden fjord system exhibits vertical cliffs with caves and pocket beaches. Low-lying areas of Broggerhalvoya in the Kongsfjord-Krossfjord system have well developed Patterned ground and sorted circles (Svendsen et al., 2002). Lagoons and single bars are common along all the coasts (e.g. Brandallaguna).

Coastal Sandur deltas are another important coastal landform. These are formed during the glacial retreat which leads to the exposure of terrigenous sediments and further transport and deposition resulting in the Sandur delta (Bourriquen et al., 2018). The sandur deltas of the Brogger peninsula are some of the examples formed by this geological process (Austre Lovénbreen, Midtre Lovénbreen, and Vestre Lovénbreen glaciers). Stone circles and patterned grounds are observed on the coastal part of Kvadehuken and in the lowland areas of Broggerhalvoya (Kumar et al., 2018).

1.3.2 Slope Landforms

Talus slopes and cones, rock falls and pro talus remparts are the important slope landforms in Kongfjorden-Krosfjorden fjord system. Rock slopes exhibit characteristic fan-shaped landforms. These landforms indicate a transition zone between coastal features and mountain ranges where debris fall and deposition is high. Periglacial processes on the slope generate talus deposits, which may form pro-talus ramparts, and rock glaciers e.g., at the foot of Zeppelinfjellet (Zeppelin Mountain) and Stuphallet (Stup Cliff) on Broggerhalvoya, and in the northern part of Blomstrandhalvoya (Svendsen et al., 2002). Debris that gets piled up to a characteristic angle of repose is called talus slope. The deposits related to rock slopes are fan-shaped landforms. Certain climatic processes favor the development of slope processes, rockfalls, debris and solifluction covers. Steep mountain slopes cut by gullies, lower gradient slope with debris covers and slopes with solifluction lobes and sometimes flattened parts dominated by shales are some of the features formed by the runoff of melting snow or/and snow avalanches and gullies formed alluvial fans (Zwoliński et al., 2013).

1.3.3 Mountain Ranges and Wide Valleys

The overall distribution of mountain ranges and wide valleys follow geological structures and the meridian of the location (Zwoliński et al., 2013). Hecahoek rocks, formed of metamorphosed crystalline rocks, quartzite, marble and slate appear as mountain ranges (Zwoliński et al., 2013). Fjords of Spitsbergen might indicate the path of the old valley system, which was reformed in Quaternary glaciations (Zwoliński et al., 2013). The downslope flow of valley glaciers creates a trough, which is called a U-shaped valley (Fredin et al.,

2013). Hence, this flat bottomed and steep-walled landform is a product of the erosive action of glaciers. This erosive action is called glacial plucking, which generally occurs at the bottom of the valleys. The mountain peaks on the southern side of inner Kongsfjorden, Nunataks surrounding the inner Kongsfjorden, the pyramidical peaks of Tre Kroner and high mountains surrounding Fjord Krossfjorden, North of Kongsfjorden are some of the examples of mountain ranges in the Krossfjorden-Kongsfjorden system. The mountain ranges in the Kaffioyra generally occur as narrow ridges with narrow crests and steep slopes (Svendsen et al., 2002).

Landforms associated with limestone terrain are also reported here. Dissolution of limestone bedrocks resulted in landscape with Karst topography. Most of these are small-scale surface features, however, some larger features like caves also occur, e.g. caves in Blomstrandhalvoya (Blümel, 1971).

1.4 SURFACE MATERIAL IN AND AROUND KONGSFJORD-KROSSFJORD REGION

The border area around Kongsfjorden-Krossfjorden fjord system is largely covered by marine deposits, slope deposits and weathering material (Kristiansen and Sollid, 1986). Glacio-fluvial, fluvial and recent moraine are also present. Patches of exposed bedrock can also be seen in the area.

Glaciers: More than 60% land area of Svalbard is covered with glaciers (Etzelmüller et al., 2000). In Svalbard, glaciers of every kind can be seen. In Spitsbergen, valley glaciers are predominate, and massive ice caps are frequent in Nordaustlandet, Edgerya, and Barentsrya. Glaciological research began in the late 19th century and became very popular in the 20th century. They include ice-core studies and research in meteorology, mass balance, glacier flow, glacial erosion, and radio-echo sounding (Liestøl, 1988). Since many of the glaciers in Svalbard are known to surge frequently (86 have surged till the beginning of 20th century), it is challenging to use fluctuations of glaciers as climatic indicators. Important morphological feature developed during the glacier advance and deglaciation is river valleys, through which the water drains from the glaciers (Svendsen et al., 2002). At the Kongfjorden fjord, the glaciers can be found directly calving into the sea. These glaciers are both subpolar and polythermal, indicating that they have both warm (above or near zero, where meltwater can exist), and cold (below zero) zones. Surging glaciers are common in the study area.

Von Postbreen, a large land-terminating glacier; Kongsvegen and Tunabreen, large tidewater glaciers; Midtre Lovénbreen, Tellbreen and Longyearbreen, small valley glaciers, Kongsvegen and Midtre Lovénbreen are examples of some of the important glaciers in the main island of Spitsbergen (Sevestre et al., 2015). Kongsfjorden and Krossfjorden are dominated by tidewater glaciers: Lilliehookbreen at the head of Krossfjorden (Lilliehookfjorden) and five other calving glaciers along its eastern coast. Kronebreen and Kongsvegan at the head, and Conwaybreen and Blomstrandbreen on northern coast of Kongsfjorden. Glacial, periglacial and hydroglacial processes influence the landscape.

The main glacierized area consists of a large glacier complex in the inner part of the fjords, which has many calving fronts at its head. These glaciers drain the large icefields of Isachsenfonna and Holtedalfonna. Blomstrandbreen on the northern side of Kongsfjorden also has a calving front. On the southern side, there are several valleys or cirque glaciers (Svendsen et al., 2002), and None of them reaches the fjord. Glaciers are associated with a different types of moraine deposits and their front portion is characterized by rivers and outwash plains (Berthling et al., 2020).

Moraines: Towards the end of glaciers, large amounts of rock debris get deposited beneath the active glacier. The size of accumulated debris ranges from meter-sized boulders to millimeter-sized dust particles. This random mixture of rock particles is called till and mounds, sheets or sinuous ridges of sediment deposits are called moraines (Elvevold et al., 2007). Austre Lovénbreen, Midtre Lovénbreen and Vestre Lovénbreen are some of the examples for terminal moraines (Bourriquen et al., 2018).

1.5 SUMMARY

Arctic regions are important in geomorphological studies as the region is characterized by the lowest sediment fluxes. However, this region has large sediment storage and a favorable topography for sediment transportation. The Svalbard archipelago in the Arctic has been widely studied in terms of its landscape, glacial dynamics, sediment transports, and paleoclimate. The Fjord system is a transition zone between the terrestrial and marine environment, acting as an important sediment archive that influenced by both the environments. These archives provide ample opportunity for high-resolution studies on paleo-environmental changes. Diverse geomorphological features developed in Svalbard are due to the interplay of active climate and surficial processes. Detailed study on these features will help to understand the sequence and mechanism of the surficial processes.

ACKNOWLEDGMENT

Authors are thankful to the Central University of Punjab, Bathinda for providing the administrative and infrastructural facilities. AK is thankful to DST-INSPIRE fellowship for providing financial support towards her PhD.

REFERENCES

Bergh, S. G., Maher, H. D., & Braathen, A. (2000). Tertiary divergent thrust directions from partitioned transpression, Brøggerhalvøya, Spitsbergen. *Norsk geologisk tidsskrift*, *80*, 63–82.

Berthling, I., Berti, C., Mancinelli, V., Stendardi, L., Piacentini, T., & Miccadei, E. (2020). Analysis of the paraglacial landscape in the Ny-Ålesund area and Blomstrandøya (Kongsfjorden, Svalbard, Norway). *Journal of Maps*, *16*(2), 818–833.

Blümel, W. (1971). Kleinkarst und Formen selektiver Erosion auf der Blomstrandhalvøya/NW Spitzbergen. (Karst and erosion forms on Blomstrandhalvøya/NW Spitsbergen). *Der Aufschluss*, *22*, 149–163.

Bourriquen, M., Mercier, D., Baltzer, A., Fournier, J., Costa, S., & Roussel, E. (2018). Paraglacial coasts responses to glacier retreat and associated shifts in river floodplains over decadal timescales (1966–2016), Kongsfjorden, Svalbard. *Land Degradation & Development*, *29*(11), 4173–4185.

Brückner, H., & Schellmann, G. (2003). Late Pleistocene and Holocene shorelines of Andréeland, Spitsbergen (Svalbard): Geomorphological evidence and palaeo-oceanographic significance. *Journal of Coastal Research*, 971–982.

Elvevold, S., Dallmann, W., & Blomeier, D. (2007). *Geology of Svalbard*. Norwegian Polar Institute, 978-82-7666-237-5

Etzelmüller, B., Ødegård, R. S., Vatne, G., Mysterud, R. S., Tonning, T., & Sollid, J. L. (2000). Glacier characteristics and sediment transfer system of Longyearbreen and Larsbreen, western Spitsbergen. *Norsk Geografisk Tidsskrift*, *54*(4), 157–168.

Evans, D. J., Strzelecki, M., Milledge, D. G., & Orton, C. (2012). Hørbyebreen polythermal glacial land system, Svalbard. *Journal of Maps*, *8*(2), 146–156.

Fredin, O., Bergstrøm, B., Eilertsen, R., Hansen, L., Longva, O., Nesje, A., & Sveian, H. (2013). Glacial landforms and Quaternary landscape development in Norway. *Quaternary Geology of Norway*, edited by: Olsen, L., Fredin, O., and Olesen, O., Geological Survey of Norway Special Publication, Geological Survey of Norway, Trondheim, 525.

Gjermundsen, E. F., Briner, J. P., Akçar, N., Salvigsen, O., Kubik, P., Gantert, N., & Hormes, A. (2013). Late Weichselian local ice dome configuration and chronology in Northwestern Svalbard: Early thinning, late retreat. *Quaternary Science Reviews*, *72*, 112–127.

Hagen, J. O., Liestøl, O., Roland, E. R. I. K., & Jørgensen, T. (1993). *Glacier atlas of svalbard and jan mayen* (Vol. 129, p. 141). Oslo: Norsk polarinstitutt.

Howe, J. A., Austin, W. E., Forwick, M., Paetzel, M., Harland, R., & Cage, A. G. (2010). Fjord systems and archives: A review. *Geological Society, London, Special Publications*, *344*(1), 5–15.

Ingvaldsen, R., Reitan, M. B., Svendsen, H., & Asplin, L. (2001). The upper layer circulation in Kongsfjorden and Krossfjorden—A complex fjord system on the west coast of Spitsbergen. *Memoirs of National Institute of Polar Research* (Special issue), *54*, 393–407.

Kristiansen, K. J., & Sollid, J.-L. (1986). *Svalbard—Glacial geology and geomorphology*. Hønefoss, Statens Kartverk: National Atlas of Norway.

Kumar, P., Pattanaik, J. K., Khare, N., & Balakrishnan, S. (2018). Geochemistry and provenance study of sediments from Krossfjorden and Kongsfjorden, Svalbard (Arctic Ocean). *Polar Science*, *18*, 72–82.

Lantuit, H., Overduin, P., Solomon, S., & Mercier, D. (2010). Coastline dynamics in polar systems using remote sensing. *Geomatic solutions for coastal environments* (pp. 163–174), New York: Nova Science Publishers.

Liestøl, O. (1988). The glaciers in the Kongsfjorden area, Spitsbergen. *Norsk Geografisk Tidsskrift-Norwegian Journal of Geography*, *42*(4), 231–238.

Miccadei, E., Piacentini, T., & Berti, C. (2016). Geomorphological features of the Kongsfjorden area: Ny-Ålesund, Blomstrandøya (NW Svalbard, Norway). *Rendiconti Lincei*, *27*(1), 217–228.

Mithan, H. T., Hales, T. C., & Cleall, P. J. (2019). Supervised classification of landforms in Arctic mountains. *Permafrost and Periglacial Processes*, *30*(3), 131–145.

Overduin, P. P., Strzelecki, M. C., Grigoriev, M. N., Couture, N., Lantuit, H., St-Hilaire-Gravel, D., . . . & Wetterich, S. (2014). Coastal changes in the Arctic. *Geological Society, London, Special Publications*, *388*(1), 103–129.

Pienitz, R., Doran, P. T., & Lamoureux, S. F. (2008). Origin and geomorphology of lakes in the polar regions. *Polar Lakes and Rivers: Limnology of Arctic and Antarctic Aquatic Ecosystems*, 25–41.

Sevestre, H., Benn, D. I., Hulton, N. R., & Bælum, K. (2015). Thermal structure of Svalbard glaciers and implications for thermal switch models of glacier surging. *Journal of Geophysical Research: Earth Surface*, *120*(10), 2220–2236.

Streuff, K. (2013). *Landform assemblages in Inner Kongsfjorden, Svalbard: Evidence of recent glacial (surge) activity. University of Tromso* (Doctoral dissertation, Master's Thesis in Geology).

Svendsen, H., Beszczynska-Møller, A., Hagen, J. O., Lefauconnier, B., Tverberg, V., Gerland, S., . . . & Azzolini, R. (2002). The physical environment of Kongsfjorden—Krossfjorden, an Arctic fjord system in Svalbard. *Polar Research, 21*(1), 133–166.

Trusel, L. D., Powell, R. D., Cumpston, R. M., & Brigham-Grette, J. (2010). Modern glacimarine processes and potential future behaviour of Kronebreen and Kongsvegen polythermal tidewater glaciers, Kongsfjorden, Svalbard. *Geological Society, London, Special Publications, 344*(1), 89–102.

Zwoliński, Z., Giżejewski, J., Karczewski, A., Kasprzak, M., Lankauf, K. R., Migoń, P., . . . & Stankowski, W. (2013). Geomorphological settings of Polish research areas on Spitsbergen. *Landform Analysis, 22.*

2 The Influence of Changing Climate on the Mass Balance of a Part of Central Dronning Maud Land, East Antarctic Ice Sheet

Pradeep Kumar,[1] Vikash Chandra,[1] A. K. Swain,[2] Deepak Y. Gajbhiye[1], S. P. Shukla,[1] and A. Dharwadkar[1]

[1] Polar Studies Division, GSI, Faridabad, Haryana

[2] SU: Odisha, GSI, Bhubaneswar, Odisha

CONTENTS

DOI: 10.1201/9781003284413-2

2.1 INTRODUCTION

East Antarctica Ice Sheet (EAIS) form the largest source of snow and ice with several basins, separated by ice drainage divides (Rignot et al., 2008) (Figure 2.1). Dronning Maud Land (DML) is hydrologically divided into many small basins. Nivlisen basin has the second-largest ice shelf in the central DML area. An ice-free landmass named, Schirmacher Oasis is located on the Princess Astrid coast in the central part of the central Dronning Maud Land (cDML) basin of East Antarctica (Figure 2.2). This ice-free area is elongated along WNW–ESE and has a maximum length of

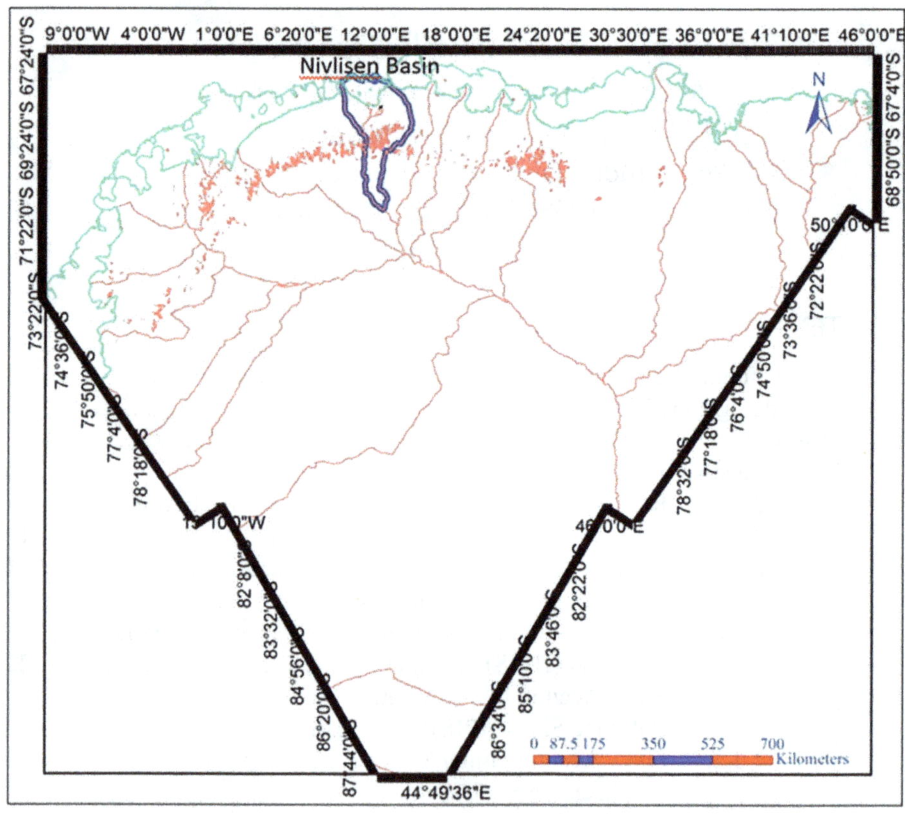

FIGURE 2.1 Catchment map of the Dronning Maud Land (DML), East Antarctica, and highlighted basin in the study area of this project.

Note: Green polygons are ice shelf. Maroon lines are ice divide. Blue-colored polygon is Nivlisen basin. Maroon-filled polygons are exposed rock mass.

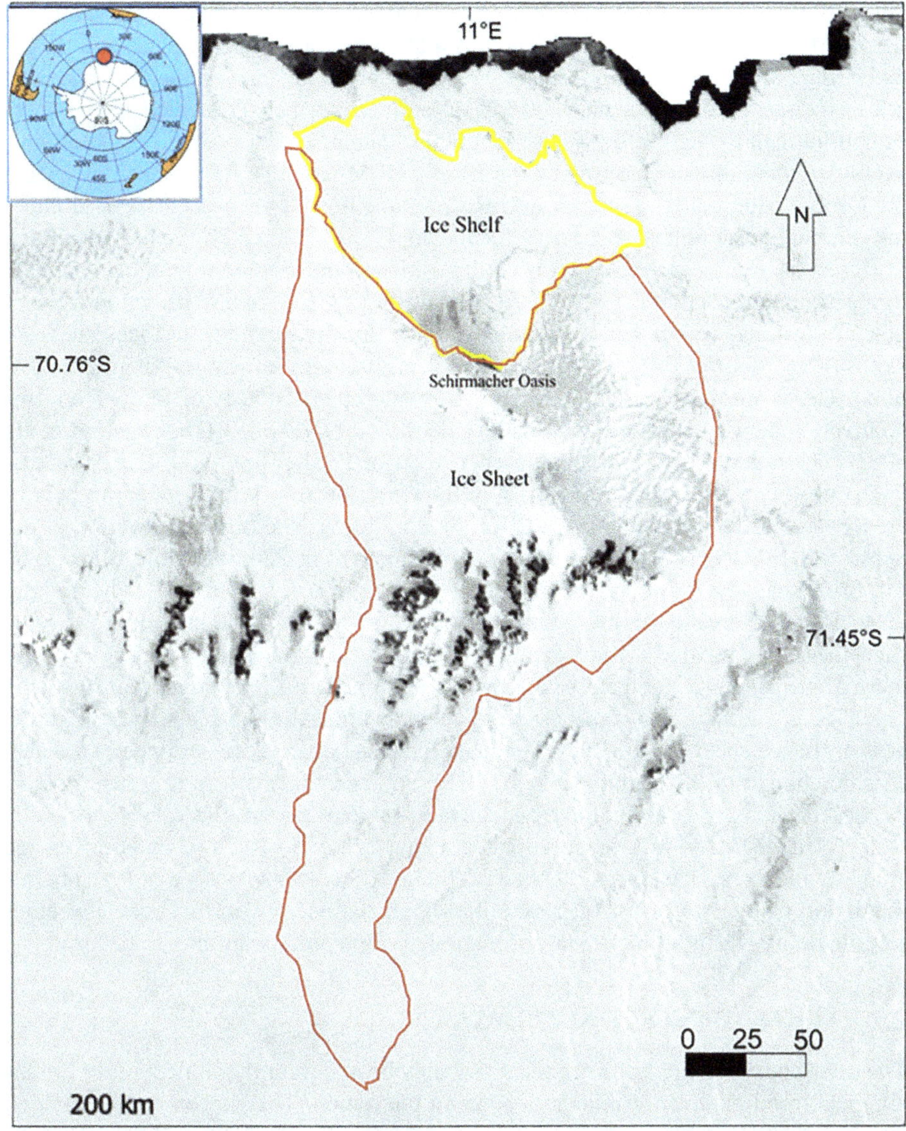

FIGURE 2.2 Map showing the location of the study area including Schirmacher Oasis in cDML, East Antarctica.

Note: Ice shelf marked in yellow polygon and ice sheet marked in a red polygon.

19.55 km and maximum width of 3.35 km covering an area of about 34 sq. km. The Schirmacher Oasis is surrounded by the Polar ice sheet to its south and the Nivlisen ice shelf to its north.

The general slope of the Polar ice sheet in the study area is towards the north. The maximum height in the study area near Wohlthat Mountains is 1597 above mean sea level (AMSL). The study area is bounded by a mountain chain towards the south and the Southern Ocean in the north. There is a 580 Km long chain of mountain ranges extending parallel to the coast from 1^0 West to 16^0 East and are visible from the open sea at a distance of more than 200 km on a clear weather day which is known as Fimbulheimen located in between Maudheim Plateau and Sør-Rondane mountains. Wohlthat Mountains consisting of the Humboldt Mountains, Peterman Ranges and the Gruber Mountains are a part of this Fimbulheimen. The area has several nunataks enclosed in a polar ice sheet. Schirmacher Oasis, situated on the northern flank of this polar ice sheet, presents varied depositional and erosional landforms related to a glacial environment. Six valleys oriented in the ENE-WSW direction, several glacial erosional (*roche moutonnée*, striations, glaciated valleys, terraces etc.) and depositional features (all types of moraines) constitute prominent physiographic features. These litho-structure controlled valleys are conspicuously different from two relatively younger tectono-glacial valleys that trend NNE-SSW (Dharwadkar et al., 2012). The latter even dissects the earlier described valleys at two different places. Field evidence suggests that the retreat of the ice sheet from Schirmacher might have been episodic (Dharwadkar et al., 2012; Swain, 2015). The highest elevation of the Schirmacher Oasis is Mount Rebristaya with a height of 228 m AMSL followed by Mt. Primetnaya (also known as Trishul) with a height of 212 m AMSL. The melting of polar ice in austral summer forms water channels in the blue ice of the polar ice sheet that flows northerly which feeds all the lakes of the area and is influenced by the thermal properties of the surrounding rocks (Swain, 2019). The stress pattern of the Polar Ice Sheet near Schirmacher Oasis depends upon the surface slope, the thickness of the ice sheet and bedrock slope and the changes in these parameters indicate climate change in the region. (Swain, 2020) and is manifested in the meteorological observations. The annual average surface air temperature has a range between –8.8 °C (in 2002) to –11.1 °C (in 2010) with an average of –10.26 °C and the average annual wind speed has varied between 7.9 and 11.3 with an average of 9.91 m/s for the period from 1995 to 2014 (Russian Research station, Novolazervaskaya weather website data). During this period, the annual average precipitation is 19.627 mm.

2.2 METHODOLOGY AND DATA USED

The cryospheric set up in the region is probably best to infer the impact of the climatological trend on each of its components in the region. This region can be divided into three major components based on the topographic and cryospheric diversity (Figure 2.3).

The first component, Polar Ice Front (PIF), is a more than 8 km long exposed ice wall at the southern margin of Schirmacher Oasis. The second component is the Polar Ice Sheet which occupies the southern part of the study area and reaches an altitude of more than 3200 m AMSL. The third component is the Nivlisen ice shelf (NIS) with an aerial extent of about 7480 sq. km towards the northern part of the study area. The methodologies opted for the observations in different regions vary according to their nature.

Assessment of annual and cumulative recession/advancement have been carried out by the observation of the PIF. To quantify the recession /advancement, various

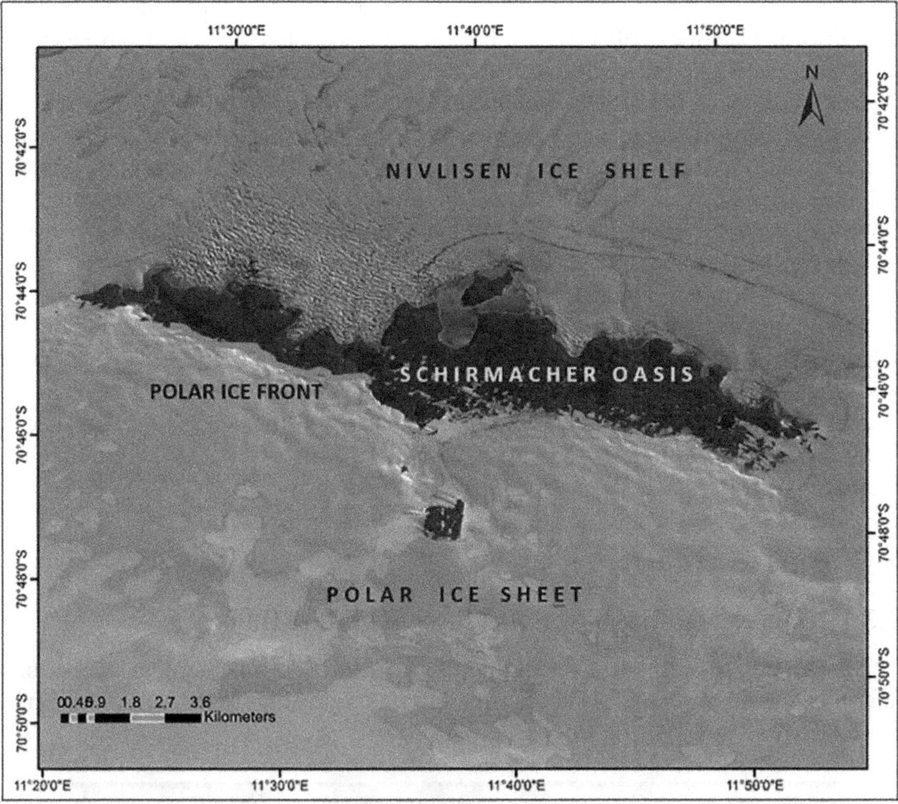

FIGURE 2.3 Satellite imagery showing the aerial view of Schirmacher Oasis with Nivlisen ice shelf to the north and Polar ice sheet to the south

Source: Courtesy: Google Earth.

methods have been adopted that get advanced with time and precision could also be achieved. With the initiation of the observation in 1983–84 plain table was used (Kaul et al., 1985; Singh and Jayaram, 1989; Ravindra and Dey, 1992; Mukerji et al., 1995; Chaturvedi et al., 1999a, b, c; 2008, 2009; Shrivastava et al., 2011; Swain and Roy, 2011; Swain and Mandal, 2011; Swain and Raghuram, 2013; Shrivastava et al., 2014 and Swain and Raghuram, 2015). Since 1996, baseline shift could be measured with the help of fixed stations over the hard ice near the margin of the PIF (Chaturvedi et al., 1999a). Both the methods were used to estimate the recession values in terms of distance and vacated area. After 2017–18, ice volume loss during summer observations could be calculated using Total Station (TS) in reflectorless (RL) mode. This method generates the point cloud of the ice surface of the wall and after processing it, change of ice loss in terms of volume could be estimated for a small part of the ice wall in a pilot study (Kumar and Habib, 2018; Chandra and Gajbhiye, 2019).

The estimation of accumulation/ablation of snow over the ice sheet and ice shelf has been achieved by installing a stake network over it. These stakes have also been replaced from time to time for continuity of the data (Swain, 2015; Kumar and Habib, 2018; Chandra and Gajbhiye, 2019). The estimation of the ice mass loss has also been calculated based on out-flux from the basin around Schirmacher Oasis (Kumar et al., 2020).

One of the most important components for the estimation of mass balance is ice thickness. The measurement of the ice sheet has been performed two times during the study period using Ground Penetrating Radar (GPR) which works on the delineation of different Earth based on the dielectric. GPR has been used with the combination of the antenna ranges from low frequency (18 MHz) to high frequency (400 MHz).

The mass balance could be calculated using the input-output method (IOM) by calculating the net volume of influx and out-flux ice across the grounding line (Shepherd et al., 2012; Schoof, 2007). This method of estimating mass balance quantifies the change between glacier mass gained through snowfall and lost by sublimation meltwater runoff and the discharge of ice into the sea. The methodology examines the change in SMB and ice dynamics separately at the scale of individual drainage basins.

2.3 OBSERVATIONS

2.3.1 Recessional Trend of Polar Ice Sheet Front (PIF)

The PIF is the ice wall at the southern margin of the Schirmacher Oasis. The total length of the observed PIF is about nine km in length (Figure 2.4). In places, part

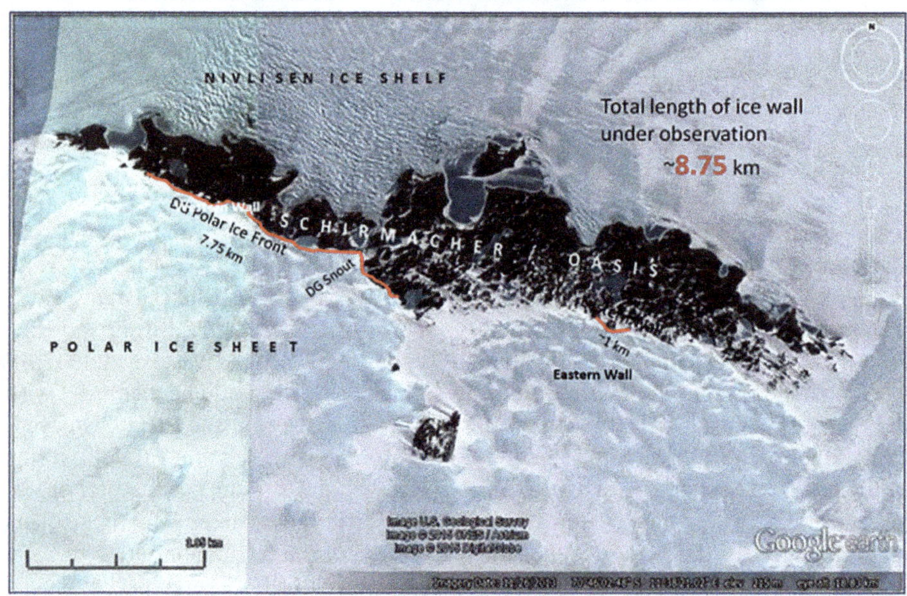

FIGURE 2.4 Observation location marked in redline at PIF to the south of Schirmacher Oasis.

of the ice wall is collapsed and forming a smooth edge, ramps structure on the ground and periglacial lake. Hence, the margin of the ice wall is not continuous for observations. The wall behind the periglacial lakes has been inferred using optical satellite data.

Due to the long stretch of the PIF and the availability of the time-series data, the PIF is divided into three parts, i.e., Eastern Wall, Dakshin Gangotri (DG) Snout and Western Wall. DG snout is a tongue-like structure hence topographically it is different from the other part of the PIF.

2.3.1.1 Recessional/Advancement Trend of the Eastern Wall

The Eastern wall is a discontinuous Polar Ice sheet front located towards the eastern part of the Schirmacher Oasis and close to the base camp at Indian Station, Maitri. Calving of ice blocks was reported in this area earlier (Swain and Chandra, 2017). For precise estimation of these changes and ice loss or gain, a Semi-Automatic Robotic Total Station was utilized (Figure 2.5) and point cloud data was generated (Figure 2.6) at definite 1-degree horizontal and vertical intervals.

During the pilot study, the TIN surface has been generated using ArcGIS and estimated the ice mass loss for the season 2017–18, which was about 65 metric tons of water equivalent.

FIGURE 2.5 Eastern Ice Wall measurement using Total Station.

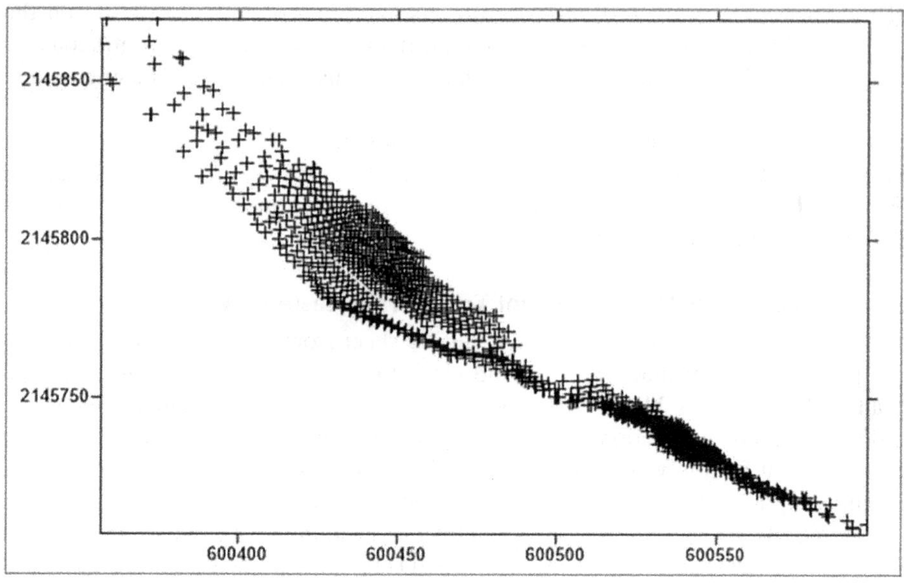

FIGURE 2.6 Plan view of point clouds of the 1326 observation points of the ice wall.

2.3.1.2 Recessional/Advancement Trend of DG Snout

Annual monitoring of the DG glacier snout is carried out since 1982 (Figure 2.7) and in recent years, it is monitored and mapped by Total Station. The distance between the fixed observation point and ice wall is interpreted in terms of recession or advancement.

During 2017–18, the average annual recession of the DG snout was found to be 25 cm. The recorded recessional pattern is the net difference between the forward displacement of the snout and the melting and breaking of the snout. If the DG snout forward displacement is more than melting and breaking, data will show advancement in the snout and vice versa. To delineate both the factor, a stake farm is installed at 1360 m south of the snout over the ice sheet (Figure 2.8) which shows the annual velocity of the surface of the ice sheet over the DG snout to be 9.9 m during 2018–19 (Kumar and Habib, 2018) (Table 2.1).

Hence the net recessional value will be 0.25 m (actual measured value) + 9.99 m (advancement due to movement of the snout) = 10.24 m for the DG snout (Kumar and Zahid, 2018).

The comparison of data collected over the years from the fixed points around the DG snout shows continuous variable fluctuation in the recessional pattern of the DG snout (Figure 2.9). The net annual recession of the DG snout during 2018–19 is 25 cm and the cumulative recession since 1996 is 13.58 m (Figure 2.10).

During 2017–18, the highest recession value is recorded at point no. 14A, which is located at the western part of the DG glacier snout and measured to be 1.41 m. The

FIGURE 2.7 Recent view of DG glacier snout (January 27, 2018). Position of DG snout in the inset image.

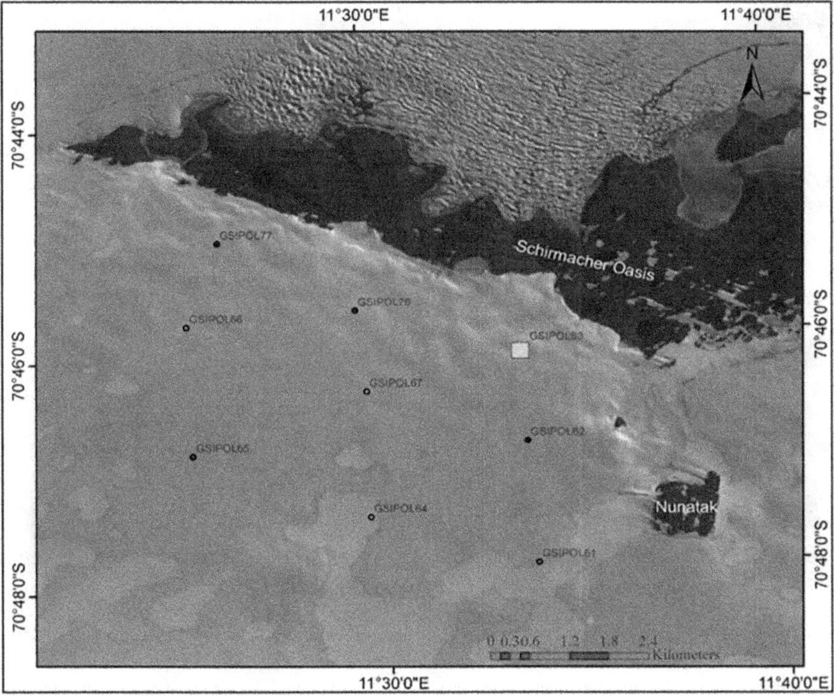

FIGURE 2.8 Location of stake farm referred by GSIPOL 63 and stakes installed in the south of ice sheet front over the ice sheet.

TABLE 2.1
Displacement Data from the Stake Farm Fixed South of DG Snout for 2018–19

Stake	Latitude (S)			Longitude (E)			Elevation	Displacement
Name	*dd*	*mm*	*ss.ss*	*dd*	*mm*	*ss.ss*	*m*	*M*
GSIPOL63 Dec. 2017	70° 46'16.89082"			11°33'51.90299"			310.46	9.989
GSIPOL63 Dec. 2018	70°46'16.74370"			11°33'51.03750"			307.85	

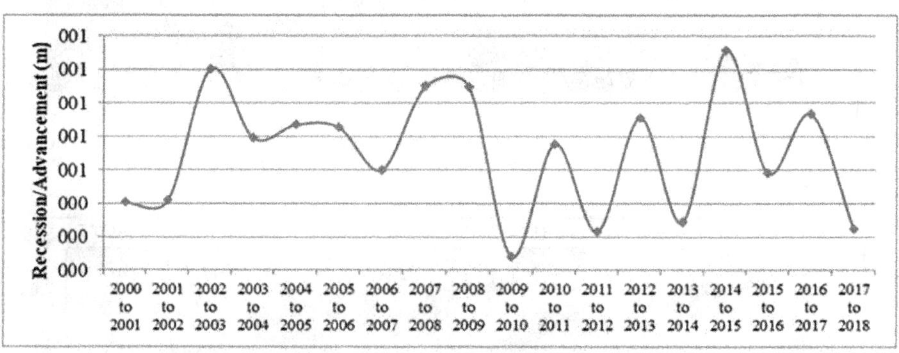

FIGURE 2.9 Annual average recessional pattern in DG snout.

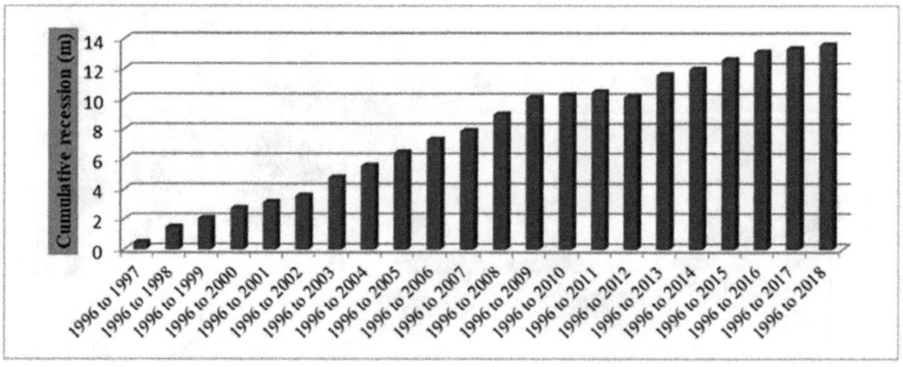

FIGURE 2.10 Cumulative recession of DG snout since 1996.

highest advancement is shown at point no. 15 located towards the western side of the DG glacier snout and measured to be 1.15 m. The shrinking continues from all sides and the eastern arm of the DG glacier snout shrinks towards the west and the northern arm of the snout shrinks towards the south.

2.3.1.3 Monitoring of Western Wall

The western wall extends from the DG snout to the eastern extremity of Schirmacher and a vertical ice escarpment (Figure 2.11). However, a snow ramp has covered the glacial at a few locations. At such locations, the front is assumed at the break in slope along the projected direction from the fixed point towards the glacier wall (Swain and Chandra, 2017; Kumar and Habib, 2018; Chandra and Gajbhiye, 2019).

The recession of the western wall was measured during the peak of austral summer. The average annual recession of the western wall during 2012–13 is 10.45 m (Table 2.2). Such a high rate of recession of more than 10 m is seen after more than a decade; the earlier maximum being 10.96 m for the period January 2002 to January 2003 was followed by another higher recession of 5.28 m for the period January 2003 to January 2004. Recessions at different observation points vary in magnitude and a higher recession was observed towards the west. During 2016–17, the extreme westernmost observation points (nos. XX2 and XX3) showed a cumulative recession of 36.92 m and 28.81 m respectively (Swain and Chandra, 2017). In recent years, a high annual average recession of 2.79 m was observed the period January 2018 to January 2019 (Chandra and Gajbhiye, 2019). From 2001 to 2019, the western wall has shown a cumulative recession of 42.08 m. (Chandra and Gajbhiye, 2019) (Table 2.2).

2.3.2 Snow Accumulation/Ablation Studies

The area selected for glaciological studies has been divided into three main parts (Figure 2.12) based on geomorphic characteristics (Kumar and Habib, 2018; Chandra and Gajbhiye, 2019). The first part of the study area is part of the Polar Ice Sheet to the south, the second part is part of the ice shelf to the north and the third part is sandwiched in between these two parts at the margin of the ice sheet. Observations

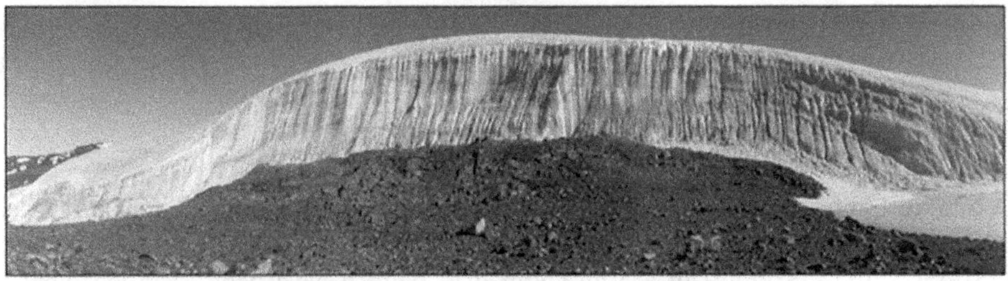

FIGURE 2.11 Part of PIF (western wall).

TABLE 2.2
Average Annual and Cumulative Retreat (+)/Advance (-) of Western PIF since 2001 (m)

Year	*Annual (m) recession (+)/advancement (-)*	*Duration*	*Cumulative recession (m)*
2001–2	2.22	*2001–2*	2.22
2002–3	10.96	*2001–3*	13.18
2003–4	5.28	*2001–4*	18.46
2004–5	1.16	*2001–5*	19.62
2005–6	0.7	*2001–6*	20.32
2006–7	1.84	*2001–7*	22.16
2007–8	(-) 1.28	*2001–8*	20.88
2008–9	0.66	*2001–9*	21.54
2009–10	0.27	*2001–10*	21.81
2010–11	1.19	*2001–11*	23.00
2011–12	(-) 3.02	*2001–12*	19.98
2012–13	10.45	*2001–13*	30.43
2013–14	(-) 0.37	*2001–14*	30.06
2014–15	2.7	*2001–15*	32.76
2015–16	0.78	*2001–16*	33.54
2016–17	3.39	*2001–17*	36.93
2017–18	2.36	*2001–18*	39.29
2018–19	2.79	*2001–19*	42.08

made during the austral summer period of November 2018 to February 2019 are summarized below:

1. One set of measurements of snow accumulation/ablation over the ice sheet between Schirmacher Oasis and Wohlthat Mountain (high Polar ice sheet, HPIS).
2. One set of measurements of snow accumulation/ablation on the ice sheet in Schirmacher Oasis-Polar ice sheet margin (SOPIM) or Impeded Polar Ice Sheet (IPIS).
3. One set of measurements of snow accumulation/ablation on the ice shelf (Nivlisen ice shelf).

2.3.2.1 Snow Accumulation/Ablation Studies on High Polar Ice Sheet (HPIS)

A total of 103 individual stakes were installed in the area between Schirmacher Oasis and to Wohlthat Mountains. Three networks of 25 stakes at each pole GSIPOL3, GSIPOL5 and GSIPOL9 in a 50mX50m grid pattern have been installed during 2016–17 and 2017–18. Accumulation/ablation observations in this area have been measured for a particular period and calculated annually. The annual average is calculated by using the following equations:

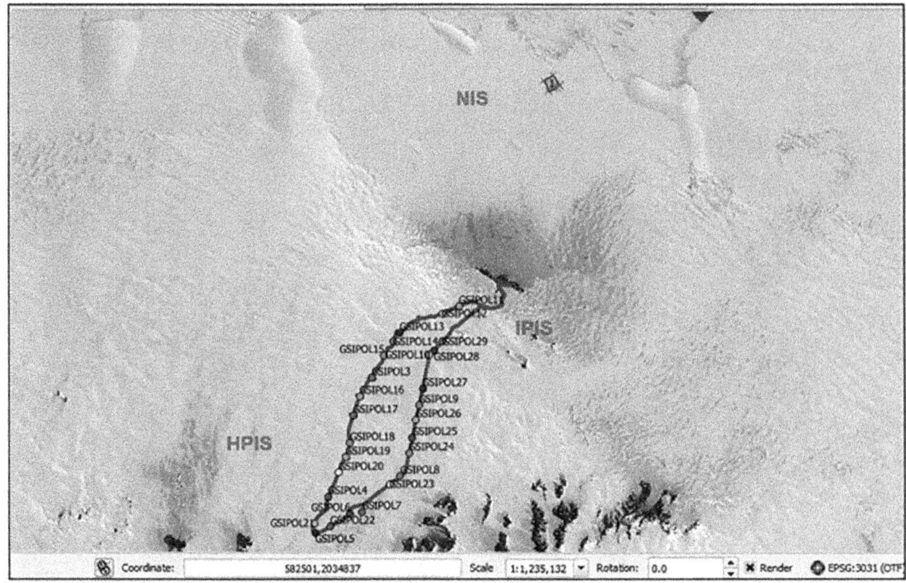

FIGURE 2.12 Location of individual stakes and stakes network on HPIS (Base Image – OLI).

Equation 1: Calculation of annual average accumulation/ablation.

(Exposed height of the stake in a particular year–Exposed height of the stake in the previous year accumulation/ablation = -- -- X 365 (Days in year)

Days between the observation dates

Data shows an average annual accumulation of 0.985 m except at pole GSIPOL7, which shows 0.07 m ablation (Figure 2.13) and (Table 2.3). The interior part close to the Wohlthat Mountains and others parts that are in the shade of nunatak and hills (Small ellipses) shows less accumulation as compared to the northern or open part (Bigger ellipse) of the HPIS area (Figure 2.14).

2.3.2.2 Snow Accumulation/Ablation Studies on Impeded Polar Ice Sheet (IPIS)

Part of the study area which is sandwiched between the Polar Ice Sheet and Nivlisen ice shelf has been christened as Impeded Polar Ice Sheet (IPIS). Field observations were taken on IPIS region, located south of the Schirmacher Oasis. To study the accumulation/ablation in the area, 24 individual stakes are installed, besides two networks, one of 25 stakes (at GSIPOL63) over the DG snout area (50 m x 50 m grid pattern) during 2016–17 (Swain and Chandra, 2017) and other of 16 stakes, 4 km south of ALCI Runway (50 m x 50 m grid pattern) during 2010–11 (Swain and Mandal, 2011) on Polar Ice sheet. The study of this region is important due to anthropogenic activity associated with two permanent scientific stations in Schirmacher

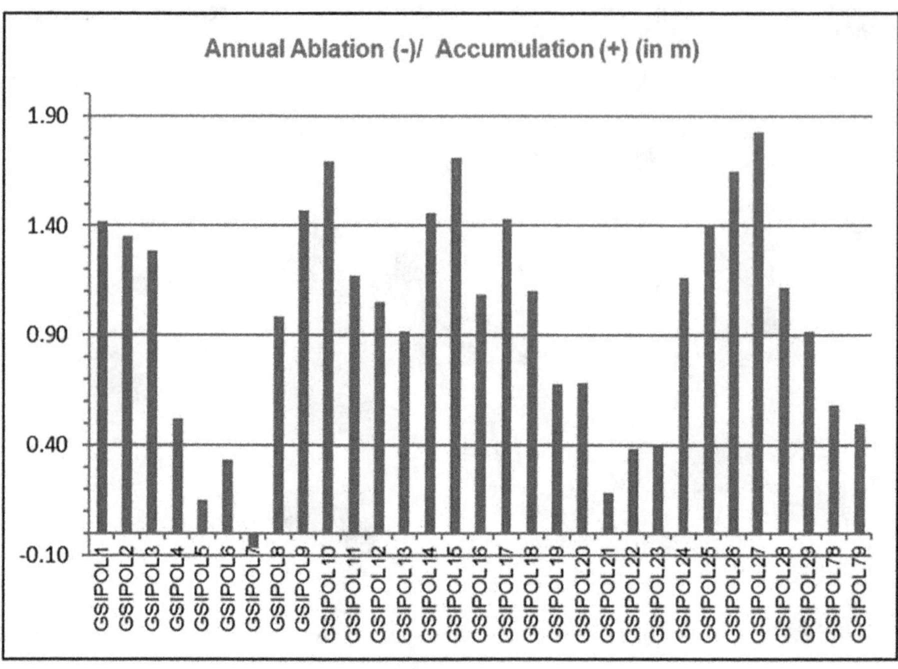

FIGURE 2.13 Annual ablation (-)/accumulation (+) of stakes installed on HPIS.

TABLE 2.3
Accumulation (+)/Ablation (-) of Individual Stakes Installed on HPIS

S. No.	Stake No. Name	Latitude South	Longitude East	Abl. (-)/Acc. (+) Annual m
1	GSIPOL1	70°55'18.42"	10°57'7.80"	1.422
2	GSIPOL2	71°5'33.16"	10°38'19.21"	1.355
3	GSIPOL3	71°7'44.20"	10°35'51.40"	1.284
4	GSIPOL4	71°30'54.67"	10°23'6.99"	0.523
5	GSIPOL5	71°37'55.14"	10°19'18.13"	0.151
6	GSIPOL6	71°33'14.58"	10°37'51.94"	0.332
7	GSIPOL7	71°31'2.15"	10°45'28.76"	-0.070
8	GSIPOL8	71°24'37.85"	11°3'6.25"	0.985
9	GSIPOL9	71°11'1.49"	11°6'11.77"	1.468
10	GSIPOL10	71°1'29.50"	11°6'52.66"	1.690
11	GSIPOL11	70°51'34.90"	11°18'0.20"	1.173
12	GSIPOL12	70°53'30.80"	11°9'24.64"	1.052
13	GSIPOL13	70°58'33.13"	10°47'16.22"	0.920
14	GSIPOL14	71°0'22.53"	10°44'43.00"	1.456
15	GSIPOL15	71°3'10.28"	10°40'52.08"	1.709
16	GSIPOL16	71°11'31.50"	10°30'51.20"	1.086
17	GSIPOL17	71°15'12.90"	10°29'11.50"	1.428

S. No.	Stake No. Name	Latitude South	Longitude East	Abl. (-)/Acc. (+) Annual m
18	GSIPOL18	71°20'22.31"	10°30'15.97"	1.099
19	GSIPOL19	71°23'6.46"	10°29'9.53"	0.676
20	GSIPOL20	71°26'9.03"	10°26'48.40"	0.686
21	GSIPOL21	71°36'13.47"	10°17'39.57"	0.181
22	GSIPOL22	71°36'17.39"	10°27'31.10"	0.382
23	GSIPOL23	71°26'29.90"	10°57'46.56"	0.402
24	GSIPOL24	71°20'5.75"	11°5'41.18"	1.163
25	GSIPOL25	71°17'12.64"	11°5'55.21"	1.398
26	GSIPOL26	71°13'56.00"	11°6'3.61"	1.649
27	GSIPOL27	71°7'54.56"	11°6'21.29"	1.825
28	GSIPOL28	71°0'32.77"	11°8'46.81"	1.120
29	GSIPOL29	70°58'37.25"	11°12'38.36"	0.920
30	GSIPOL78	70°53'56.79"	11°22'29.02"	0.582
31	GSIPOL79	70°50'47.97"	11°30'54.09"	0.492
			Average	**0.985**

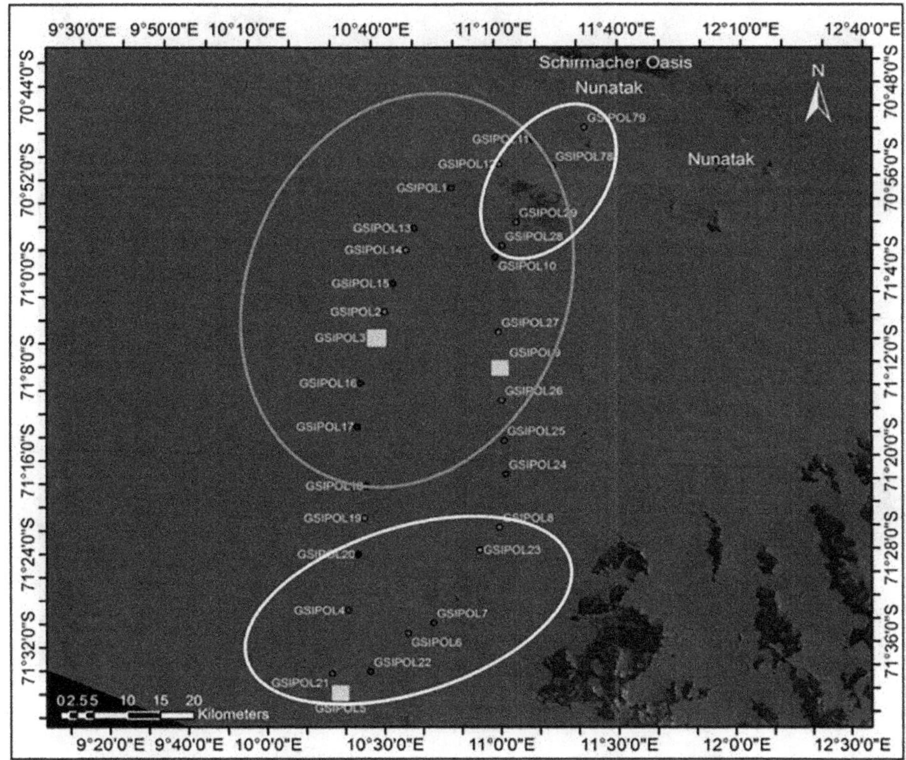

FIGURE 2.14 Map of study area showing location of the three regions.

Oasis. The anthropogenic activities include flight operations, tourism, and scientific work. Green square boxes show the locations of the stakes network (Figure 2.15). The long-term snow accumulation and ablation studies in this area around Schirmacher Oasis in central Dronning Maud Land have shown temporal as well as spatial fluctuations (Table 2.4). Stake networks on IPIS during 2018–19 show an average annual accumulation of 0.24 m whereas the network of 25 stakes over DG snout PIF (at GSIPOL63) and 16 stakes near ALCI Runway show an average annual ablation of 0.34 m and 0.46 m respectively (Figures 2.16, 2.17, & 2.18) and (Tables 2.4, 2.5, & 2.6). Snow accumulation on Antarctic ice sheet is a combined result of snow precipitation and snow drifting. Contribution to snow accumulation through precipitation is about 30–35% whereas contribution to snow accumulation through snow drifting is 65–70%. Since these two stakes networks are in the shade of nunataks or mound/hillocks or lying in the valley portion, snow accumulation received through drifting is less as compared to other open areas on IPIS. Thus, spatial variation of snow accumulation/ablation pattern even in this small area could be attributed to morphological characteristics of the region.

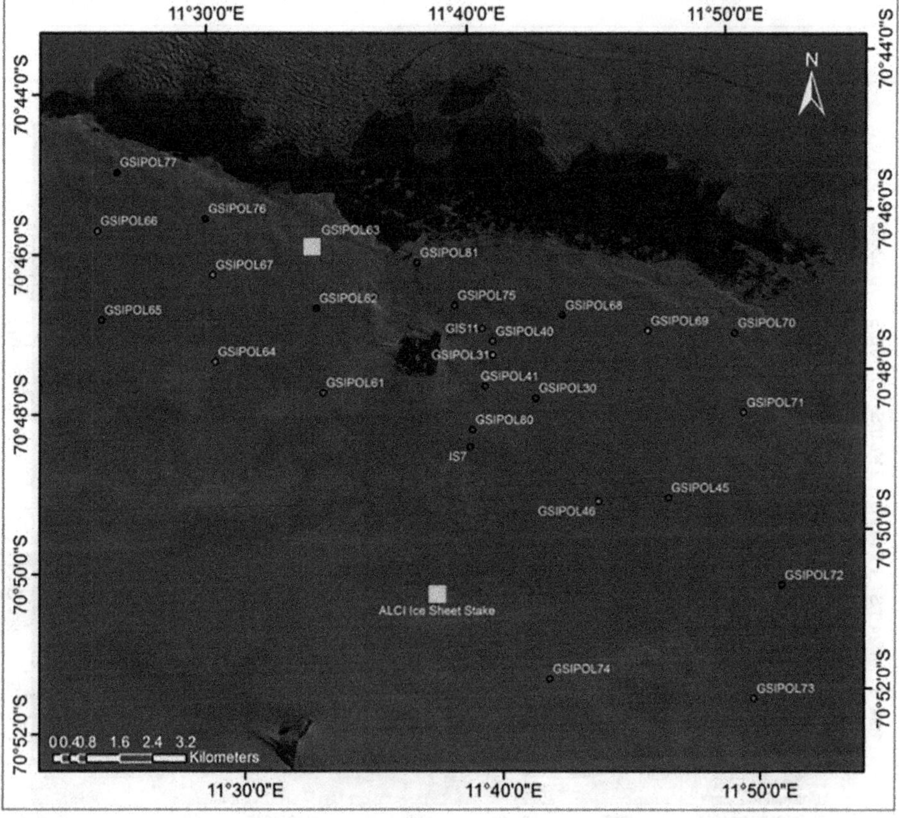

FIGURE 2.15 Location of individual stakes and stakes network (green square box) on IPIS region.

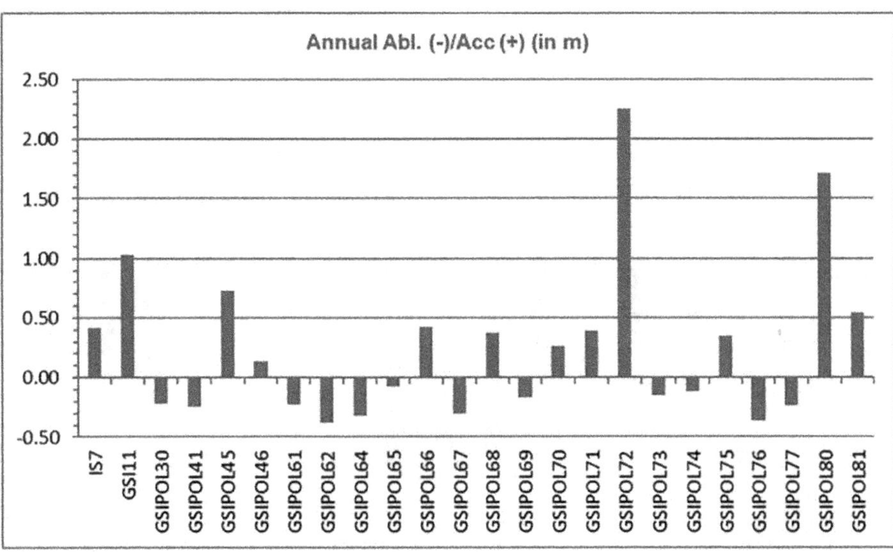

FIGURE 2.16 Annual accumulation (+)/ablation (-) pattern of individual stakes fixed in IPIS area.

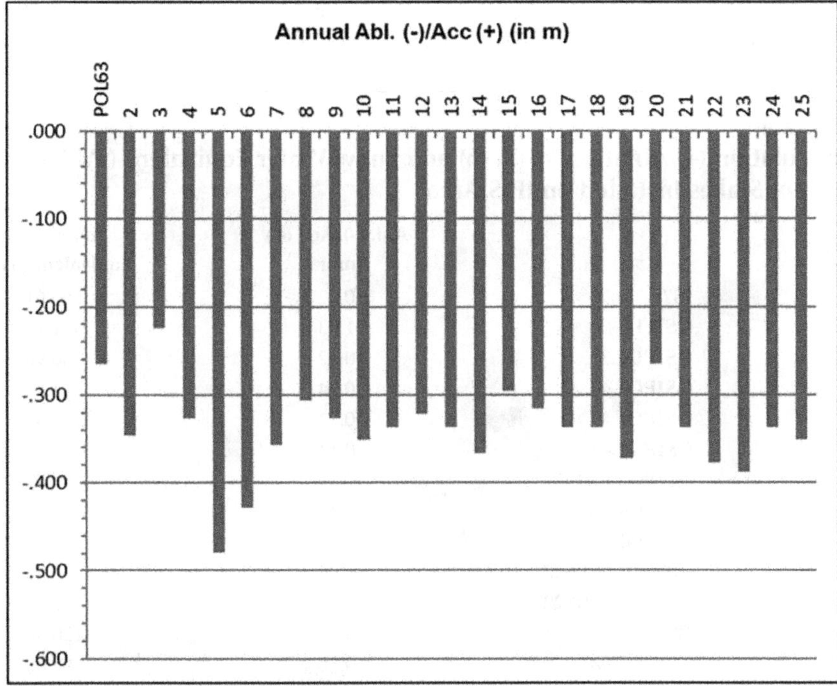

FIGURE 2.17 Annual accumulation (+)/ablation (-) pattern of stakes network GSIPOL63 on IPIS area, Dec. 17 to Dec. 18

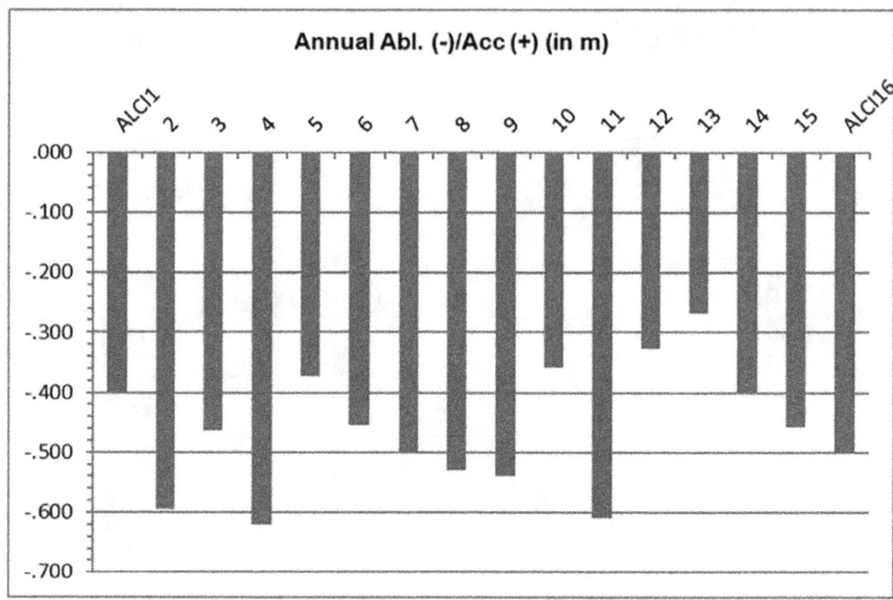

FIGURE 2.18 Annual accumulation (+)/ablation (-) of stakes network at ALCI Runway, Dec. 17 to Dec. 18

TABLE 2.4
Accumulation (+)/Ablation (-) (in m) and Snow Water Equivalent (SWE in kg/m²) of Stakes Installed on IPIS Area

S. No	Stake No.	Abl. (-)/Acc (+) Annual	Snow Water Equivalent (SWE)
1	IS7	0.42	166.61
2	GSI11	1.03	411.17
3	GSIPOL30	-0.22	-86.96
4	GSIPOL41	-0.24	-96.75
5	GSIPOL45	0.72	289.48
6	GSIPOL46	0.14	54.54
7	GSIPOL61/IS29	-0.22	-89.72
8	GSIPOL62	-0.38	-151.32
9	GSIPOL64	-0.32	-127.13
10	GSIPOL65/IS23	-0.08	-30.76
11	GSIPOL66/IS22	0.42	167.68
12	GSIPOL67	-0.30	-120.98
13	GSIPOL68	0.38	150.60

S. No	Stake No.	Abl. (-)/Acc (+) Annual	Snow Water Equivalent (SWE)
14	GSIPOL69	-0.17	-66.93
15	GSIPOL70	0.26	104.58
16	GSIPOL71	0.39	155.23
17	GSIPOL72	2.25	900.40
18	GSIPOL73	-0.15	-58.90
19	GSIPOL74	-0.12	-46.28
20	GSIPOL75	0.35	140.30
21	GSIPOL76	-0.36	-143.14
22	GSIPOL77	-0.24	-94.06
23	GSIPOL80	1.71	683.96
24	GSIPOL81	0.54	216.62
	Mean	0.24	96.83

TABLE 2.5
Accumulation (+)/Ablation (-) (in m) and Snow Water Equivalent (SWE in kg/m²) of Stakes Network (at ALCI Runway) Installed on IPIS Area

S. No.	Abl. (-)/Acc (+) Annual (in m) *365 Days	Mass of snow =d*Area* thickness kg	Snow Water Equivalent = Snow density* thickness kg/m²
1	–0.398	–395000	–158
2	–0.595	–590000	–236
3	–0.464	–460000	–184
4	–0.620	–615000	–246
5	–0.373	–370000	–148
6	–0.454	–450000	–180
7	–0.499	–495000	–198
8	–0.529	–525000	–210
9	–0.539	–535000	–214
10	–0.358	–355000	–142
11	–0.610	–605000	–242
12	–0.328	–325000	–130
13	–0.267	–265000	–106
14	–0.398	–395000	–158
15	–0.459	–455000	–182
16	–0.499	–495000	–198
	–0.462		**–183.25**

TABLE 2.6
Accumulation (+)/Ablation (-) (in m) and Snow Water Equivalent (SWE in kg/m²) of Stakes Network over DGPIF (GSIPOL63) Installed on IPIS Area

S. No.	Stake No.	Abl. (-)/Acc (+) Annual (in m)	Snow Water Equivalent
1	**GSI POL63**	–0.265	–104
2	**2**	–0.347	–136
3	**3**	–0.224	–88
4	**4**	–0.326	–128
5	**5**	–0.479	–188
6	**6**	–0.428	–168
7	**7**	–0.357	–140
8	**8**	–0.306	–120
9	**9**	–0.326	–128
10	**10**	–0.352	–138
11	**11**	–0.336	–132
12	**12**	–0.321	–126
13	**13**	–0.336	–132
14	**14**	–0.367	–144
15	**15**	–0.296	–116
16	**16**	–0.316	–124
17	**17**	–0.336	–132
18	**18**	–0.336	–132
19	**19**	–0.372	–146
20	**20**	–0.265	–104
21	**21**	–0.336	–132
22	**22**	–0.377	–148
23	**23**	–0.387	–152
24	**24**	–0.336	–132
25	**25**	–0.352	–138
	Mean	**–0.339**	**–133.12**

The annual ablation/accumulation pattern of the ALCI stake network farm shows three years of continuous accumulation from 2011 to 2013. The next three year's data show continuous ablation from 2014 to 2016 (Figure 2.19). The data show temporal fluctuation in annual accumulation/ablation with a net cumulative of 0.28 m accumulation since 2011 (Figure 2.20).

2.3.2.3 Snow Accumulation/Ablation Studies on Ice Shelf (Nivlisen Ice Shelf)

During 38th Indian Scientific Expedition, a scientific convoy to Nivlisen ice shelf was planned to continue data acquisition on ablation/accumulation and dynamics of the ice shelf, from the installed stakes network. In the area around abandoned Dakshin Gangotri station, one network consisting of 20 stakes was installed. During 2019–20, four individual bamboo stakes (GSIN1, GSIN2, GSIN3 and GSIN4) with an interval of 14 km distance (approx.) in the E-W direction were installed on the Nivlisen Ice shelf (Figure 2.21). Initial data acquisition was done with the help of DGPS and measuring tape. This data will serve as base data for future observation (Figure 2.22).

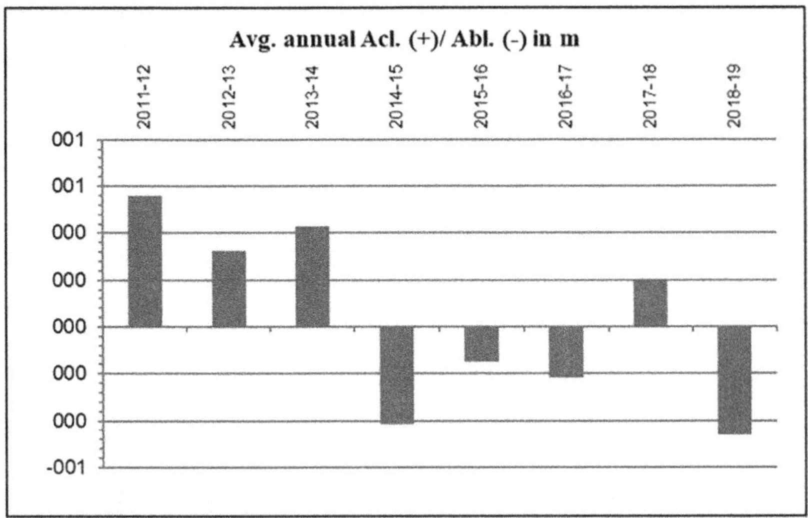

FIGURE 2.19 Annual accumulation (+)/ablation (-) pattern of stakes network at ALCI Runway on IPIS area since 2011.

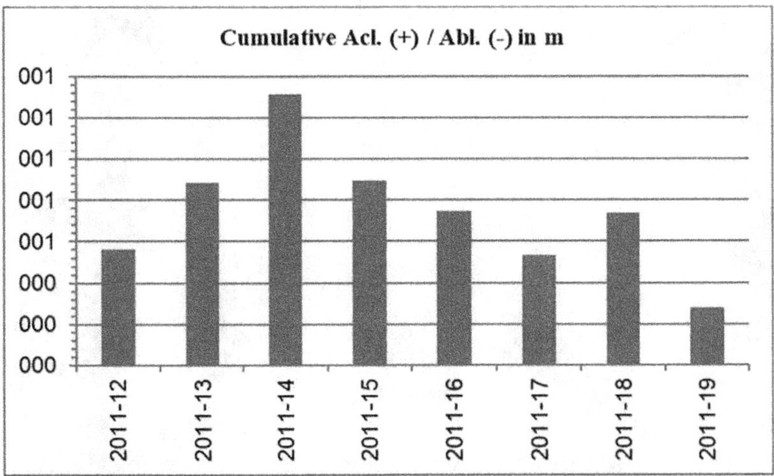

FIGURE 2.20 Cumulative accumulation (+)/ablation (-) of stakes network at ALCI Runway on IPIS area since 2011.

2.3.4 The Ice Sheet Thickness Study

To estimate the mass of ice preserved on the cDML part of Antarctica, a volumetric calculation is required. The thickness of the ice sheet was estimated for a part of the study area on the Polar Ice Sheet using the Ground Penetrating Radar (GPR) during 2016–17 (Swain and Chandra, 2017) and 2017–18 (Kumar and Habib, 2018) by using multiple low frequency (MLF) antennas with a frequency ranging from

FIGURE 2.21 Installation of stakes and initial data acquisition by DGPS on Nivlisen ice shelf.

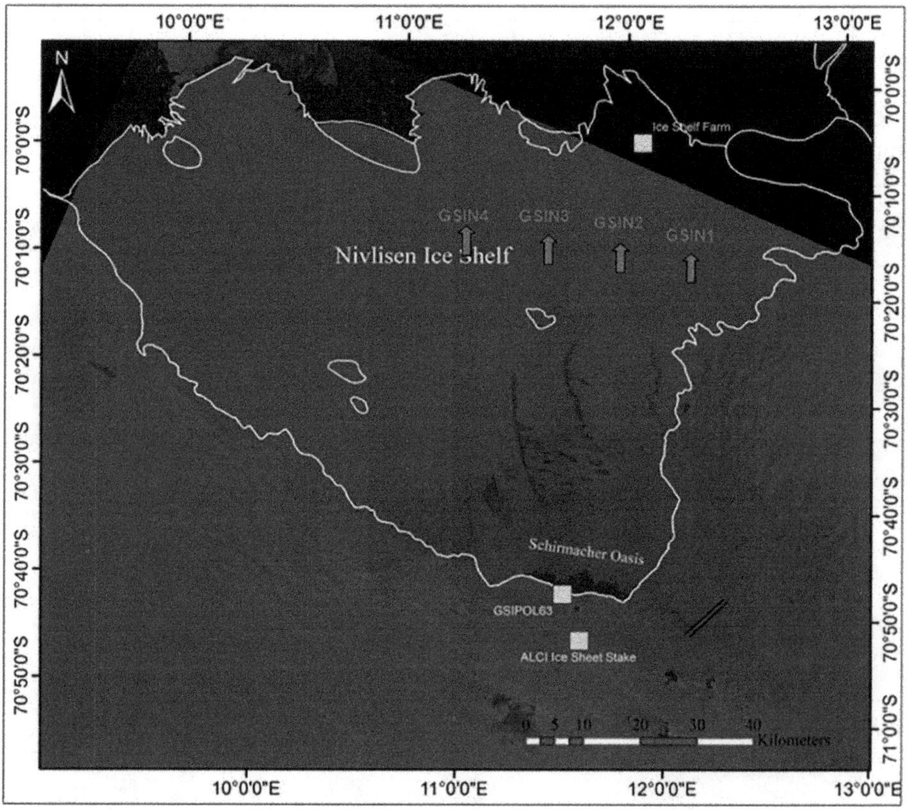

FIGURE 2.22 Location of four individual stakes installed over Nivlisen ice shelf.

16 MHz to 80 MHz. During 2018–19, a GPR survey was conducted in the interior parts of HPIS, between Schirmacher Oasis to Wohlthat Mountains, using 25 MHz and 400 MHz antenna. The average thickness of the ice sheet on HPIS is more than 500 m, which includes snow thickness of more than 50 m to 100 m in parts of the

area. Most of the route taken for the GPR survey was devoid of any barriers. The GPR survey was carried out around Vettiah nunatak for a length of 3.4 line-km in a discrete manner (Figure 2.23 & Figure 2.24) by a 400 MHz antenna showing bedrock-ice sheet interface at a depth of 90 m maximum (Figure 2.25). Firn layer

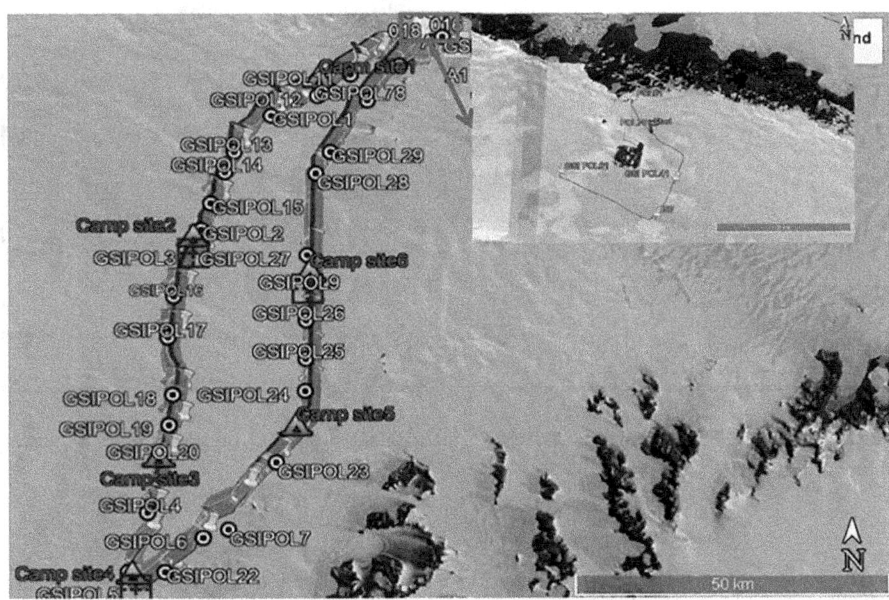

FIGURE 2.23 GPR traverse path on HPIS and on IPIS (in the inset).

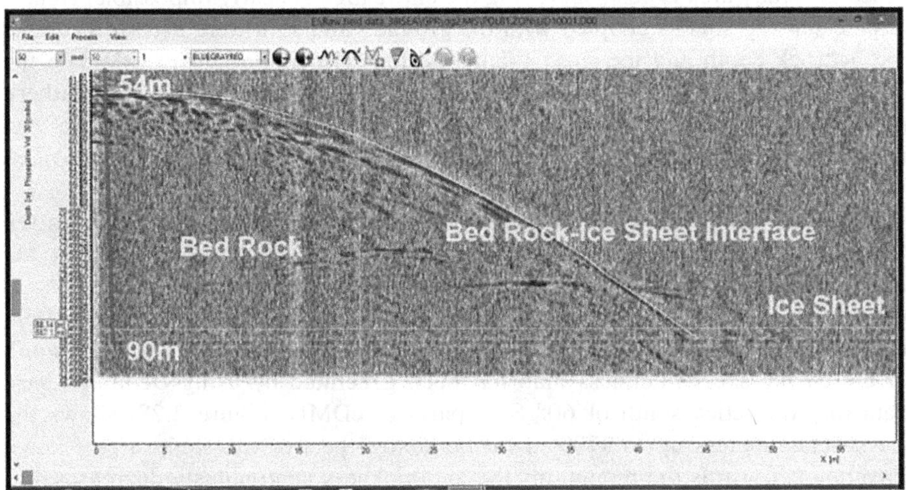

FIGURE 2.24 GPR radargram showing subsurface topography in IPIS region by using antenna of 400 MHz.

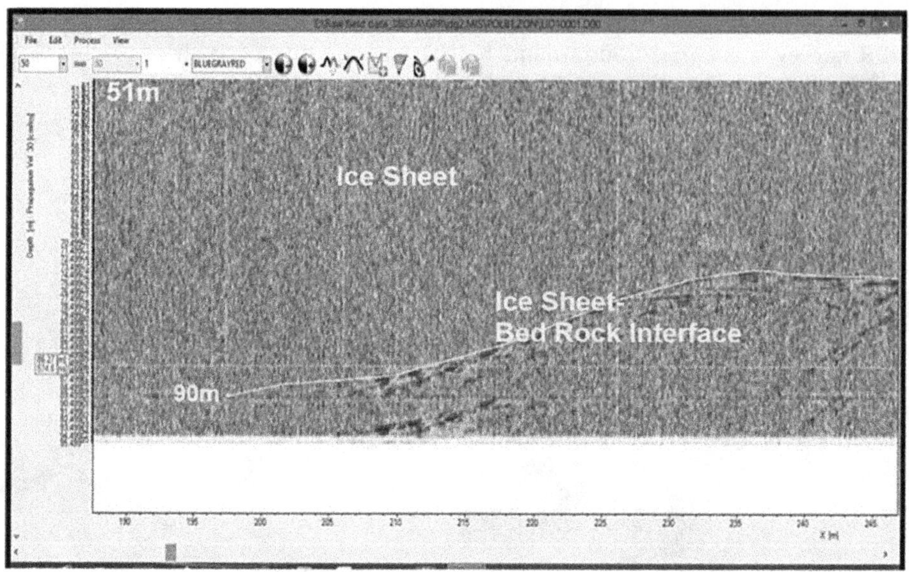

FIGURE 2.25 GPR radargram showing subsurface topography in IPIS region by using antenna of 400 MHz.

of 0–1 m depth and different types of layering into the ice sheet were also identified in this region. Hidden mound/hillocks and entrapped boulders/rock mass were also found below the ice sheet (Figure 2.26). The area is also characterized by numerous deep fractures/voids and channels filled with water into the ice sheet. Since the study area is vast, surveying the entire area is nearly impossible. Hence, remote sensing studies coupled with the ground validation were used to estimate the bedrock depth and ice sheet thickness. It is also important to calculate the in-flux and out-flux of ice mass. The in-gate is the higher reaches in the southern part of the study area and the out-gate is the boundary between the ice sheet and ice shelf towards the north, i.e., the grounding line where the ice sheet or ice mass leaves the mainland of the continent.

The grounding line (at Nivlisen ice shelf) data have been accessed from the Scientific Committee on Antarctic Research (SCAR) database (Rignot et al., 2011). The hinge line is the corresponding surface line of the grounding line at the bottom of the ice sheet (Figure 2.27). Bedmap2 describes surface elevation, ice-thickness, and the sea floor and subglacial bed elevation of the Antarctic south of 60° S. Ice sheet thickness map has been generated by using Bedmap2 base data for Antarctica south of 60° S in parts of cDML (Figure 2.28) shows the ice sheet thickness up to 1700 m in the central part of the study area (shown in yellow). Towards the mountains the ice thickness continuously decreases, but not uniformly. Some parts within the study area show an abrupt change in ice

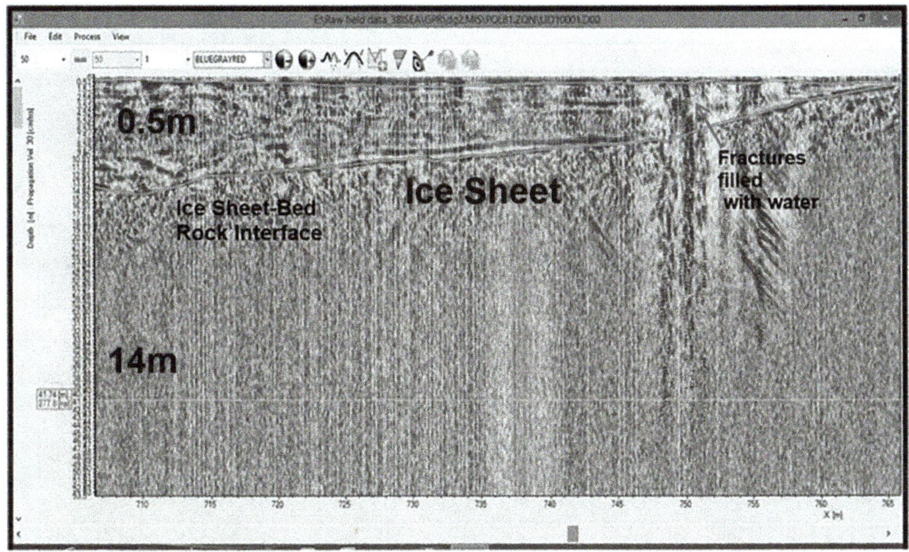

FIGURE 2.26 GPR radargram showing subsurface topography in IPIS region by using antenna of 400 MHz.

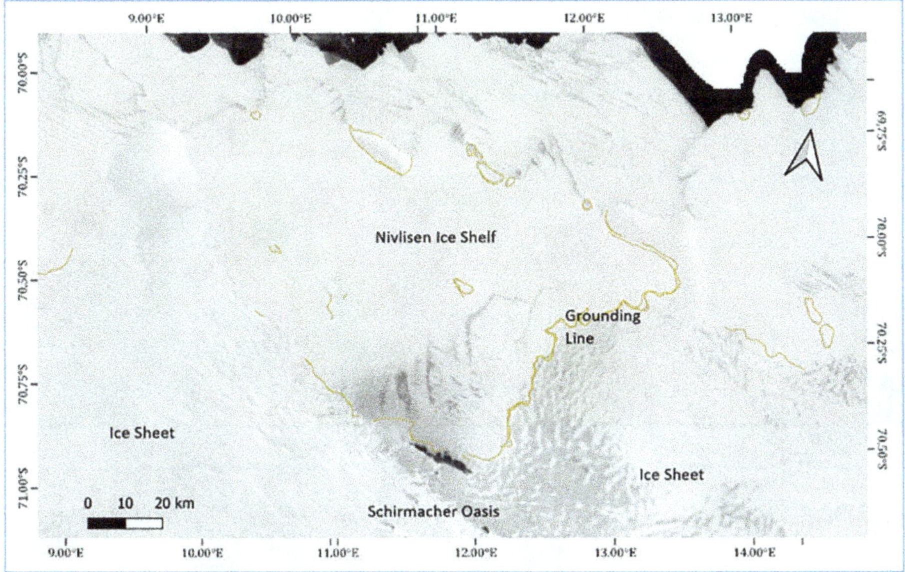

FIGURE 2.27 Map showing grounding line (brown line) in the study area.

Source: Rignot et al., 2011.

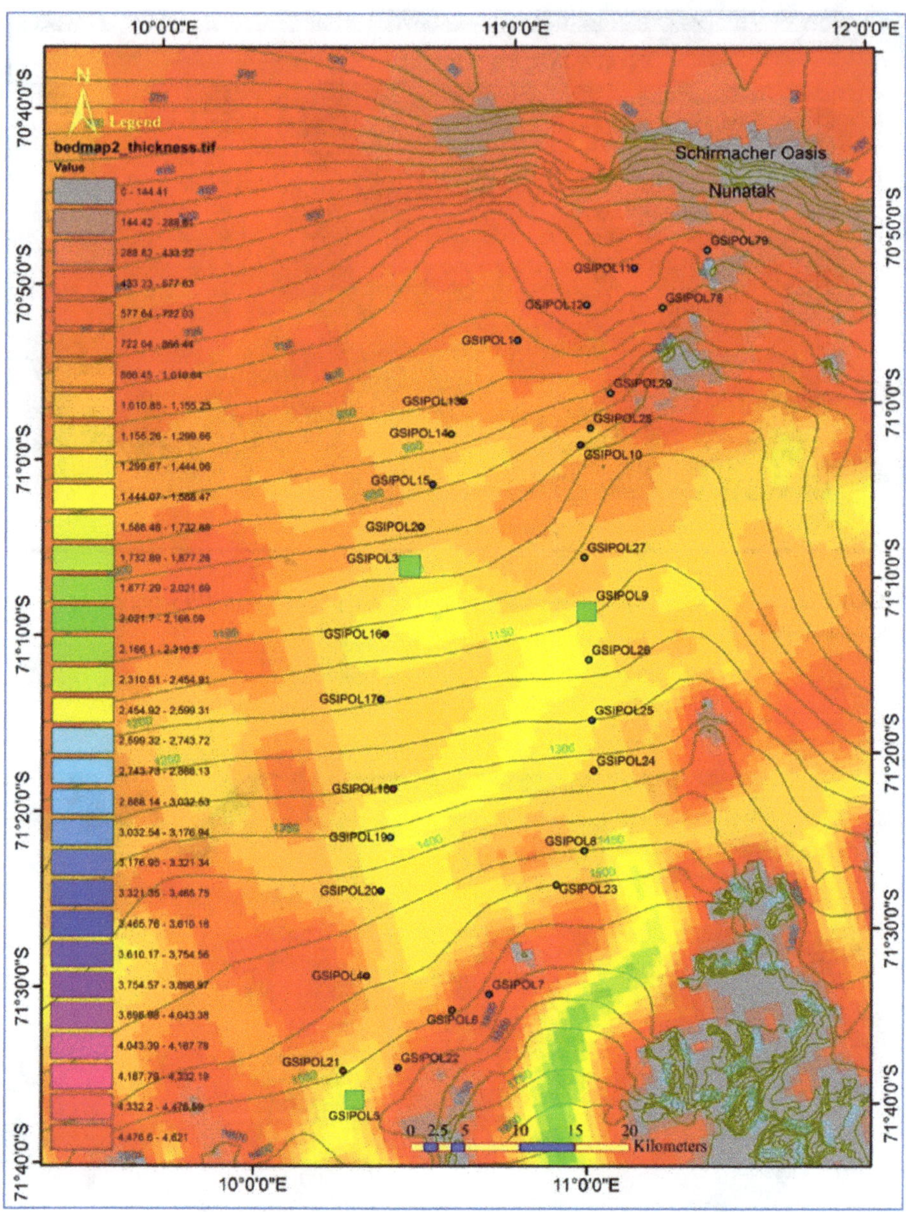

FIGURE 2.28 Thickness of the ice sheet from Bedmap2 data by British Antarctic Survey.

thickness. In the southeastern part of the study area, the ice thickness is found to be an order of 2100 m to 2300 m (green elongated patches) that forms a part of Somovken glacier, which is the fastest flowing ice stream in the study area. The topographic contour shows the elevation range between 1550 m to 1850 m for elevated Somovken Glacier from the surface.

2.4 MASS BALANCE ESTIMATION AT GROUNDING LINE (FLUX GATE)

This estimation of the mass balance is attempted to derive using the input-Output method (IOM) (Shepherd et al., 2012). Using this method, the mass loss has been estimated across the fluxgate at the grounding line in the cDML area (Figures 2.29a and 2.29b). The

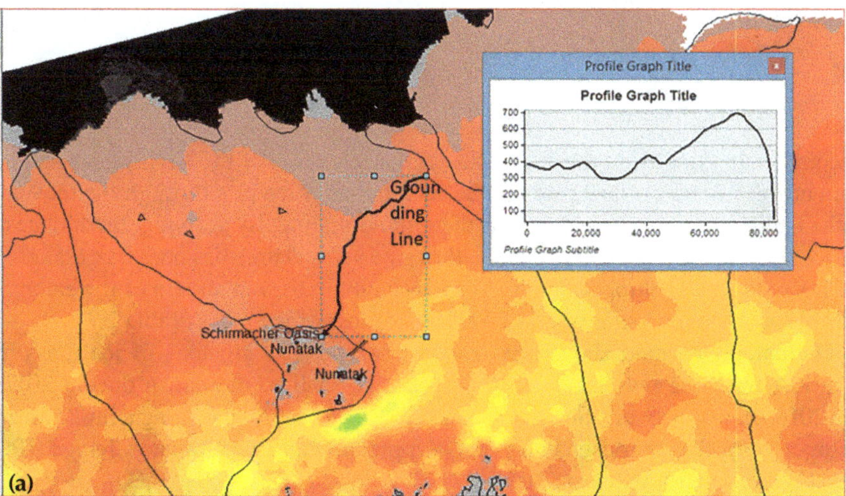

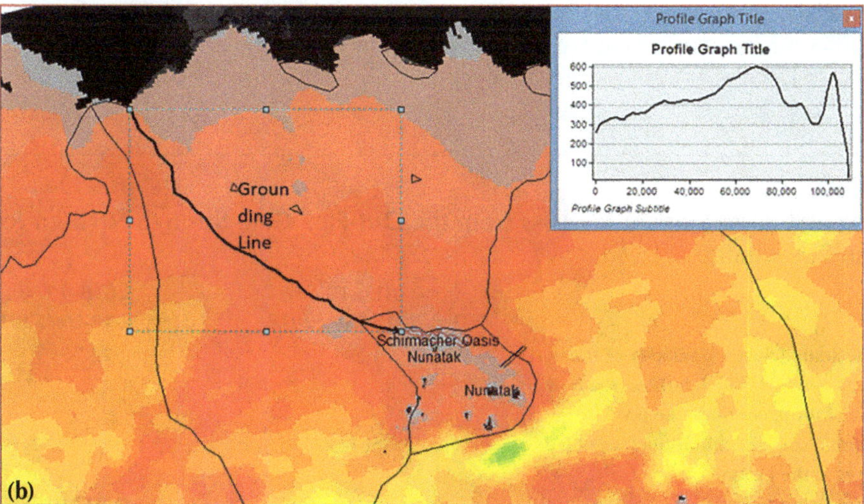

FIGURE 2.29 (a) Map showing the ice thickness across the grounding line in a part of cDML, East Antarctica. Inset: Thickness profile across the grounding line to the east of the Schirmacher Oasis, East Antarctica. (b) Map showing the ice thickness across the grounding line in a part of cDML, East Antarctica. Inset: Thickness profile across the grounding line to the west of the Schirmacher Oasis, East Antarctica.

transverse area of the ice has been calculated along the grounding line and multiplied by the velocity which is the third dimension to estimate the ice volume loss. Every year the change in the velocity and change in the thickness will change the mass loss estimation. During this calculation, it is assumed that the bottom of the ice sheet across the grounding line is flat and vertical velocity is the same across the grounding line. The velocity has been estimated using ground observations and locally averaged out for the simplicity of the calculation for the period from 2017 to 2019.

2.4.1 Out-Flux Calculation for Grounding Line at Nivlisen Ice Shelf

The grounding line from the west coast to the east coast of the Nivlisen ice shelf has been utilized as an outflux gate for the calculation of ice loss through the movement of the ice sheet into the ocean across the grounding line. By deriving the thickness of the ice sheet at the grounding line, the flux gate vertical area has been calculated. The outflux gate area multiplied by the linear velocity in the out stream of the catchment area in the ice sheet has given total out-flux or loss of ice mass (Table 2.7).

The calculated data shows that about 1.66 and 1.75 gigatons of ice mass have been lost from the out-flux gate at Nivlisen ice shelf to the ocean during 2017–18 and 2018–19 respectively. Sea Level Rise (SLR) was estimated using a multiplication factor of 1/361.8 as proposed by Intergovernmental Panel on Climate Change (IPCC) and (Haeberli and Linsbauer 2013).

SLR (mm)=mass of ice (Gt.) X (1/361.8).

Thus, the mass of ice loss would have contributed a significant 0.0045 mm and 0.0048 mm for the period from December 2016 to December 2017 and December 2017 to December 2018 respectively.

TABLE 2.7
Calculation of Mass of Ice Loss from the Out-Flux Gate through Nivlisen Ice Shelf (Annual)

	2017–18 (Dec. 2017-Dec. 2018)	2018–19 (Dec. 2018-Dec. 2019)
The annual average velocity of the stake network which falls on the flow line of the grounding line (m)	21.59	22.76
Total Area (vertical) of the face of ice sheet at the grounding line (m^2)	83895876.42	83895876.42
Total Volume (km^3)	1.811311972	1.909470147
Density of glacier ice (Gt/km^3)	0.9167	0.9167
Mass of Ice (Giga-Tonnes)	1.660429685	1.750411284

Mass of ice (Gt) = Volume of ice (km^3) x Density of ice (Gt/km^3)

2.5 DISCUSSION AND CONCLUSION

2.5.1 Discussions

The data on snow accumulation and ablation observations of the Polar ice sheet and Nivlisen ice shelf, advance and retreat of Polar ice front along the southern margin of Schirmacher Oasis, ice velocity analysis, and ice thickness were studied as a part of the advancement of glaciological research in cDML, East Antarctica in recent years for mass balance estimation.

Observation of surface snow accumulation/ablation is important to evaluate the response of the Polar ice sheet to climatic variation both locally as well regionally which could be further correlated regionally in long term. In the case of Antarctica, the contribution of snow accumulation through precipitation is only 25–32%, whereas drifted snow is 68–75%. Snow accumulation through precipitation further decreases towards the interior part of Antarctica which may be due to a decrease in moisture content of wind. Interior parts of Antarctica mostly receive snow accumulation in the form of drifted snow as per the data from stake networks across the study area. Ice shelf and marginal coastal parts receive snow accumulation mainly due to snow precipitation which is a prominent component of overall snow accumulation. Data of surface snow accumulation/ablation from the study area shows that accumulation is dominant in the HPIS, whereas ablation is dominant in the IPIS area. The Interior part of Antarctica is always snow covered even during peak summer and the resulting high albedo causes a major part of the incoming short-wavelength sun rays reflected into the atmosphere and this is one of the major factors of sub-zero temperature in the interior parts (Siegert et al., 2019). The other reason for the sub-zero temperature of interior parts is increasing latitude (high latitude) where the area receives slant sun rays and that's too for a shorter duration (Montgomery, 2006). These are attributed to higher net snow accumulation in HPIS areas or almost zero surface melting. The marginal parts of the Antarctic ice sheet which are closer to the coast experience zero to above zero temperature in summer resulting in surface melting of snow and ice. Moreover, the presence of ice-free areas and the thermal properties of rocks also influence the Polar ice sheet recession in Schirmacher Oasis (Swain, 2019). Since ice is having low albedo, higher surface melting of ice takes place.

Polar Ice Front (PIF) is being monitored since 1996 to observe the advance/retreat pattern. The advance/retreat of the ice front is the net result of the annual rate of movement of the ice sheet and the annual rate of degeneration (due to calving, breaking, and collapsing) of the ice front wall. If the annual rate of degeneration is higher than the annual rate of movement of the ice sheet, then it results in the net retreat of the PIF and vice versa. Processed data of PIF show the overall retreat of the wall during 2018–2019. Data from DGPIF shows an annual retreat of 1.32 m whereas the annual rate of movement of ice sheet over DGPIF shows an annual rate of downward movement of 5.23 m indicating a net degeneration of 6.55 m. Loss of mass of ice at DGPIF (water equivalent in kg/m^2) = density of ice (kg/m^3) × length of net degeneration (m) = 917 kg/m^3 × 6.55 m = 6006.35 kg/m^2 loss of ice mass (water equivalent) at DGPIF area.

These can be calculated as below,

I. If PIF Wall shows retreat then,

Net Annual Degeneration of PIF wall (in m) = Net Annual Movement of Ice Sheet (m) + Total Annual Retreat (m)

II. If PIF Wall shows advance then,

Net Annual Degeneration of PIF wall (in m) = Net Annual Movement of Ice Sheet (m)—Total Annual Retreat (m)

This study reveals that the average rate of movement at the HPIS is more than at the IPIS. The low movement rate of the ice sheet in the IPIS area is due to obstructions caused by the presence of numerous nunataks and the rugged terrain of Schirmacher Oasis. During the period of 2016–17 to 2018–19, the thickness of a part of the HPIS was estimated to be more than 500 m with snow thickness ranging from 50 m to 100 m. However, most of the places show uniform snow layers. The thickness of the IPIS area near the Schirmacher Oasis is about 90 m. Further analysis also reveals that the high slope along with high thickness near the IPIS also indicates high velocity. Higher velocity was also observed in some areas near the Somovken glacier.

Mass loss of the Antarctic ice sheet occurs primarily due to out-flux of ice into the ocean, and secondarily due to surface melting which is very less and restricted to more or less peripheral part and Antarctic Peninsula part. The rate of outflux is controlled by the rate of movement of the ice sheet across the grounding line at the junction of the Polar ice sheet and ice shelf. The vertical cross-section area of the ice sheet at the grounding line or thickness of the ice sheet at the grounding line and velocity of the ice sheet based on the net movement of the ice sheet for a particular observation period in between the influx and outflux gate has given the volume of the ice loss from the basin. Since this mass of ice is ultimately going into the ocean, it will contribute to the sea level.

2.5.2 Conclusions

The glaciological studies in a part of cDML near the Schirmacher Oasis, East Antarctica suggests that an average annual accumulation of 0.98 m is observed in the High Polar Ice Sheet area which is equivalent to 394.01 kg/m^2 Snow Water Equivalent (SWE). An average ablation of 0.33 m to 0.46 m annual (133.12 kg/m^2 to 183.25 kg/m^2 SWE respectively is found in Impeded Polar Ice sheet area. The Polar Ice Front at Schirmacher Oasis shows an annual retreat of 2.79 m during 2018–19 and a cumulative retreat of 42.08 m since 2001. The Dakshin Gangotri Polar Ice Front (DGPIF) shows an annual retreat of 1.32 m during 2018–19 and a cumulative retreat of 14.9 m since 1996. During 2018–19, the ice sheet velocity varies from 1.98 m/y to 51.65 m/y with an average velocity of 21.16 m/y in parts of the High Polar Ice sheet area whereas the ice dynamics was 4 m/y to 97 m/y for the same period during 2017–18. The ice sheet movement of 1.57 m/y to 13.71 m/y with an average of 6.78 m/y in parts Impeded Polar Ice sheet area has been recorded, whereas the Ice sheet movement over DGPIF is found to be of the order of 5.23 m/y which reveals a total 6.55 m annual degeneration of DGPIF wall during 2018–2019, which is equivalent to 6006.35 kg/m^2 loss of ice mass at DGPIF area. Out flux of ice mass is calculated at the grounding line at Nivlisen ice shelf, east and west of Schirmacher Oasis. The calculated data shows that 1.7504 gigatonnes of ice mass has been lost from the out-flux gate at Nivlisen ice shelf during 2017–2018 through ice sheet movement towards the periphery of the continent. which was 1.66 gigatonnes for the same time during 2016–2017. It was because of the faster rate of movement of the ice sheets during that period. The mass of ice loss would have contributed to the Sea Level Rise by about 0.0045 mm and 0.0048 mm during 2017–18 and 2018–19 respectively.

REFERENCES

Chandra, V. and Gajbhiye, D. (2019) *Ice Sheet Dynamics from Schirmacher Oasis to Wohlthat Mountains, cDML, East Antarctica and Their Stress Pattern.* Unpublished Report of Geological Survey of India, pp. 14–68.

Chaturvedi, A., Asthana, R., Kachroo, K., Oberoi, L.K. and Singh, A. (2008) Glaciological Observations during the 22nd Indian Antarctic Expedition. *Scientific Report of Twenty-Second Indian Expedition to Antarctica, Scientific Report, Ministry of Earth Sciences, Technical Publication No. 20*, pp. 191–197. ISBN: 978-81-906526-2-9.

Chaturvedi, A., Shrivastava P.K., Kaul, M.K., Chakraborty, S.K. and Mukerji, S. (2009) Recession of Dakshin Gangotri Glacier Snout in Schirmacher Oasis, East Antarctica. *Ind. Jour. Geo. Sci.* Vol. 63, No. 2, pp. 229–234.

Chaturvedi, A., Singh, A. and Beg, M.J. (1999a) Trend of Depositional Patterns on Ice Shelf near Dakshin Gangotri Station. *Scientific Report of Fifteenth Indian Expedition to Antarctica, Tech. Pub. No. 13*, Department of Ocean Development, Govt. of India, New Delhi, pp. 313–320.

Chaturvedi, A., Singh, A., Gaur, M.P., Krishnamurthy, K.V. and Beg, M.J. (1999b) A Confirmation of Polar Glacial Recession by monitoring the Snout of Dakshin Gangotri Glacier in Schirmacher Range. *Scientific Report on Fifteenth Indian Expedition to Antarctica, Tech. Pub No. 13*, Department of Ocean Development, Govt. of India, New Delhi, pp. 321–336.

Chaturvedi, A., Singh, A., Jayapaul, D., Asthana, R. and Ravikant, V. (1999c) Glaciological Studies Carried Out During the Wintering Period of XV Expedition and the Summer Period of the XVI Indian Expedition to Antarctica. *Geol. Surv. Ind., Rec.* Vol. 131, pt 2, pp. 71–74.

Dharwadkar, A., Roy, S.K. and Kumar, P. (2012) *Glaciological Observations During the 31st Indian Antarctic Expedition.* Unpublished Report of Geological Survey of India, pp. 11–30.

Haeberli, W. and Linsbauer, A. (2013) Global Glacier Volumes and Sea Level–Small but Systematic Effects of Ice below the Surface of the Ocean and of New Local Lakes on Land. *Cryosphere.* Vol. 7, No. 3, pp. 817–821.

Kaul, M.K., Chakraborty, S.K. and Raina, V.K. (1985) Ablation on the Antarctica Shelf Ice Sci. Rep. of Second Indian Expedition to Antarctica. *Tech. Pub. No.2, Department of Ocean Development*, Govt. of India, New Delhi, pp. 81–86.

Kumar, P., Gajbhiye, D., Swain, A.K., Shukla, S.P., Chandra, V. and Habib, Z. (2020) Estimation of the Ice Mass Loss from the Out-flux Gate of the Basin AroundSchirmacher Oasis, East Antarctica. *Geosciences: The Basic Science for a Sustainable Future, 36th International Geological Congress* Geological Society of India. Abstract No. 1489–1371. p. 898.

Kumar, P. and Habib, Z. (2018) *Ice Sheet Dynamics from Schirmacher Oasis to Wohlthat Mountains, cDML, East Antarctica and Their Stress Pattern*. Unpublished report of Geological Survey of India, pp. 16–116.

Montgomery, K. (2006) Variation in Temperature with Altitude and Latitude. *J. Geogr.* Vol. 105, No. 3, pp. 133–135, http://doi.org/10.1080/00221340608978675.

Mukerji, S., Ravikanth, V., Bejarniya, B.R., Oberoi, L.K. and Nautiyal, S.C. (1995) A Note on the Glaciological Studies. Carried Out During Eleventh Indian Expedition at Antarctica. *Sci. Rep. of Eleventh Indian Expedition to Antarctica, Tech. Pub. No. 9, Department of Ocean Development*, Govt. of India, New Delhi, pp. 153–162.

Ravindra, R. and Dey, A. (1992) Geological, Geomorphological and Glaciological Studies Carried out During Austral Winter of 1990 at Antarctica. *Geol. Surv. Ind.* Vol. 125, pt 2, pp. 99–101.

Rignot, E., Bamber, J.L., Van Den Broeke, M.R., Davis, C., Li, Y., Van De Berg, W.J. and Van Meijgaard, E. (2008) Recent Antarctic Ice Mass Loss from Radar Interferometry and Regional Climate Modelling. *Nat. Geosci.* Vol. 1, No. 2, pp. 106–110.

Rignot, E., Mouginot, J. and Scheuchl, B. (2011) Antarctic Grounding Line Mapping from Differential Satellite Radar Interferometry. *Geophys. Res. Lett.* Vol. 38, p. L10504.

Schoof, C. (2007) Ice Sheet Grounding Line Dynamics: Steady States, Stability, and Hysteresis. *J. Geophys. Res. Earth Surf.* Vol. 112, No. F3.

Shepherd, A., Ivins, E., Aa, G., Barletta, V., Bentley, M., Bettadpur, S., Briggs, K., Bromwich, D., Forsberg, R., Galin, N., Horwath, M., Jacobs, S., Joughin, I., King, M., Lenaerts, J., Li, J., Ligtenberg, S., Luckman, A., Luthcke, S. and Zwally, H. (2012) A Reconciled Estimate of Ice-Sheet Mass Balance. *Science.* Vol. 338, No. 6111, pp. 1183–1189. http://doi.org/10.1126/science.1228102

Shrivastava, P.K., Asthana, R. and Roy, S.K. (2011) The Ice Sheet Dynamics around Dakshin Gangotri Glacier, Schirmacher Oasis, East Antarctica vis-à-vis Topography and Meteorological Parameters, Jour. *Geol. Soc. Ind.* Vol. 78, pp. 117–123.

Shrivastava, P.K., Roy, S.K. and Mallik, R.K. (2014) *Glaciological Studies in Central Dronning Maud Land, East Antarctica during the 33rd Indian Antarctic Expedition.* Unpublished Report of Geological Survey of India, pp. 1–12.

Siegert, M.J., Kingslake, J., Ross, N., Whitehouse, P.L., Woodward, J., Jamieson, S.S.R., et al. (2019). Major ice sheet change in the Weddell Sea Sector of West Antarctica over the last 5,000 years. *Rev. Geophys.* Vol. 57. https://doi.org/10.1029/2019RG000651.

Singh, A. and Jayaram, S. (1989) Secular Movements of the Dakshin Gangotri Glacier Snout, Antarctica. *Geol. Surv. Ind., Rec.* Vol. 122, pt. 2, p. 200.

Swain, A.K. (2015) *Geomorphological Evolution of Schirmacher Oasis, East Antarctica*, Unpublished Ph. D. Thesis, Ravenshaw University, Cuttack.

Swain, A.K. (2019) Influence of Thermal Conductivity of Rocks on Polar Ice Sheet Recession Near Schirmacher Oasis, East Antarctica. *J. Geol. Soc. India.* Vol. 93, No. 4, pp. 455–465. http://doi.org/10.1007/s12594-019-1200-2.

Swain, A.K. (2020) Glacier Stress Pattern as an Indicator for Climate Change. *Climate Change and the White World*, pp. 119–138. Springer, Cham. https://doi.org/10.1007/978-3-030-21679-5_9.

Swain, A.K. and Chandra, V. (2017) *Ice Sheet Dynamics in and Around Schirmacher Oasis, cDML, East Antarctica*. Unpublished Report of Geological Survey of India, pp. 1–73.

Swain, A.K. and Mandal, A. (2011) *Glaciological Observations during the 30th Indian Antarctic Expedition*. Unpublished report of Geological Survey of India, pp. 1–12.

Swain, A.K. and Raghuram (2013) *Glaciological Studies in Central Dronning Maud Land, East Antarctica During the 32nd Indian Antarctic Expedition (Nov 2012–March 2013)*. Unpublished report of Geological Survey of India, pp. 1–14

Swain, A.K. and Raghuram (2015) *Glaciological Studies in Central Dronning Maud Land, East Antarctica during the 34th Indian Antarctic Expedition*. Unpublished report of Geological Survey of India, pp. 1–45.

Swain, A.K. and Roy, S.K. (2011) *Glaciological Work Carried Out During 27th Indian Antarctic Expedition*. Unpublished report of Geological Survey of India, pp. 1–10.

3 Antarctic Climate History and Its Relationship with Global Climate Changes

Evidence from Ice Core Records

Ashutosh K. Singh,[1] *Devesh K. Sinha,*[1*]
Ankush Shrivastava,[2] *Vikram Pratap Singh,*[3]
Kirtiranjan Mallick,[4] *and Tushar Kaushik*[5]

[1] Department of Geology, Centre of Advanced Studies, University of Delhi, Delhi

[2] Department of Geology, Mohanlal Sukhadia University, Udaipur, Rajasthan

[3] Department of Geology, Indira Gandhi National Tribal University, Amarkantak

[4] Department of Geology, Utkal University, Vani Vihar, Bhubaneswar, Odisha

[5] Biodiversity and Palaeobiology, Agharkar Research Institute, Pune

CONTENTS

DOI: 10.1201/9781003284413-3

3.1 INTRODUCTION

The long and continuous ice cores obtained from drilling thick ice sheets of Antarctica provide us with one of the best high-resolution records of climate change during the last 800 kilo years (kyr). Away from anthropogenic activities, the ice core records are relatively uncorrupted and well preserved. The climatic history of our planet is studied from three major types of archives including continental sediments, marine sediments, and ice sheets (Figure 3.1). All three types of archives have their advantages and constraints in interpreting paleoclimate. Not only does the climate forcing factors in these records vary but also the nature of the responses of the continental land mass, ocean water and ice sheets to the climate forcing factors is different. Out of the three, the Ice cores have many advantages over the other two in terms of their completeness, and the ability to calibrate actual and measured past temperature data through proxies, though the oldest age that an ice core can go

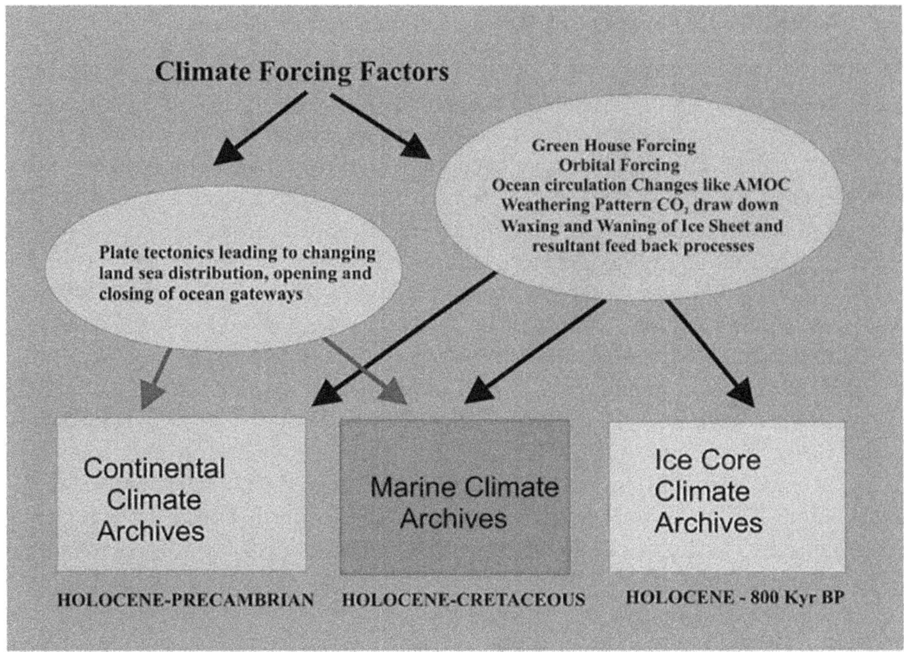

FIGURE 3.1 The three main types of climate archives, are their applicability in time and forcing factors.

up to is ~800 kyr before present (BP) in Antarctica and 160 kyr BP in Greenland. This constrains the correlatability of the ice core records from other marine and continental global sedimentary records that go up to millions of years in the past (Figure 3.1). Though the segment of geological time represented by the ice cores is very small as compared to marine and continental records, the ice cores have contributed immensely to our understanding of the climate forcing and resultant climate response in the recent past which can be applied even to time beyond 800 kyr (Figure 3.1).

Ice sheets have attracted the curiosity of scientists for a long. A desire to know the history of the Ice sheet and the information they preserve about past climate started as early as 1930 when Dr Hans Wilhelmsson Ahlmann, Dr Harald Ulrik Sverdrup and Dr Ernst Sorge made initial studies of the ice sheet by making pits and came out with initial data about the snow accumulation, density and other physical properties over the pit profile from the Greenland Ice sheet (Sorge, 1933; Ahlmann, 1935; Sverdrup, 1935). After nearly twenty years, the serious scientific research on the Antarctic ice core began during 1949–52 by members of the Norwegian-British-Swedish Antarctic Expedition (NBSAE) (Swithinbank, 1957). The ice core drilling was done to nearly 100 meters on the polar ice shelf along the Queen Maud Land Coast, Antarctica (Schytt, 1958). Since then, the drilling techniques gradually improved and many more cores were recovered from various parts of the Antarctic ice sheet going up to the last 800 kyr BP.

3.2 MARINE ISOTOPIC STAGES

As our ongoing discussions are focused on glacial/interglacial stages it is worthwhile to give a brief introduction to Marine Isotopic Stages (MIS). MIS are intervals of depleted/enriched ocean water in heavier oxygen isotope ^{18}O as recorded in the marine carbonate fossils. They have been numbered from MIS-1 (present Holocene Interglacial) to MIS-2, MIS-3, etc. with odd number stages representing relatively warmer intervals than the even-numbered stages which represent cold intervals (Lisiecki and Raymo, 2005). As more and more high-resolution data came from the marine sediments it was found that individual stages do not remain either warm or cold throughout and there are small perturbations e.g., an odd number MIS may have some mild cold intervals while an even-numbered MIS may have some mild warm intervals. Later, many authors used substages and divided the individual MIS into substages a, b, c, etc. For example, widely used subdivisions of MIS 5 have been MIS5a, 5b, 5c, 5d, and 5e. Though the subdivisions of the MIS 5 have been followed quite uniformly across the scientific community there has been inconsistency amongst various workers so far as subdivisions of other MIS are concerned. Sometimes they have been designated as a, b, and sometimes putting a decimal after the numbered MIS. Railsback et al. (2015) presented a rationalized way of subdividing the Marine Isotope Stages to lettered marine isotope substages for the last 1 million years. However, for the older stages, no such uniform scheme is available and the nomenclature of Lisiecki and Raymo (2005) is followed.

With the unprecedented rate of rising greenhouse gases concentration in recent years, leading to an unavoidable global warming scenario, maximum attention is paid to the reducing ice cover at the poles. The ice core data provide us with analogues in the past when the climate and the greenhouse forcing were not very different from pre-industrial times. As in most geological studies, we apply the Huttonian concept of "Present is the key to the past," for climate change study can we apply "past is key to the future?" Few of the interglacial intervals detected based on proxy data like MIS 19 (790–61 kyr BP) (Vavrus et al., 2018), MIS 11 (424–374 kyr BP) (McManus et al., 2003) and MIS 5e (123–9 kyr BP) (Yin and Berger, 2010), have been comparable to warming in the pre-Industrial times.

3.3 ICE CORE RECORDS

The proxy records provide us clues about the duration of the warming events and the Earth's natural mechanism of returning to cooler conditions which are very crucial information in understanding current climate change and its consequences. Ice core records have truly preserved the signatures of variations in the rate of snowfall, and the temperature of the region where they are located. In addition, the wind-blown dust and sea salt record, and the trapped air bubbles containing major and trace gases of the prevailing atmosphere, all provide us clues to the climate changes through ages and over wide regions (Alley, 2000). One of the main issues with the present climate change is understanding the nature of thresholds because the climate system is non-linear and abrupt changes in the past can be possibly repeated in the future with even more vigor and amplitude. Evidence of such abrupt climate changes of the past has been very well preserved in the ice cores. As the climate itself is a net result of several forcing factors including greenhouse forcing and orbital forcing besides others, the measurement of greenhouse gas concentrations (GHC) helps in understanding the relative role of greenhouse gas forcing and orbital forcing (insolation changes) in glacial/interglacial cycles (Yin et al., 2009; Yin and Berger, 2010). We do not have any control over the orbital forcing and thus the only concern of mankind is the forcing factor which is augmented by our activities and also can be moderated if we take preventive measures. We can never make out when the next threshold is going to be crossed in terms of GHC so that the climate changes from one mode to the other mode in decades (e.g., Dansgaard-Oeschger cycles). When one tries to look at a parallel for the Holocene interglacial, we are slowly approaching the end of the 400 kyr eccentricity cycle. With the anthropogenic and natural CO_2 rise we can expect our interglacial to extend and be longest.

Thus, estimating GHC is an important aspect of understanding the dynamics of the past climate and the ice cores provide a direct measurement of the GHC trapped in the air bubbles. Several proxies can be utilized to understand past climate with the help of ice cores (Figure 3.2).

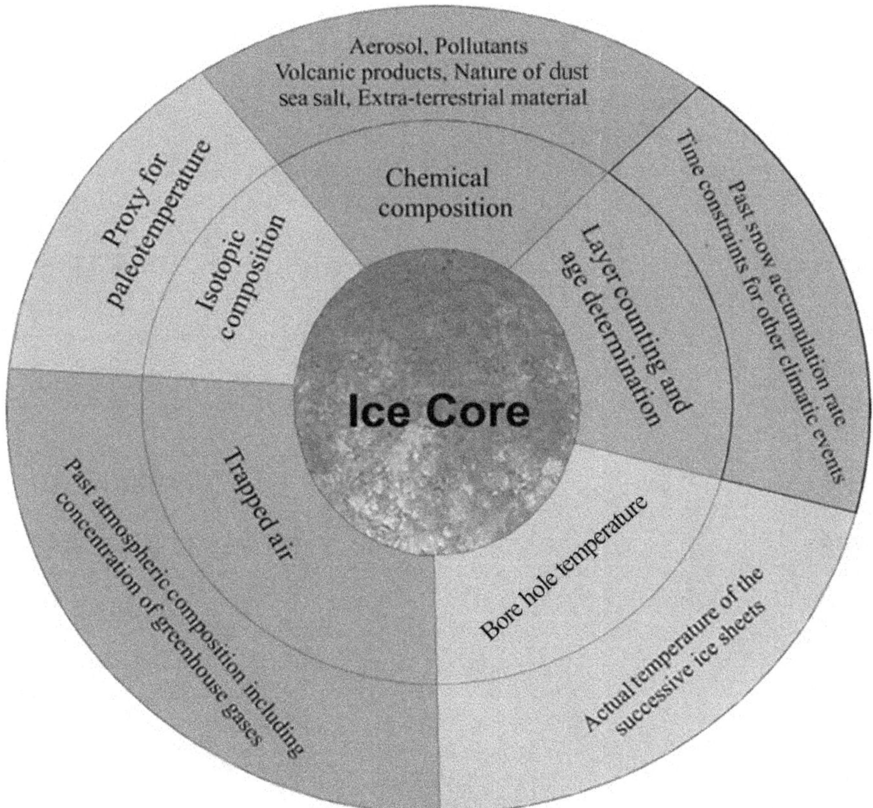

FIGURE 3.2 The diverse studies made from ice cores. The center shows the cross-section of an ice core. The first concentric ring shows the methods adopted and the outer ring indicates the interpretation drawn.

One of the most important results obtained from the ice core records is that not all interglacials have the same forcing factors influencing or triggering them, nor do the same forcing factors cause similar climate changes. This is natural as a large number of forcing factors work in concert producing a net result. Also, different permutations and combinations of the forcing factors play a key role in causing glacial/interglacial events. Ice cores have provided many conceptual advancements to our understanding of climate forcing and climate response.

3.4 LOCATION OF THE ANTARCTIC ICE CORES

To better understand and validate the mechanisms and hypotheses proposed to explain the long and short-term climate changes in the Antarctic region and their global connectivity, the locations of the ice cores are strategically planned (Figure 3.3, Table 3.1).

For example, the ice cores drilled from the interior of Antarctica (Like Dome C), where precipitation (in the form of snow) is rather low, truly record the changes in atmospheric composition, precipitation, and temperature, least influenced by the Southern Ocean temperature variability. On the other hand, the ice cores taken from areas near the coast (e.g., Law Dome), have recorded the Southern Ocean temperature variability. Also due to the high snow accumulation rate, the Law dome core has recorded historical temperature which has been compared with the instrumental record and has helped in validating the importance of ice cores as faithful archives of paleotemperature (Etheridge et al., 1996; Rubino et al., 2013).

The Core from the Kohnen Station in the Dronning Maud Land sector of Antarctica was selected because this sector faces the Atlantic Ocean (Figure 3.3) and will be directly influenced by the meridional atmospheric and ocean circulation, especially for correlating with the Greenland Ice core record. The long and

TABLE 3.1
Some Important Ice Cores and Their Importance in Antarctica Discussed in the Chapter

Ice Core	Importance (Important Reference)
Vostok Ice Core East Antarctica	Recorded events that can be correlated to the larger millennial events of Greenland, including the Younger Dryas (Sowers and Bender, 1995; Petit et al., 1999)
Dome Fuji East Antarctica	720 kyr of CO_2 record, (Dome Fuji Ice Core Project Members, 2017; Watanabe et al., 2003; Kawamura et al., 2017)
EPICA Dome C East Antarctica	Longest record of CO_2_up to MIS 20, (EPICA Community Members, 2004; Lüthi et al., 2008; Jouzel et al., 2007; Loulergue et al., 2008)
Law Dome Core East Antarctica	High snow accumulation rate, Historical record of temperature (Rubino et al., 2013)
Kohnen, Dronning Maud Land, East Antarctica	Facing the Atlantic Ocean and therefore reveals the best data for a comparison with the Greenland ice cores, accumulation rate twice that of EPICA Dome C. (Oerter et al., 2009)
Byrd Core, West Antarctica	A detailed record of CO_2, (Neftel et al., 1988)

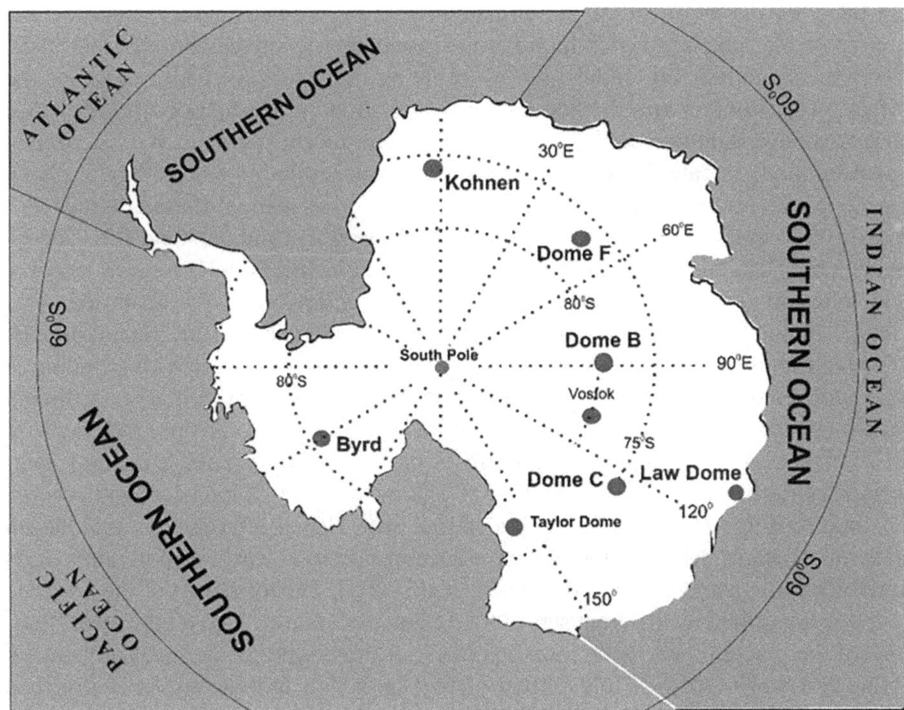

FIGURE 3.3 Location of some important ice cores in Antarctica discussed in the present chapter.

continuous ice cores in the dry ice sheets from the interior of the icy continent generally do not have stratigraphic breaks through they have very low accumulation rates and extend to 800 kyr whereas stratigraphically younger ice cores from the coastal sites have higher snowfall rates.

3.5 KNOWING THE RECENT PAST AND CLIMATE FORCING FACTORS

The ice core record has a unique attribute of having preserved signatures of climate change from a period of no anthropogenic activity to the historical times and thus has the potential to understand the sensitivity of the climate system and thresholds when climate changes from cold to warm or warm to cold conditions. Also, the ice cores add conceptual advancements to our understanding of the climate system because the cores have preserved climate records as a proxy that can be compared with actually measured climate parameter records of the pre-industrial periods and over the transition to the industrial period (Etheridge et al., 1996; Rubino et al., 2013). Another important aspect of the ice core is the nature of atmospheric and oceanic linkages between the two poles and understanding their teleconnectivity through a very high-resolution record of climate change in the last few thousand years. Understanding the teleconnections of the global climate system necessitates a very precise chronology of the samples from which the measurements of various proxies are made. The teleconnections are understood by applying the principle "cause precedes the effect." The problem arises when the error in estimating the age of closely spaced successive events at two different places is large. In recent years much advancement has been made to date the ice cores and a very high-resolution chronology has been developed at both the Antarctic and Greenland Ice cores using different dating techniques discussed in a later section. This has allowed us to understand the teleconnection between north and south and surprising results like Bipolar Seesaw (Stocker and Johnsen, 2003) have been revealed. The paleoclimatic studies from ice cores (Jouzel et al., 2007) revealed that in the last four climactic cycles the glacial periods are much longer than the interglacials (Lang and Wolff, 2011) and this finding is also supported by the Land records (Tzedakis et al., 2006) and marine sediment record (Lisiecki and Raymo, 2005). What may be the reason for this? Is it because on a longer time scale from the beginning of the Cenozoic we are still facing a long-term cooling trend (Miller et al., 1987)? The availability of long ice core records in Antarctica up to 800 kyr, has revealed that each glacial-interglacial cycle has an individual pattern, which in the Antarctic ice cores is most clearly seen in interglacial variability (Lang and Wolff, 2011; Jouzel et al., 2007). The formation of the polar ice sheets is controlled by different sets of forcing factors (Figure 3.1), yet the major findings of the glacial-interglacial intervals from all three types of archives are similar (Lang and Wolff, 2011). While climate change in the low latitude region is governed by a multitude of factors including land use, anthropogenic activities, ocean circulation, greenhouse gas emission and external forcing like sun's radiation, the climate change at the poles, particularly Antarctica are mainly controlled by greenhouse concentration, orbital forcing, and feedback processes. The Antarctic amplification, which is the feeling of a climate signal two to three times the global average, must be understood correctly. For a clear understanding of the Antarctic amplification, the ice core records provide an excellent opportunity. One can analyze the contributions made by atmospheric circulation and heat transport, ocean circulation and the greenhouse forcing so far as recent climate change is concerned. However, going back to the past one can understand how the changes were made in the climate of Antarctica at a time when no anthropogenic activities were present. One of the most significant findings from the long Antarctic ice core has been to understand the character and profile of the glacial and interglacial intervals. The first common attribute is that the intervals (glacial or interglacials) have duration but all intervals are not of equal duration. The second is their amplitude and the third is the profile. The amplitude of the glacial and interglacial intervals have changed with time and the major change occurred after the Mid Pleistocene transition (Clark et al., 2006). Profile means how does an interval start and end. For example, how does an interglacial start and end, how many small cold events an interglacial has like the MIS 5 which is divided into 5a-e. Similarly, a glacial interval may have several short warm intervals throughout its duration (Masson-Delmotte et al., 2010). Another interesting feature is knowing whether each glacial or interglacial has been a response to the same or combination of external forcing (Yin and Berger, 2010)? Why are they different? The difference they owe to the forcing or to the response due to changes in ocean circulation and related heat transport (Broecker et al., 1990)? Does each glacial and interglacial interval have a similar global spatial signature (Lang and Wolff, 2011)? Such questions to a great extent have been answered by the study of the ice cores. In recent years the growing emphasis has been on the model-data comparison and the long ice cores have facilitated this to a great extent. Given that the few interglacials had comparable warming as in present during which the external forcing like orbital and greenhouse is known, one can assess how our present system may return to a colder event based on the study of past interglacials. Ice core data and its correlation with global climate change from high, mid and low latitude regions have added to our understanding of the active vs. passive roles of the ice sheet in causing gradual/abrupt climate changes. Understanding abrupt climate changes as recorded in the ice core is essential to our understanding of the behavior of the climate system because of the recent global warming. The role of high vs low latitude is to be understood at various time scales. Modelling studies have also revealed that even the cause of abrupt events like Younger Dryas can be a change in the tropical Pacific ENSO variation (Clement et al., 2001).

3.6 FORMATION OF THE ICE SHEET AND ITS MOVEMENT

Ice cores originate from atmospheric precipitation of ice when the moisture-laden winds from low latitudes reach the poles and precipitate the snow which turns into the ice with time (Figure 3.4). The Antarctic Ice cores store a plethora of information about

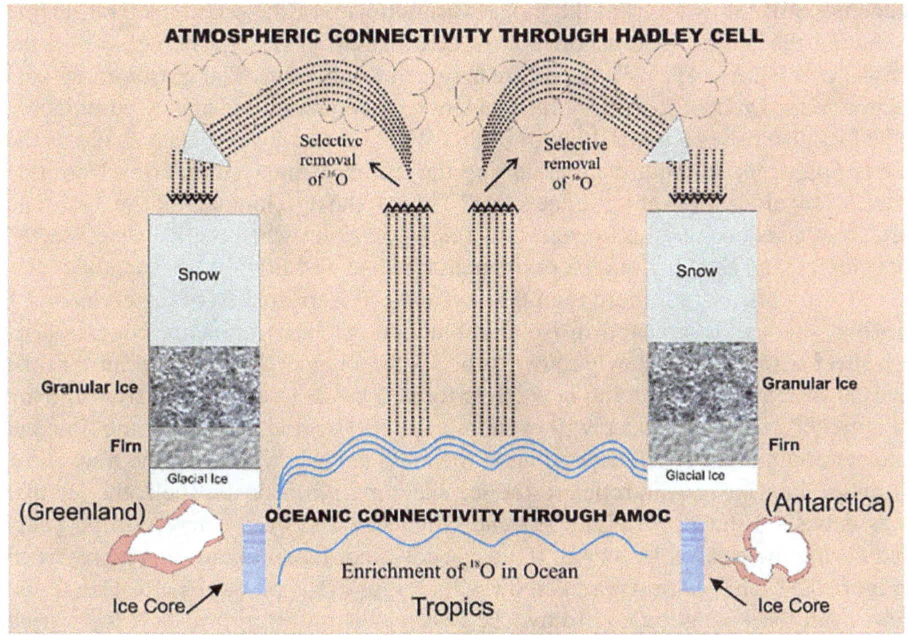

FIGURE 3.4 Conceptual diagram showing deposition of snow in Greenland and Antarctica and signals of oxygen isotope preserved in the glacial ice and oceanic sediments.

the Antarctic environment and its connection with the global climate. While the isotopic composition of the ice cores reflects the local temperature (Jouzel et al., 1997), the chemical composition of the ice cores provides information about the input of dust, sea salt, volcanic material, pollutants, other aerosol material, extra-terrestrial material, etc. (Wolff et al., 2009; Schuepbach et al., 2013; McConnell et al., 2014).

The layer counting of the ice cores gives us information about the past snowfall rate. Owing to the flow of ice from the interior to the ice sheet margin, ice deposited during the earliest history of the ice sheet is not likely to have been preserved, but the maximum age of extant ice is unknown (Brook and Buizert, 2018).

The icy continent Antarctica, the largest desert on Earth, contains 70% of its freshwater, in the form of snow, enough to raise the global sea level by 200 ft. if completely melted. At first impression, the above facts appear to contradict each other because the usual picture of dessert seems to be a place with no or little water. However, the Antarctic continent qualifies for being classed as dessert as the average annual precipitation in coastal areas is ~166 mm with precipitation at the South Pole less than 5 mm per year. In general, the precipitation in Antarctica follows the temperature belts. The reasons for being so dry are many. First is the fact that the cold air has little capacity to hold moisture and the Antarctica interior is characterized by sinking air at and near the South Pole forming the falling arm of the polar cell. This high-pressure zone, similar to other high-pressure zones of the Earth is dry. The other strong reason is that during the austral winter the ice expands to an area that itself is larger than the continent. The interior of Antarctica becomes still far away from the moist warm climate of the coastal region further reducing the moisture transport due to continentality. Despite being so dry, the desert continent holds 70% of the freshwater of the Earth. The reason is its cold temperature which allows very little ablation as compared to snow accumulation. The fact also raises a natural question that should then we get the ice as old as the Early Oligocene in Antarctica assuming the ice has been accumulating from that time? How old is the Antarctic Ice sheet? Though the development of the Antarctic ice sheet goes back to the Eocene-Oligocene transition when the ice sheet started developing and establishing the permanent ice sheet in Middle Miocene, the oldest ice in Antarctica is not more than 800 kyr old as discovered from ice cores or 2.7 million years as discovered in isolated blue ice patches from some regions. The ice sheet always slides slowly towards the shelf and older ice melts because of the crustal heat and pressure and new ice is formed by the precipitation from the top (Figure 3.2). That means we will even lose the 800 kyr old ice after some time but the net accumulation will depend upon the mass balance. Owing to the flow of ice from the interior of Antarctica to the ice sheet margin near the shelf, the ice that was deposited during the Late Eocene-Oligocene and even a few million years old during the earliest history of the ice sheet is not likely to have been preserved (Figure 3.5). The ice in Antarctica owes its origin to the poleward moisture transport from the low latitude (Bromwich, 1990). This moisture transport has varied in the geological past as a result of climate change during the glacial and interglacial ages. This fact also gives us clues as to how we can infer the connectivity of Antarctic climate to low latitude regions.

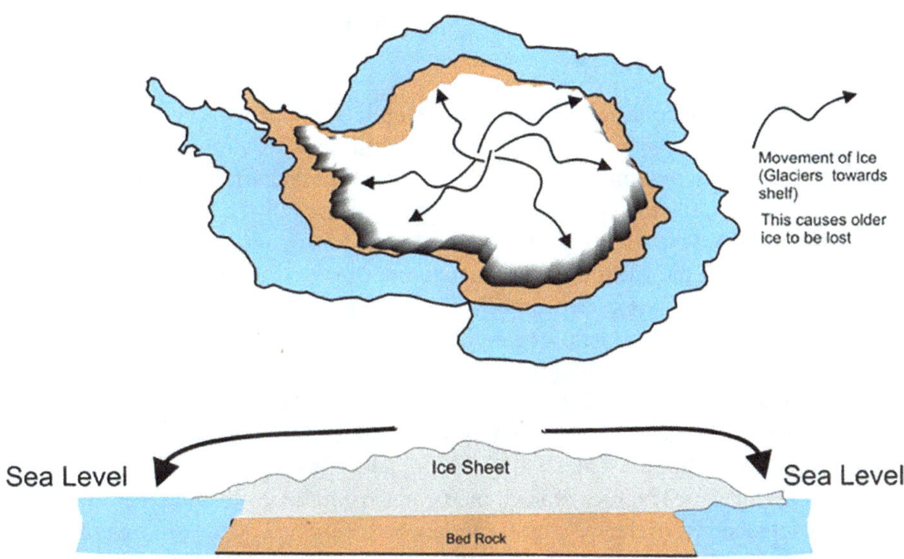

FIGURE 3.5 Cartoon showing the movement of ice sheet towards margin (shelf) of the Antarctic ice sheet.

This transport mostly precipitates near the coastal areas where the average annual precipitation is much more than the interior of Antarctica. The second factor is the surface temperature of the interior of Antarctica which also affects the ice sheet growth and mass balance. Some of the studies made based on the difference in age of the trapped air and surrounding ice concluded that the LGM temperatures were lower by 8–15 degrees Celsius than present in the interior of Antarctica (Blunier et al., 2004). The rate of movement of the ice sheet towards its margin must have varied in the geological past though no reliable data exists except the indirect effect of the melting on the oxygen isotope record of the benthic foraminifera. In addition, the concentration of Ice Rafted detritus in oceanic sediments is also an indication of ice calving near the shelf. It is not that the rate of ice sheet melting has always been due to climate change but Seroussi et al. (2017) showed that mantle plumes have an important local impact on the ice sheet, with basal melting rates reaching several centimeters per year directly above the hotspot. Altogether Antarctica is affected by and affects the climate of the world. Much of the early history of the Antarctic ice sheet is revealed not from Antarctica but from the marine sedimentary record in the Southern Ocean surrounding Antarctica. However, for the last few thousand years, several ice cores reveal a direct record of the Antarctic climate. As shown in Figure 3.2, a multitude of studies can be made from the ice cores but it is worthwhile first to consider the dynamics of the ice sheet which will help understand the information derived from the ice cores about past climate change. The first factor is what controls the accumulation, thickness, extent, and sustainability of the ice sheet.

3.7 THICKNESS, VOLUME, EXTENT, AND SUSTAINABILITY OF THE ICE SHEETS

At first look, the vast continental ice sheet of Antarctica appears very stable and gives an impression of a never melting entity through time with temperatures far below freezing point. A layman might even question the so-called effect of global warming of less than 1 °C in the last several decades on ice melting as the simple laws of physics do not allow ice melting if the temperature remains in a minus regime. But what appears is far from the truth. Recently a network of subglacial lakes was discovered beneath the ice sheet (Siegfried and Fricker, 2021). A change in surface topography was observed with the help of satellite data when such lakes drain out water or accumulate water due to the melting of ice at the bed rock ice interface. Though the presence of subglacial lakes is known for a long, like Vostok, the most conspicuous aspect of these subglacial lakes is that they seem to be interconnected and drain the water gradually to the sea. Thus, what appears above is different from what is going on in the subsurface. The rate of such meltwater from the lakes and large chunks of ice being detached from the ice shelves and iceberg calving and the amount of annual snowfall (precipitation) all contribute to the net accumulation/ablation of ice from the ice sheets (Jacobs et al., 1992).

The overall balance amongst the above-mentioned factors is generally maintained for a long time and equilibrium conditions persist through several years. The loss through iceberg calving is one of the dominant ways in which ice is lost by Antarctica. On a longer time scale, one can say that the development of the

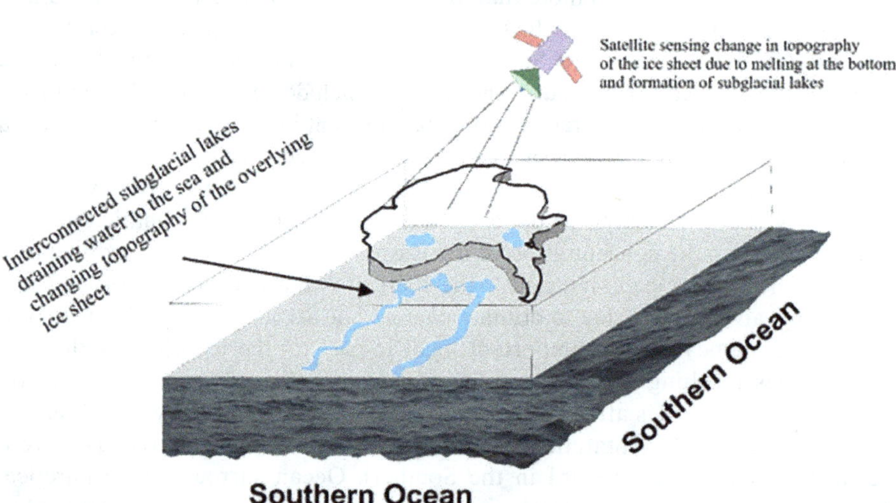

FIGURE 3.6 Cartoon depicting subglacial lakes below Antarctic ice sheet which are formed due to overlying pressure, geothermal heat flux, the friction generated heat near the ice-bed rock interface. The lakes are interconnected and drain water into the surrounding oceans. A change in topography is observed at the ice sheet surface when such large lakes cause the sinking of the ice sheet.

ice sheet itself indicates an equilibrium condition where the gain has always been more than the loss. This is the reason for the thickest ice sheet present in desert-like Antarctica where the accumulation has always exceeded the ablation since the initiation of the Antarctic ice sheet at the Eocene-Oligocene transition (EOT). The history of the Antarctic ice sheet as revealed from the stable isotope data of marine carbonate microfossils documents the overall positive growth of the ice sheet from EOT to the present (Miller et al., 1987). Under equilibrium conditions, the mass lost is balanced by the mass gained but under the present climate deteriorating scenario, the main concern is about a disturbance in this equilibrium. Much has been talked about in recent years about the melting of the polar ice sheets, particularly Antarctica which has the potential to raise the global sea level and cause displacement of millions of people living in or near the coastal area. However, this issue has several facets and we have been accustomed to discussing only the loss of ice and not the gain. This requires detailed mass balance studies of the Antarctic ice sheet. The mass balance studies include a detailed inventory of precipitation (in the form of snow) over the ice sheet, evaporation from the snow surface, iceberg calving at the coast, melting beneath ice shelves and runoff of meltwater (Connolley and King, 1993). Throughout Antarctica, the surface runoff is not prominent because the temperatures remain sub-zero throughout the year. Thus, the gain is mainly in terms of precipitation while the loss is mostly caused by iceberg calving, melting of the ice from the base near the shelf as the ice sheet very slowly slides towards the coast (Figure 3.5). The difference between the precipitation and loss of ice due to the above-listed causes is termed the net accumulation or accumulation which indicates the mass balance. The mass balance determined locally for a large number of stations can be integrated to get a comprehensive picture of the entire Antarctic continent. However, the local mass balance determination may not include all the causes of loss of ice listed above and if it is the interior area, the loss may be mainly due to evaporation. Another problem arises due to drifted snow and windy weather during the time of precipitation in the coastal areas. Due to this the clear sky precipitation and drifted snow together can give a false idea about the actual precipitation. However, in the interior of Antarctica, the precipitation of snow is mostly from the sky. Snow gauges pose problems and the estimation of net snowfall is done from the stakes set out along traverse routes and from snow-pit stratigraphy.

There have been major concerns about the possible melting of the ice sheet in Antarctica and the resultant global sea-level rise. The mass balance studies so far made are very few and the net poleward moisture transport from the coastal areas towards the interior of Antarctica is a key issue. The estimations can have considerable errors yet the data will be useful and seeing the variation during the last few decades will throw light on past variations. This transport will have also implications for our predictions of global sea-level rise. The moisture transport would also depend on the net evaporation from the Southern Ocean and a sea surface temperature rise would be beneficial for moisture transport. As proposed by Gildor (2003) a strong hydrological cycle is needed for glaciation to occur. On a short time scale, we need to have quantitative estimates on the net moisture transport and implications on the thickness of the ice sheet. The enhanced moisture transport towards the ice sheet, in some cases is defined as atmospheric rivers (Connolley and King, 1993). Jouzel et al.

(1982) based on a continuous deuterium excess profile (*d*) measured from the EPICA Dome C concluded that *d* was significantly lower (4.5‰) during the Last Glacial Maximum (LGM) compared to the recent 7,500 yr. This they interpreted to represent moisture supply for precipitation in Antarctica.

The precipitation measurements in Antarctica are marred by very few measuring stations and thus have been calculated using atmospheric measurements like the vapor or moisture transport (moisture advection) by the wind and measurement of the Moisture flux convergence (MFC). Some important studies estimated MFC over Antarctica at different heights in the troposphere ranging from 100 meters above the surface to the upper boundary of the troposphere (Bryan and Oort, 1984). They estimated the precipitation minus evaporation over Antarctica to be close to corresponding to 44 ± 21 mm per year contrary to the larger estimate by Giovinetto and Bentley (1985) who estimated it to be around 124 mm per year from the surface estimate. Another estimate of 148 mm per year was given by Masuda (1990) based on calculations of heat and moisture budgets for the polar caps from the First GARP Global Experiment (FGGE) data near the south polar cap. Though the studies are not conclusive due to various errors involved yet they give a rough estimate of the precipitation over Antarctica. These data become extremely crucial in the case of enhanced global warming as we will have to see whether global warming which can cause iceberg calving and possible sea-level rise can be compensated by the enhanced moisture transport to the Antarctic Ice sheet in the inner part? Recent work using both simplified and comprehensive Global Circulation Models (GCMs) has shown that poleward moisture transport across mid-latitudes follows Clausius-Clapeyron scaling at temperatures close to modern, but that it reaches a maximum at sufficiently elevated temperatures and then decreases with further warming (Caballero and Hanley, 2012).

3.8 WIND PATTERN IN ANTARCTICA AND ITS RECORD IN ICE CORES

The interior of Antarctica is extremely cold and this creates a high-pressure zone where huge air mass sinks. This sinking air mass is replaced by Air mass moving towards the interior in the upper reaches of the Atmosphere. The sinking air flows towards the Antarctic shelves area and due to Coriolis deflection creates Polar easterlies. The polar easterlies are the dry, cold prevailing winds that blow from the high-pressure areas of the Antarctic polar highs towards low-pressure areas within the Westerlies. These are dry airs and do not contribute to the precipitation in Antarctica. The temperature of the continent increases gradually towards the margin due to the presence of the ocean. Thus, on the continent the high-pressure zones gradually are encircled by concentric low-pressure belts and the dry polar easterlies dominate the continent.

Snowfall occurs throughout the year in Antarctica making thick ice sheets. During summer there is always sunlight present on the pole and during winter it's dark. Thus, the nature of the summer ice and winter ice differ. Each layer of snow is thus distinguished from the layer above and the layer below. Also, the thickness of the winter and summer snow differs owing to the difference in precipitation amount. The winds also bring continental dust from the southern continental areas which also impart color to the snow layers. Each layer of snow stores immense information regarding the climate and other parameters prevailing at that time as shown in Figure 3.2. The successive layers of snow provide valuable time-series data to scientists. The prevailing wind directions are also deduced by analyzing the geochemistry of the dust and comparing it with the continental land record. The annual bands remain preserved though due to overlying load the underlying layers are compressed. The top few meters of the fresh snow are granular crystals and are known as Fern. With time the Fern becomes ice. Just like sedimentary beds the ice sheets are lying over one another.

3.9 MEASURING PALEOTEMPERATURE OF ICE CORES WITH OXYGEN AND HYDROGEN ISOTOPE

The past temperature of the ice samples from the ice cores can be measured with the help of isotopic ratios of oxygen ($^{18}O/^{16}O$) and hydrogen ($^{2}H/^{1}H$) of the water molecules released by the melting of the ice cores. Isotopes of an element have the same atomic number (proton number) but different mass numbers. Oxygen has mainly two isotopes ^{16}O and ^{18}O (^{17}O is present in a very insignificant amount). ^{16}O constitutes 99.76% of the oxygen in the water. During evaporation, the lighter isotopes get into the vapor phase easily and through the Hadley cell, the moisture is precipitated at the poles in form of snow (converted to ice through a series of processes) (Figure 3.4). The result is that the ice becomes depleted in heavier isotopes while the ocean water becomes enriched in the heavier isotope. Thus, in the ice cores, the cold (glacial) intervals will be marked by less $^{18}O/^{16}O$ ratio (expressed as $\delta^{18}O$ in parts per thousand, ‰, when compared with standard) compared to warm interglacial intervals when the ratio will be more. The opposite happens in ocean water. During such intervals, the calcareous micro-organism, foraminifera which secretes its calcium carbonate test in isotopic equilibrium with ambient sea water, arrests the $^{18}O/^{16}O$ of the water in its $CaCO_3$ test and the stable isotope analyses of their test can reveal the prevailing isotopic composition of sea water. Thus, the foraminiferal ($CaCO_3$) calcite precipitated during glacial/interglacial intervals shows higher/lower values of $\delta^{18}O$ in contrast to ice core which shows lower/higher values of $\delta^{18}O$ during glacial/interglacial intervals.

The hydrogen (^{1}H) and deuterium (^{2}H) ratio is also utilized similarly and confirms the measurements done with the oxygen isotope ratio. The hydrogen (^{1}H) constitutes 99.985 of hydrogen in the water. The other isotope is deuterium (^{2}H) with one proton and one neutron. Stable isotopes ratios of Oxygen ($^{18}O/^{16}O$) and Hydrogen ($^{2}H/^{1}H$) are generally expressed as deviations, $\delta^{18}O$ and δD, from Vienna Standard Mean Ocean Water (VSMOW) in the following way:

$$\delta = \frac{R_{\text{sample}} - R_{\text{standard}}}{R_{\text{standard}}}$$

where R_{sample} is the ratio in the sample and R_{standard} is the ratio in VSMOW

As in the case of Oxygen isotopes, during glacial/interglacial intervals the ice samples become depleted/enriched in deuterium isotopes. For several years, scientists have measured the isotopic ratios of oxygen and hydrogen in snowfall and related them with the actual temperature. Such exercise has resulted in calibrating isotopic ratios with the temperatures though there are constraints in measuring past temperatures and necessary corrections are done. The water isotope ratios (in the snow) at the precipitation site and the surface temperature measurements are done and an empirical relationship is worked out between the two (Petit et al., 1999; Jouzel et al., 1997). In a recent study, Markle and Steig (2021) discussed in detail the non-linearities associated with simple models correlating isotope ratios with temperature. These authors provided a refined and precise model taking cognizance of the absolute surface temperature, condensation temperature, and source-region evaporation temperature for all long Antarctic ice-core records (EPICA Community Members, 2006; Markle and Steig, 2021).

3.10 HOW DO WE DIRECTLY MEASURE THE TEMPERATURE OF AN OLD ICE SHEET?

The temperature of an ice sheet layer can be directly measured by lowering a thermometer into the borehole drilled to retrieve the ice core (borehole thermometry). The basic presumption is that the ice has preserved the temperature of each successive layer of snow, indicating general contemporary atmospheric temperatures when the layer accumulated. In other words, we can say that Ice Sheet has a temperature memory. The borehole thermometry data has been used for paleoclimate reconstruction for a few tens of kyr in the East Antarctic Ice Cores (Salamatin, 2000). As one reaches the base of the ice sheet near the surface of the bedrock, the heat of the Earth warms the lowest layers of the ice. That also causes melting and flows of the melted water according to the slope. These physical temperature measurements help calibrate the temperature record scientists obtain from oxygen isotopes (Figure 3.7). Ice Sheets have a very good borehole temperature memory if the accumulation rate is high. Successive layers are quickly buried before any temperature change is forced from the successive climatic variation in the atmosphere. Borehole-based temperature reconstructions at the West Antarctic Ice Sheet (WAIS) Divide site in East Antarctica (Cuffey et al., 2016) indicate a glacial-interglacial temperature change of 11.3 ± 1.8 °C for the last termination, consistent with estimates based on stable water isotopes alone (Brook and Buizert, 2018). Based on borehole temperature measurements in the upper 300 m of Rutford Ice Stream, West Antarctica, Barrett et al. (2009) found recent warming of 0.17 ± 0.07 °C $(\text{decade})^{-1}$ since 1930. At the global level, the variation in glacial/interglacial temperature on average has been around 3.5 degrees (Shakun et al., 2012) but due to polar amplification, the variation in glacial/interglacial temperature in the Antarctic ice cores is quite high (Cuffey et al., 2016).

3.11 HOW DO WE KNOW ATMOSPHERIC COMPOSITION FROM ICE CORE?

Air bubbles trapped in the ice cores provide a record of past atmospheric composition (Figure 3.7). When snowfall occurs at high latitude the layers of snow contain almost 90% of the air. Slowly with time the compaction occurs and the snow turns

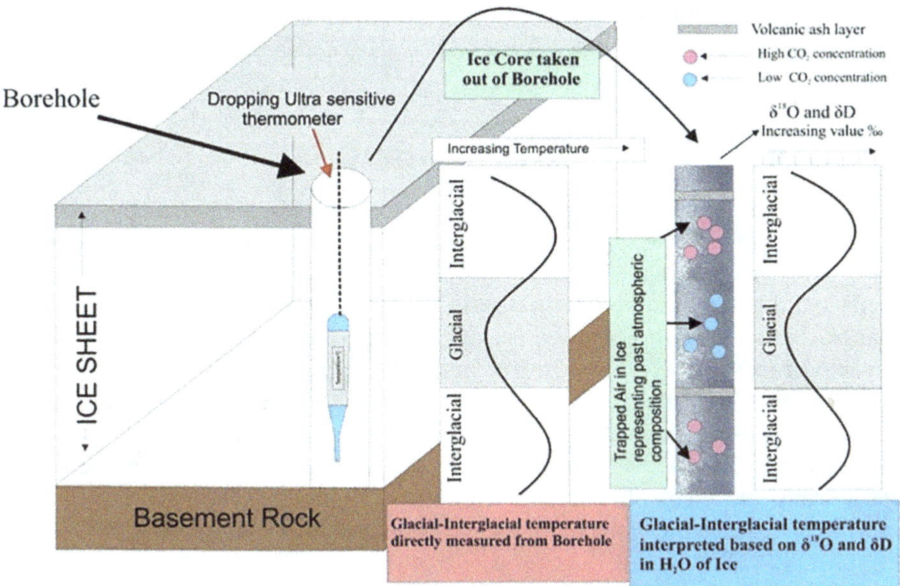

FIGURE 3.7 Cartoon depicting the method of directly measuring Borehole temperature of the ice sheet and comparison with temperatures measured with stable isotopes of oxygen and hydrogen. Also shown are the air bubbles trapped in the ice core which tell us about the past atmospheric composition. Two layers of volcanic ash are also shown.

to granular ice but still with lots of interstitial space filled with air but gradually the volume of air becomes less (Figure 3.4). Further due to compaction, the granular ice turns into firn and the entrapped air volume is further reduced. The firn-ice transition depth varies with latitude and is quite shallow in the polar region (Benson, 1962). The necessary time for ice formation in the firn aquifer, estimated from the empirical formula, decreases exponentially as the pressure increases, showing that the overburden pressure acting on the water-saturated firn is the most important factor in determining the depth of firn-ice transition (Kawashima and Yamada, 1997). With further compaction, the full transformation of the snow to ice via granular ice and firn is completed (Figure 3.4). The Ice becomes airtight and preserves the signatures of the prevailing atmosphere in the form of air bubbles (Figure 3.7). In the Ice cores recovered from this glacial ice, these air bubbles provide us with valuable information about the composition of the atmosphere including greenhouse gas concentration. The comparison of the ice core records from Greenland and Antarctica has not always provided synchronous signals of climate change but many times the signals are connected via meridional circulation both from the atmosphere and ocean (Figure 3.4). Further, the gases above the ice sheet very freely mix and thus their ages (when trapped as an air bubble) are quite younger than the ice which encloses them, especially in the interior of Antarctica where the snow accumulation is quite slow this age difference can be in thousands of years (e.g., in Vostock) but where snow accumulation is fast (close to the coast) the age difference will be smaller. Thus the methods adopted to date for each ice core may vary depending upon the snow accumulation rate (White and Steig, 1998).

3.12 VOLCANIC ASH IN THE ICE SHEETS

The role of volcanic eruption and volcanic aerosols has a profound effect on the climate as they inhibit solar insolation. The Volcanic eruption throws tons of volcanic ash high into the atmosphere and such ash is also carried to the poles to atmospheric shells and deposited in the ice sheets. As Ice cores record the signature of climate change, the causative link between volcanic eruption and climate can be studied with the help of ice cores (Figure 3.7). The atmospheric pathways of the preserved volcanic ash give an idea of prevailing wind directions (Basile et al., 2001). In general, the cold periods in Antarctica are characterized by much greater dust fallout than is found during interglacials (Delmonte et al., 2002). Due to an increase in the flux of dust during glacial conditions the growth of the ice grains is inhibited and thus the general decrease in the diameter of the ice grain also indicates glacial conditions.

3.13 DUST RECORDS FROM ANTARCTIC ICE CORES

Delmonte et al. (2008) observed the presence of Antarctic dust of dominantly South American origin during several glacial intervals, MIS-8, MIS-10, MIS-12, and back to MIS 16 and 20 as deduced from EDC core. Their data provided evidence for a long-time westerly circulation transporting dust from South America to the interior of Antarctica during these intervals. The study was based on the strontium and neodymium isotopes as tracers. Thus, the dust record in the ice cores particularly during the cold intervals gives us clues about the prevailing wind directions.

3.14 MELT LAYERS IN ICE SHEETS

As the accumulation of snow and then conversion to ice through granular ice and firm is a continuous process and may take years, the upper layer of snow also produces meltwater whenever the summer prolongs and summer temperatures are quite high. Such meltwater percolates downwards and again freezes into solid ice which is often devoid of bubbles. Melt layers appear in bubbly ice cores as near-horizontal, typically irregular or discontinuous bands of ice with few, oddly shaped or sized, air bubbles, typically 1–100 mm thick (Langway and Shoji, 1990; Das and Alley, 2005). Physical identification of melt layers is possible up to a few hundred meters in depth but beyond that due to compaction, the physical identification becomes difficult. For older melt layers chemical identification is used. The melt layers are related to summer temperatures and thus provide important information about the prevailing summer temperatures. If the summer air temperatures are high that will be producing more melt layers. The melt layers are bubble-free and thus can be easily distinguished while examining an ice core. Various melt layers are found separated by ice core thickness (time) and indicate the paleoclimate. Melting at the ice sheet surface depends upon the temperature of the ambient air. Due to positive energy balance, the surface air is warmed up and causes the surface ice to melt. However, the Antarctic ice sheet surface melting is not so common as it often maintains sub-zero temperatures, in the areas located near the coasts and in the areas of the Antarctic Peninsula the surface melting is more due to high heat supply (van Wessem et al., 2016). Thus the surface melt produced is directly related to surface temperature and was used for paleotemperature reconstruction and paleoclimate in the Ice Cores from Antarctica (Abram et al., 2013). However, the paleoclimate proxies can be obscured if the surface melting is excessive (Thomas et al., 2021).

3.15 ESTABLISHING CHRONOLOGY OF THE ANTARCTIC ICE CORE

Developing a reliable chronology of the Ice Core is an important issue for any interpretation. As the Ice cores preserve the evidence of climate change for the past few thousand years, the chronology needs to be of very high resolution. There are several methods used to establish the chronology of the Antarctic Ice Core. For the upper part of the core where Ice compaction has not been so high, counting the annual layers is a good and reliable method (Figure 3.8). If the ice accumulation rate is quite low then the layer counting for age determination has a limitation. Another method is based on the Deuterium composition of the ice cores. An age-depth model can be constructed using δD combined with a glaciological flow model (Mulvaney et al., 2012). Though difficult, some workers successfully applied the $^{234}U/^{238}U$ dating method to date the

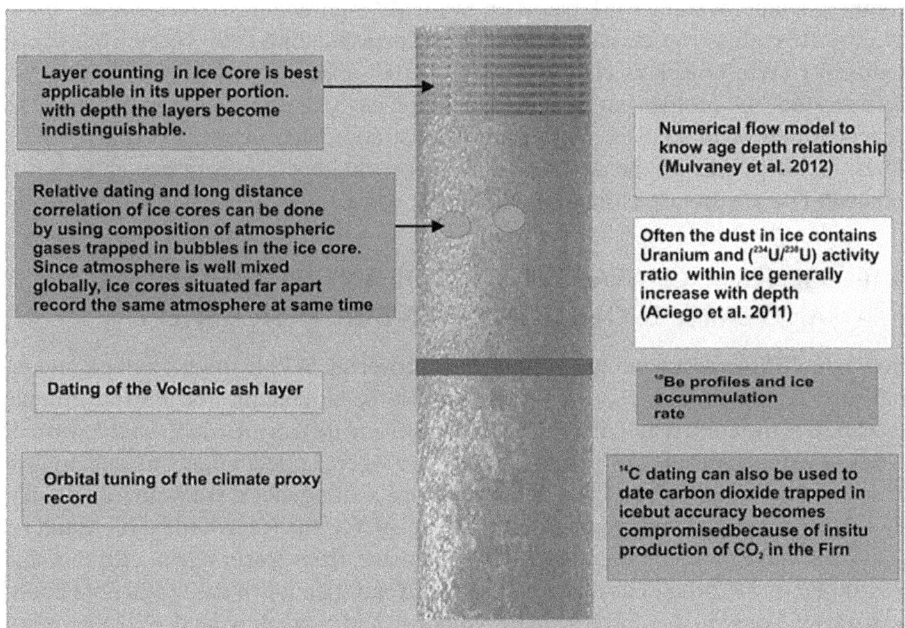

FIGURE 3.8 Different methods used for ice core dating.

EPICA Dome C Ice Core in Antarctica for the duration of 80 ka and 870 ka. The dust particles found in the ice matrix contained Uranium and measured ($^{234}U/^{238}U$) activity ratios within the ice generally increased with depth (Fireman, 1986; Aciego et al., 2011). ^{14}C dating can also be used to date carbon dioxide trapped in the ice but the accuracy becomes compromised because of the in-situ production of ^{14}C in the firn. Earth's orbital variations directly affect the surface layers of Earth and the insolation variations can leave signatures on the Ice. Ice core climatic records can be orbitally tuned and age determination is possible (Berger, 1978; Laskar et al., 2004). This approach has been used in the case of the long, 420-kyr Vostok record. However, as the climate system is non-linear, the orbitally tuned methods have limitations too. Due to this nonlinearity, the uncertainty in chronology with the orbital tuning approaches reaches up to 5 kyr, however, it has been suggested that ice core properties recording local insolation directly may overcome this limitation (Bender, 2002; Raynaud et al., 2007).

Another approach is to employ methane concentrations to develop relative chronology. The problem with methane stratigraphy occurs because during densification and compaction of the snow the upper layer exchanges air due to firn porosity before getting trapped as an air bubble. Therefore, in any ice core, contemporaneous gas and ice signals are not recorded at the same depth and air occluded in air bubbles is younger than its surrounding ice. However, the methane concentrations can still be employed for long-distance ice core correlations because the errors are similar in the two places. Blunier et al. (1998) and Blunier and Brook (2001) employed methane concentration in Antarctic and Greenland ice cores to the phase relationship (in leads or lags) of the temperature variations. Another way of ice core dating is to establish a common temporal framework based on SPECMAP chronology (Imbrie et al., 1984) by correlating ice-core climate records with corresponding time series from ocean sediments (Waelbroeck et al., 1995). Yet another approach is to establish the link between the ^{10}Be profiles and past accumulation changes and the use of peaks in the concentration of this cosmogenic isotope as stratigraphic markers (Lorius, 1990). Thus, ice cores may not be dated by a single method but a multi-proxy approach is needed to have a precise chronology of the ice cores.

3.16 GENERAL CHARACTERISTICS OF ANTARCTIC CLIMATE AND EARLY EVOLUTION OF ANTARCTIC ICE SHEET

Being the fifth largest continent of the world, covering 98% by highest albedo material, the ice, Antarctica plays a significant role in modulating the Earth's climate system. It is the coldest and driest place on Earth and differs greatly from the warmer Arctic (Gettelman et al., 2006) and it is a region where large amounts of heat are lost to space, due to high albedo which helps drive Earth's general atmospheric circulation. The cold Antarctic Circum Polar Current (ACC) which surrounds the Antarctic continent makes the continent thermally insulated from warm ocean currents and this is one of the prime reasons which make Antarctica much colder than its counterpart the Arctic in the north which is an ocean surrounded by land. What goes out from Antarctica is in the form of Polar easterlies towards the margin (shelf area), the iceberg calving and what enters the Antarctic region is part of the warmer westerlies and occasional poleward heat transport through the ocean currents. Besides the above two, the Antarctic climate is affected by the Atlantic Meridional Ocean Circulation (AMOC) changes during the geological past. The atmospheric Tele connectivity with the Indian Monsoon, El Nino Southern Oscillation, Indonesian throughflow and Arctic climate has also been documented. Changes in the North Atlantic Meridional circulation may affect and also get affected by the Antarctic climate. Based on the average annual precipitation, Antarctica can be called a Polar dessert. A dessert is defined by the average precipitation in the form of rain, snow, fog, or mist. The relative humidity of air at the South Pole is often as low as 0.03%. The average elevation of ice sheets in Antarctica is quite high (~2000 meters) and thus it further gets distanced itself from the surrounding ocean adding to its being the coldest place (Shepherd et al., 2019). Despite having such low precipitation, the thick ice sheets in Antarctica owe their presence to low temperatures, which cause less ablation than accumulation.

The origin, evolution, and establishment of a permanent Ice Sheet in Antarctica have been mainly understood from the marine sedimentary records and date back to the Eocene-Oligocene transition (EOT). In several pelagic sequences, the first major positive +excursion in benthic foraminiferal stable isotope $\delta^{18}O$ of 1.5‰ indicated initiation of the ice sheet on the Antarctic Continent with its peak reaching in the Early Oligocene designated as Oi-1 by Miller et al. (1991). Based on their study of a benthic foraminiferal stable isotope of oxygen, Zachos et al. (1996) constrained the timing and duration of the Oi-1 Event. They found that the event did not occur in a single step but in two steps with the first being at 34.0 Ma to 33.8 Ma with a 0.3–0.5 ‰ rise in $\delta^{18}O$ and the second phase was rapid with a 0.8‰ increase in 50 kyr (Zachos et al., 1996). Such studies provided deep insight into the rate of development of the Antarctic ice sheet and are of great relevance even today. Another evidence for the presence of a large ice sheet, extensive enough to reach the coastal shelf is the ice-rafted debris which also dates back to EOT providing supportive evidence for the Antarctic ice sheet (Carter et al., 2017).

Various forcing factors including orbital forcing and insolation changes, and carbon dioxide levels were attributed as the cause for the initiation of the ice sheet. The threshold values for the CO_2 were 600 ppm around 32. 8 Ma (Galeotti et al., 2016). The history of origin and subsequent permanent establishment of the Antarctic ice sheet has been well constrained by the oxygen isotope record of the benthic foraminifera from the marine sediments (Miller et al., 1987). Thus before 800 kyr BP the history of the Antarctic Ice sheet is mainly known through the marine sediments. For the last 800 kyr, the direct measurements of the Antarctic climate in terms of greenhouse gas concentration, temperature, and the response of the ice sheet to orbital and other forcing factors have been possible only with the study of the Ice Cores. The history of Antarctic climate has been revealed by a large number of cores drilled in the Eastern and Western Antarctic ice sheet but the longest record so far has been from the European Project for Ice Coring in Antarctica (EPICA) Dome C (EDC, 75^0 S, 123^0 E, 3,233 m.a.s.l. (meters above sea level) Ice Core (EPICA Members, 2004), East Antarctic Dome Fuji (Dome F, or Valkyrie Dome) Ice Core; EDML, in the interior of Dronning Maud Land at 75^0 S, 0^0 E, 2,892 m.a.s.l. (EPICA Members, 2006), and Vostok Ice Core (Petit et al., 1999) (Figure 3). Several proxies have been studied from these ice cores and the detailed methodologies can be found in the relevant literature cited here and the references therein. Here the major findings of the climatic history of Antarctica as revealed from these ice cores and the relation with global climate have been discussed.

3.17 EARLIER ANTARCTIC HISTORY FROM ICE CORES

The longest Ice Core drilled so far representing the oldest climate record goes up to 800 kyr BP. Thus, considering the marine isotopic stages this ice core represents MIS 1 to MIS 20. Some of the basic issues connected to the relation between Antarctic climate and global climate can be understood if we first understand the Antarctic climate history through these long ice cores. The climatic history deduced from the ice core begins after Mid Pleistocene Transition (MPT) which occurred around 900 kyr BP. It is worthwhile to mention the ice core record starts at least 100 Ky later than MPT. MPT changed the climate system to a 100 ky (Eccentricity driven) cyclicity from 41 Ky (Obliquity driven) cyclicity.

3.17.1 Mid Pleistocene Transition (MPT)

The Mid Pleistocene transition is an enigmatic event when the frequency of the glacial/interglacial cycles changed from 41 kyr (Obliquity) to 100 kyr (Eccentricity) cycle. This event occurred around 0.9 Ma (or 900 kyr BP) which is 100 kyr before the earliest age from the Antarctic Ice core, though based on benthic oxygen isotope record from the eastern Mediterranean ODP Sites 967 and 968 and an astronomically tuned age model employing benthic foraminiferal titanium versus aluminum (Ti/Al) ratio, Konijnendijk et al. (2015) pointed towards an early (>1200 kyr BP) onset of the Mid Pleistocene Transition and that the transition lasted till at least MIS 15 (~600 kyr BP). Thus, this MPT event has not been recorded in the Ice cores. Another unique character of the climatic signals after MPT is the increase in the amplitude of the climate variation as shown by many workers (Clark et al., 2006).

3.17.2 Post MPT Climatic History of Antarctica Revealed from Ice Cores

3.17.2.1 General Characteristics

The Antarctic climate history for the last 800 kyr is well established with the well-dated long ice cores drilled at the Vostok, Dome Fuji (Dome F) and by the EPICA project. There have been several cores drilled in the Antarctic Ice from the interior to the shelf areas in East and West Antarctica, however, the longest core drilled so far reaches up to 800 kyr corresponding to Marine Isotope Stage 20 (EPICA DOME C) and has been studied by EPICA project members. Various authors have measured a multitude of proxies from these well-dated ice cores including carbon dioxide concentration (Lüthi et al., 2008), methane concentration (Loulergue et al., 2008), water isotopes and paleotemperature (Jouzel et al., 1997; Markle and Steig, 2021), Aeolian dust (Delmonte et al., 2008) and volcanic aerosols (Basile et al., 2001). The Antarctic ice core records also are matched by other marine proxy records like oxygen isotopic ratios of benthic foraminifera from the deep ocean sediments which are indicative of major changes in the ice volume (Brook and Buizert, 2018). All the climatic records from the ice cores in Antarctica show major, synchronous variations across the glacial cycles, which have an average duration of about 100 kyr, however, at millennial scale the Antarctic and Greenland ice core records show a phase difference. The direct measurements of the GHC from the ice cores with paleotemperature data have revealed that in general, the glacial-interglacial stages as recorded from the Ice core are coupled with greenhouse gas concentrations. Similarly, the orbital forcing factor has also played a significant role in affecting the Antarctic climate. Deep ice cores from the Antarctic interior show that on longer timescales, the Antarctic climate closely follows variations in solar insolation that drive climate change globally and that it exhibits major temperature changes at the termination of ice ages (Brook and Buizert, 2018). In general, the climate of the last 500 kyr years was characterized by extremely strong 100-kyr cyclicity, as seen particularly in the Vostok ice-core (Petit et al., 1999) and is also matched by marine-sediment records (Imbrie et al., 1993; Bassinot et al., 1994b). Like all other high-resolution glacial—Interglacial records, the Antarctic records also show a characteristic sawtooth pattern (though less prominent than those in the north) on these timescales, with a gradual cooling trend from glacial inception to peak glacial conditions, followed by a relatively fast glacial termination (Brook and Buizert, 2018). The Antarctic temperature records as deduced from the δD or $\delta^{18}O$ ratios of ice show relatively low amplitude of variation before 450 kyr and after that the amplitude is rather high. On average the amplitude ranges from 6 to 13 degrees Celsius. There is a remarkable coherence of Antarctic temperature variability and global climate change on orbital time scales, although orbital forcing due to precession acts with opposite effects in each hemisphere (Broecker and Denton, 1990). There have been many studies to understand the phase difference between climate changes occurring in the North and South. Such studies have utilized the climate records obtained from well-dated ice cores from Greenland and Antarctica. Model studies comparing GRIP (Greenland) and Byrd (Antarctic) Ice Core data indicate coupling of the climate signal between north and south, though with a phase difference. The coupling is on a millennial-scale mostly around 1000–2000 years. However, the temporal characteristics of the climate records from Antarctic Ice Cores are much different from those of the Greenland Ice cores. Explaining the causative factors for the phase difference has been attempted by several workers and some agreement has been reached regarding the physical mechanism which makes this phase difference. One of the earlier attempts to understand the phase difference was explained by Broecker (1998), Sowers and Bender (1995) based on ^{14}C measurements, $^{18}O/^{16}O$ and methane concentration showed that Antarctica warmed during the Younger Dryas cold event recorded in the North Atlantic. This phase difference has been termed a bipolar seesaw. Stocker and Johnsen (2003) while discussing the bipolar seesaw, showed that the models point towards storage of heat in the south Atlantic in case of a collapse of Thermohaline Circulation (THC) in the North Atlantic but so far as warming of the Antarctic interior is concerned (because of storage of heat in south Atlantic), the different models show opposite results. One shows that under such conditions the Antarctic will warm while the other shows it will cool. Thus, this points towards our poor understanding of the physicality of these mathematical relations. Such problems need high-resolution well-dated marine sedimentary records tied with ice core records to understand the causative factors. However, the study by Stocker and Johnsen (2003) to some extent resolved the time scale of the bipolar see-saw. Wunsch (2003) however did not see any evidence of a 'bipolar seesaw' in either frequency band. The ice cores climate record provides us with a process understanding of the interhemispheric coupling of climate change from the conceptual models (Pedro, 2018).

3.18 IMPORTANT ISSUES REGARDING FINDINGS FROM ICE CORE RECORD

The Antarctic ice core provides a maximum paleoclimatic record spanning the last 800 kyr. One would expect that the 20 MIS representing glacial-interglacial intervals should be reflected in the ice core record and mimic the record from marine carbonate oxygen isotope. Then what is the additional information one gets from the ice core record? How do they add to our knowledge of the paleoclimate for the last 800 kyr? To answer these questions let us look at the data which is provided by the ice cores. One of the most valuable data is an actual measurement of the Carbon-dioxide and methane from the trapped air bubble in the ice cores. Second is the ability to compare the Greenland ice core record for the last 127 kyr (the oldest Greenland ice core record goes up to 127.5 kyr). The actual temperature measurements from the borehole and comparison with measured temperature from the proxy like oxygen and hydrogen isotopes show that the ice core has faithfully recorded the past temperatures. Another interesting feature is the connection between the Antarctic and tropical climate. The tropical regions of the world have witnessed intense changes in the variation in monsoonal intensities. How is Antarctic climate related to the monsoonal variation? Does it provide inputs into causative factors? Many of the studies rely on finding coeval changes in the tropical climatic parameters including Indian and African monsoon and changes observed in the CO_2 and CH_4 in Antarctic and Greenland ice cores. The debatable issues are the series of physical linkages that connect the variations in polar climate to the tropical climate. The cause-and-effect relationship has been explained in varied ways. The typical enhancement of the African and Indian Monsoon during MIS-13 has been based on productivity proxies in the Arabian sea which have also been related to the onset of an intensive meridional overturning circulation in the Atlantic Ocean at the end of the Mid-Pleistocene transition (Ziegler et al., 2010). Modelling studies have also pointed towards a fully alternate hypothesis of the tropics causing changes in the polar climate (Clement et al., 2001). Though comparison between Antarctic and Greenland ice core data has resulted in important concepts like a bipolar see-saw while at the same times issues like whether north leads south or south lead north in terms of causative relationships have been debated. Another interesting issue is whether the bipolar seesaw works in the same way through the ocean and atmosphere. How have the various forcing factors like orbital forcing, greenhouse gas concentrations, and Ocean circulation changes, affected the Antarctic climate during different marine isotopic stages in concert with one another? How are individual glacial or interglacial intervals (Marine Isotopic Stages) different from each other? What is the nature of the profile of marine isotopic stages? Whether the thick ice sheet of Antarctica has been a passive witness to the climate changes or has also affected the climate changes? High-resolution well-dated ice core records from Antarctica have indeed advanced our conceptual understanding of the ocean and climate dynamics operating through the last several hundred kyr. Antarctic Ice cores have a unique setup because they have also arrested the effect of feedback processes operating through changes in albedo due to the waxing and waning of the ice sheet at a longer and shorter time scale. This has resulted in what we call Antarctic Amplification which is less pronounced than Arctic amplifications probably due to contrasting differences in land-sea distribution. In the following section, the detailed record of various parameters from the Antarctic ice cores has been discussed together with comments on the Antarctic climate and its global connectivity. For the sake of simplicity, the discussion has been farmed in chronological order starting from the oldest record from the ice core to the recent past.

3.19 OLDEST ICE RECENTLY DISCOVERED FROM ANTARCTICA AND IMPLICATIONS ON CLIMATE

Apart from the long Ice cores discussed above, a very significant discovery of 2.7 million-year-old ice was made from the Blue Ice Area (BIA) of the Allan Hills region of Antarctica by a team of Scientists (Yan et al., 2019) from ALHIC1502 and ALHIC1503 ice cores. The glaciers slowly move towards the ocean, and their ice turns blue because all its air bubbles are squeezed out due to compaction. This dense ice absorbs the red, and orange wavelengths and a general blue color is reflected. Where the flow of the glaciers is impeded, for example, by the Antarctic Mountains the blue ice may be thrust to the surface and thus this old blue ice is exhumed as a result of erosion, and ablation of the overlying ice by the wind. This is just like exposed old cratonic rocks on continents when the overlying younger rocks are eroded. Blue ice is not layered by age as we observe in ice cores but occurs exposed in areas where younger snow accumulation is either absent or compensated by erosion, ablation, or sublimation. The deepest sample lies within about 1 m of bedrock and has the oldest ^{40}Ar atm age, 2.7 ± 0.3 (1σ) myr. This ice was dated using argon and potassium contained within a piece of ice with an error of plus-minus 100 kyr (Yan et al., 2019). In their initial investigations (Yan et al., 2019) detected a very low concentration of CO_2 (300 ppm) in this 2.7 my old blue ice. This finding is very significant because this was the time of initiation of Northern Hemisphere glaciations. Though the authors interpreted the paleoclimate records from the Allan Hills BIA as discrete 'snapshots'—discontinuous but accurate portraits—of the climate state, rather than continuous time series (Yan et al., 2019).

There has not been any finding of such old blue ice from Greenland to corroborate this discovery of such low CO_2 levels. Also, the CO_2 concentration does not match with findings from other proxies of the same age. However, this discovery by Yan et al. (2019) has opened up a new line of thinking of discovering more and more old blue ice to get an idea of direct measurement of atmospheric concentration of greenhouse gases.

3.20 CARBON DIOXIDE RECORD IN ANTARCTIC ICE CORE AND MARINE ISOTOPIC STAGES

As introduced in a previous section the Marine Isotope Stages MIS) are intervals of depleted/enriched ocean water in heavier oxygen isotope ^{18}O as recorded in the marine carbonate fossils. One of the best contributions of the Antarctic Ice core records has been to make us better understand the nature of various glacial and interglacial intervals up to MIS 20. Thus, it is worthwhile to discuss the climatic record chronologically revealed from Antarctic ice cores from MIS-20 to the Holocene (MIS-1). The following discussions are based on our understanding of the CO_2 record during various MIS as documented in the Antarctic Ice Cores drilled from different strategic locations. For the sake of simplicity, the odd number MIS has been compared with the MIS-1 (Present-Holocene) while the even-numbered have been compared with MIS-2 (The Last Glacial Maximum) besides comparing the various odd and even MIS amongst themselves. Jouzel et al. (2007) provided a detailed record of the deuterium profile from EPICA Dome C for the last 800 kyr and inferred that the change in surface temperature, based on 100-year mean values, was ~15 °C over the past 800 ky, from –10.3 °C for the coldest 100-year interval of MIS 2 to +4.5 °C for the warmest of MIS 5.5 (5e). Jouzel et al. (2007) stated that the proxy records from three long cores of East Antarctica, Dome C, Vostok (Petit et al., 1999; Raynaud et al., 2005) and Dome F (Watanabe et al., 2003) are very similar for the interval of overlapping time and thus the Dome C temperature record can be considered as representative of East Antarctica. The temperature record provided another interesting fact about the nature of glacial periods 430 kyr BP. Jouzel et al. (2007) observed that all the even number Marine Isotopic Stages representing glacial intervals were warmer than the MIS-2 during the last 430 kyr. It was estimated that MIS 14 was the warmest among them with a temperature difference of ~2 °C whereas MIS-12, MIS-16 and MIS-18 were warmer by 1 °C.

Another important issue is the orbital forcing. If we look at the present orbital parameters we have 23 and a half degree obliquity which is high while the precision is such that extreme northern winters occur when the Earth is closest to the Sun (precision minimum effect). Milankovitch (1941) proposed that Earth resides in an interglacial state when its spin axis both tilts to a high obliquity and precesses to align the Northern Hemisphere summer with Earth's nearest approach to the Sun (Huybers, 2011). This general concept has been discussed and generally accepted by many workers. The obliquity maximum and precision in phases can trigger deglaciation. The importance of the Northern hemisphere comes from the fact that nearly 68% of the land lies in the northern hemisphere while 32% lies in the southern hemisphere of which a major part is Antarctica covered 98% by the highest albedo material, the ice. The triggering of deglaciation and later residing in an interglacial period may be different phasing of the orbital parameters. (Hays et al., 1976; Liu, 1998; Huybers and Wunsch, 2005;

Imbrie and Imbrie, 1980; Paillard, 1998; Berger et al., 1999; Tziperman et al., 2006; Berger, 1988; Saltzman, 2002) .

MIS 20 has been a cold interval with 4 substages, MIS 20a, b, c, and d (Railsback et al., 2015). The most detailed record of the CO_2 concentration in the EPICA DOME C was provided by Lüthi et al. (2008). Lüthi et al. (2008) recorded CO_2 concentrations level of 188 ppmv in MIS 20 (814–790 kyr) from EPICA Dome C ice core. MIS 20 lasted for 24 kyr. The next 9 kyr witnessed a rise in CO_2 concentrations by 70 ppmv (= 258 ppmv) during the earlier part of MIS-19 which is an important analog for the present interglacial because of its similar orbital configuration, especially the phasing of the obliquity maximum to precession minimum (Suganuma et al., 2018).

Lüthi et al. (2008) found that in the five glacial-interglacial transitions of the past 450 kyr, similar slopes were present. At the beginning of MIS 19 (~790 ky), the CO_2 concentrations reached 261 ppmv, similar to later MIS 13 and MIS 15 but this value is not as high as the values during the last four recent interglacials (Lüthi et al., 2008). When we compare the orbital pattern of the MIS-5, MIS-9 and MIS-19, they are similar to MIS 11 (424–374 Ky) and Holocene (11700- present). However, both the Holocene and MIS 11 do not show such a decrease rather the CO_2 increases slowly. Thus, we have a case where these odd numbers MIS (1, 5,9,11, and 19) have the same orbital parameters yet, two of them namely MIS-1 and MIS-11 differ in the profile of CO_2 increase. In the interglacial MIS Stage 5 (130–71 Ky), MIS Stage 9 (337–300 Ky), and 19 (790–61 Ky) a similar pattern was observed by Lüthi et al. (2008), where in about 10 kyr the CO_2 levels decreased by 15 ppmv before attaining a saw tooth pattern to glacial intervals. An intriguing question and a fact that emerge from such values is that every interglacial is not the same but many interglacials share the same values of carbon dioxide. Also, another issue that emerges is the beginning of interglacial and sustained warming in an interglacial. The present value of carbon dioxide (410 ppmv) is scary as it points towards unprecedented warming in years to come given the average residence time of carbon dioxide in the atmosphere as ~200 years.

The data of Luthi et al. (2008) for MIS 18 (761–12 kyr) shows two phases in terms of carbon dioxide concentration. In the earlier phase, a value of 177 ppmv is reached and then it rapidly increases to a plateau at a constant value of between 205–212 ppmv lasting for 20 kyr. The beginning of the MIS 17 (Termination VIII) is marked by the decline in carbon dioxide levels (~30 ppmv), thereafter rapidly increasing by 40 ppmv. Though considering carbon dioxide we have left the orbital forcing discussion which also might have played an important role. Parrenin and Paillard (2012) based on their modelling studies found that precession played a fundamental role in the triggering of termination VIII (~720 kyr ago) while obliquity played a vital role in the triggering of termination VI (~530 kyr BP).

Thus, can we say that the orbital forcing triggers deglaciation while the greenhouse concentration maintains the interglacial warm periods? Another important concept that one gets from the study of the ice core is that every interglacial is different, some may be similar. During 40 kyr of MIS 17, the CO_2 levels range between 215 and 240 ppmv (Lüthi et al., 2008) which is considerably less as compared to other interglacials during the past 800 kyr. Lüthi et al. (2008) noted very low values of CO_2 during MIS 16 (180 ppmv for 3 kyr) and they postulated that it may be due to more pronounced glacial carbon storage in the ocean. Significantly during MIS 16, the CO_2 levels were the lowest ever found in ice cores, 172 ppmv (667 kyr BP). This is contrary to the generally accepted range of CO_2 for the late Quaternary (170 to 300 ppmv). However, the precise measurement of CO_2 from the ice core raised some important questions. A comparison of carbon dioxide levels in MIS17 with those of MIS 13, 15 and 19 shows that the values are significantly lower in MIS17. This is not supported by general carbon cycling models (Kohler and Fischer, 2006), nor by similar temperature variations. That means some additional force was playing a role. Similarly, a difference is noted in the glacial carbon dioxide concentration in MIS 14, MIS 16 and MIS 18, where a significantly low concentration of CO_2 occurs as compared with other glacial. Lüthi et al. (2008) tried to answer this difference in the CO_2/temperature relationship. They suggested that the difference in the CO_2/ temperature relationship is probably due to a slow increase in the CO_2 levels of about 25 ppmv from 800 to 400 kyr and a slow decrease of 15 ppmv during the last four glacial cycles. The idea that the changes in CO_2 levels occur at several kyr cycles due to changes in the weathering (Kump et al., 2000), or by major reorganizations in the carbon reservoir of the global ocean is supported by Wang et al., (2004) who reported a 500 kyr cycle in the Quaternary carbon cycle found in marine records. Thus, we see that the odd and even MIS, their CO_2 concentrations, temperature records and orbital similarities all play a resultant role by various combinations making every MIS different.

3.21 MARINE ISOTOPE STAGE 3, ANTARCTIC ISOTOPIC MAXIMA (AIM EVENTS), AND BIPOLAR SEESAW

Though Marine Isotope Stages are characterized by enriched and depleted benthic foraminiferal stable oxygen isotope and numbered from 1 (recent Holocene) interglacial in descending stratigraphic order yet not all the odd number Stages represent true interglacial. These intervals depict the ice volume effect and thus they are also intervals of relatively higher sea levels. Similarly, MIS 3 was also defined based on oxygen isotopic records from marine sediments (foraminifera) (e.g. Imbrie et al., 1984; Bassinot et al., 1994a,b; Waelbroeck et al., 2002). However, MIS 3 has been an exception to the general rule of designating all interglacial by an odd number of Stages, and, during MIS 3, the sea levels ranged between 60 and 90 m below the present mean sea level (Chappell, 2002; Waelbroeck et al., 2002; Siddall et al., 2003; Siddall et al., 2008), and thus, therefore, cannot be described as an truly interglacial. Because the difference between glacial and cooling must be clear. Siddall et al. (2008) gave further justification of MIS 3 should not be considered a true interglacial because its duration is between 60 and 25 Ka and this would not conform with the change in frequency of 100 ky of Milankovitch forcing of glacial/interglacial after Mid Pleistocene (Lisiecki and Raymo, 2005).

During MIS 3, the Antarctic climate is marked by some warming events when the temperature has gone up to 1 to 3 degrees Celsius and a high value of the oxygen isotope ($\delta^{18}O$). These warming events are known as Antarctic isotopic Maxima (AIM) events and are characterized by gradual warming and gradual cooling. This is in contrast to the Dansgaard–Oeschger events in the North where rapid warming (8–16 degrees) is followed by gradual cooling to Greenland stadials. This shows a distinct difference in the pattern of the warming events in the north and south. Now the question is whether AIM events and D-O events are linked? Many workers have explained the physicality of this linkage by a Southern Ocean heat reservoir. In the normal AMOC (Atlantic Meridional Ocean Circulation), the warm water flows to the north, cools become more saline, and sinks. But during the DO events, the AMOC becomes slow and the northward heat transport is reduced. This results in a southern heat reservoir causing warming of the subsurface and surface waters of the South Atlantic and Southern Oceans and is known as the thermal see-saw (Stocker and Johnsen, 2003; Pedro et al., 2018).

3.22 COMPARISON OF GREENLAND AND ANTARCTIC ICE CORE RECORD: EPICA

In paleoclimate studies, precise chronology is essential to understand the cause-and-effect relationship (Figure 3.9). There are two important issues. The first is that the cause will always precede the effect. The second issue is the lag and lead. If two events are separated by a few hundreds or thousands of years, the older one is likely the cause of the latter event. But the "pair repeatability" at least for more than two-three times

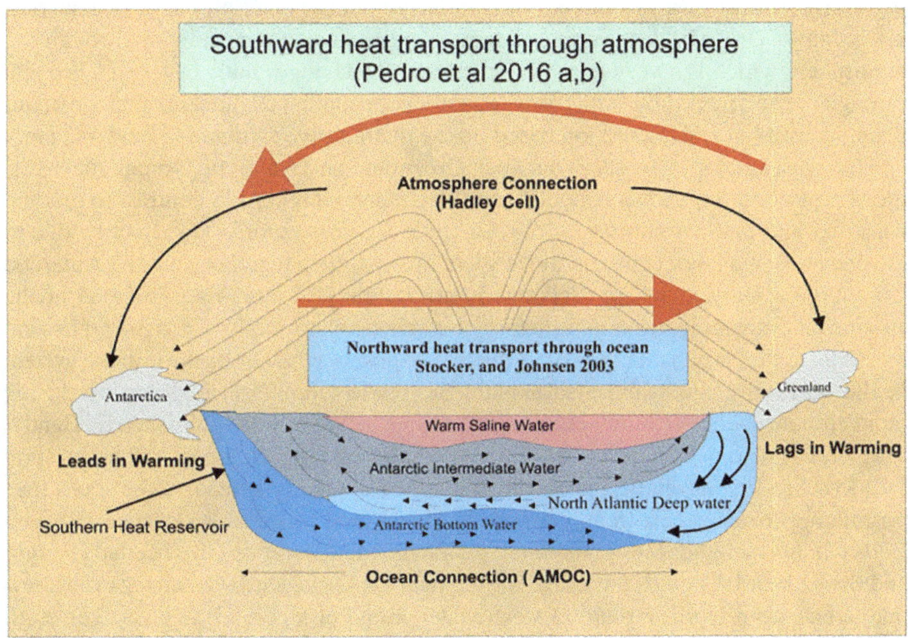

FIGURE 3.9 Teleconnectivity of Antarctic and Greenland ice core record through ocean and atmosphere.

needs to be established. All events cannot become a cause for later events and the difference between coincidence and correlation must be explained by understanding the physicality of the processes. Thus, explaining a viable physical mechanism through which the cause and effect are related is most crucial. The whole idea is to avoid coincidence and concentrate on correlation. Establishing the teleconnections in the climate needs a thorough understanding of the leads and lags and the physical connectivity of the events.

Antarctica and Greenland climate connectivity are classical examples. The (dis)connectivity has been studied since the beginning of the Antarctic ice sheet and important studies have been made on the (dis)connectivity between the North Atlantic and South Atlantic reflecting changes in Greenland Ice and Antarctic Ice sheet. Many advocated even the initiation of the Northern Component Water to the development of Antarctic Circumpolar Current while few advocated that the NADW is independently developed on its own due to changes in depth of the Greenland-Scotland ridge (Abelson and Erez, 2017). Thus, this debate is going on for various time scales right from the initiation of the northern hemisphere glaciations to the last of the Dansgaard–Oeschger events. An important study was made by Blunier et al. (1998) in this regard. Blunier et al. (1998) measured methane from the Greenland and Antarctic ice cores and established a precise chronology. They found that the Greenland warming at 47, 36, and 47 kyr is preceded by Antarctic warming. The lead of the Antarctic warming was approximately 1 kyr. They also argued about the physical mechanism through which the Antarctic warming leads Greenland warming by 1 kyr. Based on the amount of lead/lag Blunier et al. (1998) argued that since 1 kyr lag of Greenland warming cannot be attributed to connectivity through the atmosphere which is a very quick mixing time, the changes must have been brought through ocean circulation. Thus, this is another conceptual advancement made in our understanding of teleconnection based on the quantitative estimate of lead and lag.

One of the major differences between Greenland and Antarctic climate records is the abrupt warming and slow cooling in the Greenland record in contrast to gradual warming and gradual cooling in the Antarctic ice core record. This may be due to the presence of a large continental ice sheet in Antarctica together with the Antarctic Circumpolar Current where a change is not so quick. Thus this poses one of the significant intriguing questions before the scientists to explain a viable physical mechanism that is capable of explaining this opposite behavior of the climate system in the north and south and also explain the phase (antiphase) relationship during rapid climate change (Blunier et al., 1998). Opposite views were published by Bender et al. (1994) who believed that those interstadial events which are of longer duration (> 2 kyr) are first developed in the north and subsequently spread to the south thus postulating that northern temperature variations lead those in Antarctica. So who leads whom is a big question and more data from more locations are needed (Wright and Steig, 1998). Though the cause for the heat for Dansgaard–Oeschger events initiate come from North Atlantic Ocean water, Bond et al. (1993) and Stocker et al. (1992) believed that this heat is drawn to the north Atlantic from the Southern Ocean which results in a cooling of surface air temperatures in the south and corresponding warming in the north. Another believer of the south leading north was Charles et al. (1996) who identified a 1.5 kyr lead of the southern hemisphere climatic fluctuation over the northern hemisphere based on stable oxygen isotope of benthic foraminifera from the Southern Ocean. They also believed that the climate of the two hemispheres is not linked through the deep-water variability. Another important illustrative example of the south leading north is the Antarctic Cold Reversal's 1.8 kyr lead to Younger Dryas (Chappellaz et al., 1993; Chappellaz et al., 1997). Whether a lead is a cause for the lag can only be understood by the repeated events. Is Antarctic cold reversal responsible for the Younger Dryas? One thing is clear, there was no contemporary cold event in Antarctica during Younger Dryas but rather the climate of Antarctica warmed during Younger Dryas.

3.23 OCEAN VS. ATMOSPHERE TELECONNECTIONS

There have been considerable debates over the Bipolar seesaw but the relative role of the atmospheric and ocean heat transport remains enigmatic. The ice core record of the Antarctic Cold Reversal (ACR) and its comparison with the Greenland Ice core record does show that the ACR corresponds to 14.7–13 kyr BP which is the timing of the Bølling-Allerød event in the North Atlantic. The drawdown of the heat from the southern hemisphere causing Bølling-Allerød warming in the north and ice expansion in the south has been proposed by many and is a typical example of the bipolar seesaw (Stocker and Johnsen 2003), but on the other hand, Pedro et al. (2016a, 2016b) show that the Bipolar seesaw works oppositely in the atmosphere and ocean (Figure 3.8). They compared several proxy records of paleoclimate from the continental climate archives from Australia, Patagonia, and Amazon (speleothems) and concluded that during northward heat transport of the ocean, there is a southward heat transport via the atmosphere (Figure 3.8). The study revealed that despite the strengthened southward heat transport via the Hadley cell, the southern high latitudes cool because the northward anomaly in ocean heat transport exceeds this southward anomaly in the atmosphere. During the ACR the AMOC strengthens but in concert with that the Hadley cell southward heat transport also strengthens the evidence which is preserved in the dry and warm climate record from the continents. Thus, the debate over whether the bipolar sea seesaw through the ocean is compensated by opposing behavior in the atmosphere is resolved to some extent through more records are needed. The compensation does occur with the ultimate dominance of the ocean heat transport.

3.24 UNDERSTANDING LATE PREINDUSTRIAL HOLOCENE CO_2 VARIATIONS FROM ICE CORES

The Law Dome ice core has a high snow accumulation rate and has documented the carbon dioxide concentration during historical times. The coupled carbon-climate system (Arora et al., 2013) needs to be understood from a robust record of CO_2 including a time representing no anthropogenic activities to pre-industrial to industrial time. All our future predictions of average global temperatures change due to greenhouse gas emissions are based on the understanding of the coupling of the GHG and climate. As the climate system depends on a large number of factors, filtering out the effect of GHG can be done if we have a record. The Paris agreement has targeted a substantial reduction of GHG emissions by the end of 2021 leading to limiting global warming to well below 2 °C and pursuing efforts to limit it to 1.5 °C. All estimates of future global warming require an additional understanding of the carbon cycle processes and feedback, together with the relationship between CO_2 emissions and CO_2 concentration (Friedlingstein et al. (2014). Aerosol forcing is another important aspect in understanding coupled carbon-climate systems. Feedback processes particularly at the polar ice sheets do modify the effect of CO_2 forcing on climate. Law Dome, Antarctica, provides the best time-resolved ice core records due to the very high accumulation rate and relatively quick bubble close-off time at this site (Etheridge et al., 1996; Goodwin, 1990). Quick bubble close-off time ensures that the air trapped in the ice cores is truly representative of the overlying atmosphere and the age difference is also reduced. If the bubble closes off time is large, as in the case of cores having a very slow accumulation rate the difference in the age of the ice and that of the trapped air is quite large.

3.25 THE HISTORICAL RECORD OF CO_2 FROM ICE CORE (LITTLE ICE AGE: PREINDUSTRIAL TO INDUSTRIAL TIMES)

Etheridge et al. (1996) provided a record of atmospheric CO_2 mixing ratio for the period 1006 AD to 1978 AD by analyses of the air enclosed in three ice cores from Law Dome, Antarctica. The age resolution was excellent due to the high accumulation rate at the Law dome. The CO_2 data from the ice core at least overlapped with direct instrument measurement for 20 years. They determined CO_2 mixing ratios between 275–84 ppmv during the pre-industrial period with lower values between 1540–1800 AD (Little Ice Age). Etheridge et al. (1996) demonstrated major growth in CO_2 after the Industrial period except between 1935–45 AD. The Law Dome core is presently isolated from the ice flow of the main inland ice sheet by the drainage effect of the adjacent glaciers and the top layers are free from ice deformation. These factors make the Law dome ice core ideal for comparison of instrumental records and the transition from no anthropogenic activity to preindustrial to industrial times. Snow surface melting is also almost absent. The record of the low CO_2 during 1540–1800 AD in the Law Dome was argued by Etheridge et al. (1996) to represent a global signature recorded in the Law Dome ice core. CO_2 concentrations are often altered and influenced by local factors including particulate impurities, land surfaces, forest fires or volcanoes, etc. The Law Dome ice core record is naturally averaged over a decade which is many times the atmospheric mixing time between two hemispheres and thus they represent a global signature. The debate persists whether the LIA was caused by the Green House forcing or there were other causes. The empirical temperature CO_2 relationship requires a drop in temperature of .094 °C per ppmv. Raynaude et al. (1993) estimated that the temperature changes in the Vostok ice core are 50% due to radiative forcing and out of this about 70 % was attributed to CO_2 forcing. Keeping this estimate in view the Law dome Core which has many similarities with Vostok, Ethridhe et al. (1996) gave a 0.2 °C cooling estimate corresponding to a 6 ppmv lower CO_2 during LIA. However, the magnitude of the LIA cooling as estimated by several workers is 1–2 °C (Grove, 1988). Thus the CO_2 forcing might have been a contributing factor but not the only cause of the LIA (Etheridge et al., 1996). Though the Maunder minimum (MM), a period of extremely low solar activity also occurred during this period approximately AD 1650 to 1715, the timing of the features is not suggestive of causation and should not, in isolation, be used as evidence of significant solar forcing of climate during LIA (Owens et al., 2017). The record of the Little Ice age has been documented in many parts of the world (Bradley and Jonest, 1993) and is thus considered a global climatic event. The Antarctic Ice core data from the Law Dome does give clues about one forcing factor i.e., the greenhouse forcing though the entire cooling cannot be attributed to GHF. Another forcing factor for the initiation of the Little Ice Age was postulated by Miller et al. (2012) who attributed the LIA to an unusual 50-year-long episode with four large sulfur-rich explosive eruptions, each with global sulfate loading >60 Tg. Thus, we have three probable forcing mechanisms for the LIA including low CO_2 concentration as recorded in the Law Dome Ice Core, explosive volcanism, and Maunder Minimum.

3.26 ANTARCTICA-TROPICS CONNECTIONS

Studies have been made to understand the linkages of the Antarctic climate with the Asian Monsoon. $\delta^{18}O$ record from a well-dated (^{230} Th) stalagmite from Yongxing Cave in central China preserves an excellent record of variation in Asian Monsoon strength for the period 88 to 22 kyr (Chen et al., 2016). This record was analyzed by removing the effect of the 65 °N insolation signal to see the relationship between the residual $\delta^{18}O$ record and Antarctic temperature variability. It was observed that the residual $\delta^{18}O$ record is not in phase with Antarctic temperature variability on sub-orbital time scales during the Marine Isotope Stage (MIS) 3 (Chen et al., 2016). The linkages between north and south have been possible either with ocean circulation (Stocker and Johnsen 2003) or with the help of atmospheric circulation (Pedro et al., 2016a, 2016b). The stalagmite record from China also demonstrates that the Antarctic climate is linked to the northern hemisphere's monsoonal changes through the atmosphere. Chen et al., 2016) found that the periods of reduced monsoon strength are closely associated with the Antarctic warm climate and occur during the Terminations.

During MIS-13 interglacial, widespread enhancement of Monsoon was reported by several workers like Yin and Guo (2008) but Ziegler et al. (2010) argued against it doubting the ages of the planktic foraminiferal isotope excursion in the equatorial Indian Ocean and, Sapropel layers in the Mediterranean (Bassinot et al., 1994b; Rossignol-Strick et al., 1998) during MIS-13. These authors provided an alternative explanation for the productivity increase in the Arabian Sea during MIS-13 also relating to low methane concentration in the EPICA dome C ice core in Antarctica. The argument was that heavy precipitation worldwide would have increased the wetland production of methane and should have caused the EPICA Dome methane record to shoot up during MIS-13 which is otherwise. The EPICA record shows a low concentration of methane during MIS-13.

Another interesting teleconnection of the Indian Monsoon and Antarctic climate has been through Mascarene High. This high-pressure region is located between 25 °S–35 °S and 40 °E–90 °E near the Mascarene Islands in the southern Indian Ocean. This is the engine that powers the Indian monsoon (Kumar et al., 2021; Han et al., 2017). Azhar et al. (2020) analyzed the relationship between sea ice extent in Antarctica from June-July-August-September from 1979 to 2013 and its relationship with Indian Monsoon. These authors observed a positive relationship between the sea ice extent (SIE) and Mascarene High. The high ice phase corresponds respectively to the strengthening of the Mascarene High as well as an increase in Indian summer monsoon rainfall. However, Prabhu et al. (2009) observed a connection between the patterns of the western Pacific Ocean sector's SIE of the Antarctic in March with that of the Indian monsoon rainfall. In both the studies the sea ice extent in Antarctica is positively correlated with Indian Summer Monsoon rainfall. If we extend this observation to the past then can we say that long-term Antarctic cold events led to enhanced summer monsoon? And as the bipolar seesaw, the North Atlantic warm events would be the interval of the enhanced summer monsoon. Thus, as Kutbach (1981) observed, the monsoon intensity is governed by insolation changes at 65 °N, and Northern hemisphere insolation changes are related to Indian monsoon strength. Guo et al. (2009), based on a comparison of the Chinese loess records with temperature records from Antarctica suggested that the climates of both hemispheres were different during MIS 13, with the northern hemisphere much warmer. Another support for this interpretation comes from Antarctic EPICA Dome C temperature records (Jouzel et al., 2007) which show that MIS 13 was relatively cold interglacial as compared to following interglacial periods. Thus we see that the Antarctic Ice core records provide us with a great understanding of the climate system, feedback, forcing factors, response system, and global teleconnectivity from north to south and from polar to tropical latitudes. Though the oldest record goes back only to 800 kyr, it provides conceptual advancements in our understanding of the cause and effect relationship and physical processes of linkages, that can be applied for time beyond 800 kyr.

3.27 CONCLUSIONS

The Antarctic climate and its variability in the recent past are best studied through the long ice core records obtained from a few important localities in Antarctica. Ice cores have many advantages over marine sedimentary records in terms of their completeness; ability to calibrate actual and measured temperature data through proxies. Ice core records have faithfully preserved the signatures of variations in the rate of snowfall and temperature of the region where they are located. In addition, the windblown dust and sea salt record, and the trapped air bubbles containing major and trace gases of the prevailing atmosphere, all provide us clues to the climate changes through ages and over wide regions. (Alley, 2000). Also, the ice cores add conceptual advancements to our understanding of the forcing factors, their response, and the thresholds in the climate system when it changes from cold to warm and vice versa. While climate change in the low latitude region is governed by a multitude of factors including land use, anthropogenic activities, ocean circulation, greenhouse gas emission and external forcing like sun's radiation, the climate change at the poles, particularly Antarctica are mainly controlled by greenhouse concentration, orbital forcings, and feedback processes. The feedback process probably is a key factor for Antarctic amplification, which is a manifestation of a climate signal two to three times the global average. For a clear understanding of the Antarctic amplification, the ice core records provide an excellent opportunity. One of the most significant findings from the long Antarctic ice core has been to understand the nature of the glacial and interglacial intervals including their amplitude and profile. The comparison of the paleoclimate record of the Antarctic ice cores with mid-latitude marine sedimentary records reveals teleconnections in the climate through the ocean and atmospheric circulation. Also, the abruptness of the climate change in the north pole is not so evident in Antarctica where the changes are rather gradual. The Antarctic Ice core record compared with the Greenland Ice Core record enables us to understand the interhemispheric coupling. The ice core studies have revealed that during the last glacial period, the Dansgaard–Oeschger events in Greenland and millennial-scale warm events in Antarctica are strongly coupled with a time lag through the Atlantic meridional overturning Circulation showing teleconnections between north and south. The bipolar seesaw is to a great extent proved by comparing Byrd (Antarctica) and Greenland Ice core records. The Antarctic Sea ice extent is related to the Indian and African monsoon. The Antarctic ice expansion and sea ice extension are related to enhanced summer monsoon. Well, dated stalagmite δ^{18}Orecord between 88 and 22 kyr BP from Yongxing Cave in central China characterizes changes in Asian monsoon (AM) strength and the studies have shown that the record is strongly anti-phased with Antarctic temperature variability on sub-orbital timescales during the Marine Isotope Stage (MIS) 3, thus establishing teleconnections between monsoon and Antarctic climate. Another major contribution of the Ice core record has been the identification of Antarctic Isotope Maxima Events (AIM). During MIS 3, the Antarctic climate is marked by some warming events when the temperature has gone up to 1 to 3 ^{0}C and a high value of the oxygen isotope. These warming events are known as AIM events and are characterized by gradual warming and gradual cooling. This is in contrast to the Dansgaard–Oeschger events in the North where rapid warming (8–16 degrees) is followed by gradual cooling to Greenland Stadials. This shows a distinct difference in the pattern of the warming events in the north and south. Several theories have emerged to explain the bipolar seesaw including the development of a southern heat reservoir during Greenland stadials. With further studies of high-resolution paleoclimate covering diverse geographical regions of the world with more precise chronology, we will be able to understand more about the Antarctic and global climate teleconnections. Thus, we see that the Antarctic Ice core records provide us with a great understanding of the climate system, feedback, forcing factors, response system, and global teleconnectivity from north to south and from polar to tropical latitudes. Though the oldest record goes back only to 800 kyr, it provides conceptual advancements in our understanding of the cause-and-effect relationship and physical processes of linkages that can be applied for time beyond 800 kyr.

ACKNOWLEDGMENTS

AKS & DKS thank the Delhi School of Climate Change and Sustainability (DSCCS) of the Institution of Eminence (IoE) and the Department of Geology, University of Delhi, for providing infrastructural support. Financial assistance from the Ministry of Earth Sciences, Govt of India (Sanction No. MoES/CCR/Paleo-4/2019) is thankfully acknowledged. Other authors (KM, VPS, AS, TK) thank their respective institutions for logistic help.

REFERENCES

Abelson, M., & Erez, J. (2017). The onset of modern-like Atlantic meridional overturning circulation at the Eocene-Oligocene transition: Evidence, causes, and possible implications for global cooling. *Geochemistry, Geophysics, Geosystems*, 18(6), 2177–2199. http://doi.org/10.1002/2017gc006826

Abram, N., Mulvaney, R., Wolff, E. et al. Acceleration of snowmelt in an Antarctic Peninsula ice core during the twentieth century. *Nature Geosci*, 6, 404–411 (2013). https://doi.org/10.1038/ngeo1787

Aciego, S., Bourdon, B., Schwander, J., Baur, H., & Forieri, A. (2011). Toward a radiometric ice clock: Uranium ages of the Dome C ice core. *Quaternary Science Reviews*, 30(19–20), 2389–2397. http://doi.org/10.1016/j.quascirev.2011.06.008

Ahlmann, H. W. Son, 1935. The stratification of the snow and firn on Isachsen's Plateau. In *Scientific Results of the Norwegian and Swedish Spitzbergen Expedition*. Geografiska Annaler, Taylor and Francis. ISSN: 2001-4422,

Alley, R. B. (2000). Ice-core evidence of abrupt climate changes. *Proceedings of the National Academy of Sciences*, 97(4), 1331–1334. http://doi.org/10.1073/pnas.97.4.1331

Arora, V. K., Boer, G. J., Friedlingstein, P., Eby, M., Jones, C. D., Christian, J. R., . . . Wu, T. (2013). Carbon—concentration and carbon—climate feedbacks in CMIP5 earth system models. *Journal of Climate*, 26(15), 5289–5314. http://doi.org/10.1175/jcli-d-12-00494.1

Azhar, A., Chenoli, S. N., Samah, A. A., & Kim, S.-J. (2020). The linkages between Antarctic sea ice extent and Indian summer monsoon rainfall. *Polar Science*, 100537. http://doi.org/10.1016/j.polar.2020.100537

Barrett, B. E., Nicholls, K. W., Murray, T., Smith, A. M., & Vaughan, D. G. (2009). Rapid recent warming on Rutford Ice Stream, West Antarctica, from borehole thermometry. *Geophysical Research Letters*, 36(2), n/a—n/a. http://doi.org/10.1029/2008gl036369

Basile, I., Petit, J. R., Touron, S., Grousset, F. E., & Barkov, N. (2001). Volcanic layers in Antarctic (Vostok) ice cores: Source identification and atmospheric implications. *Journal of Geophysical Research: Atmospheres*, 106(D23), 31915–31931. http://doi.org/10.1029/2000jd000102

Bassinot, F. C., Beaufort, L., Vincent, E., Labeyrie, L. D., Rostek, F., Müller, P. J., Quidelleur, X., & Lancelot, Y. (1994a). Coarse fraction fluctuations in pelagic carbonate sediments from the tropical Indian Ocean: A 1500 kyr record of carbonate dissolution. *Paleoceanography*, 9(4), 579–609.

Bassinot, F. C., Labeyrie, L. D., Vincent, E., Quidelleur, X., Shackleton, N. J., & Lancelot, Y. (1994b). The astronomical theory of climate and the age of the Brunhes-Matuyama magnetic reversal. *Earth and Planetary Science Letters*, 126(1–3), 91–108. http://doi.org/10.1016/0012-821x(94)90244-5

Bender, M. (2002). Orbital tuning chronology for the Vostok climate record supported by trapped gas composition. *Earth and Planetary Science Letters*, 204, 275–289.

Bender, M., Sowers, T., Dickson, M. L., et al. (1994). Climate correlations between Greenland and Antarctica during the past 100,000 years. *Nature*, 372, 663–666. https://doi.org/10.1038/372663a0

Benson, C. S. (1962). Stratigraphic studies in the snow and firn of the Greenland ice sheet. *SIPRE Research Report*, 70, 76–83.

Berger, A. (1978). Long-term variations of daily insolation and Quaternary climatic changes. *Journal of Atmospheric Sciences*, 35, 2362–2367.

Berger, A. (1988). Milankovitch theory and climate. *Reviews of Geophysics*, 26, 624–657.

Berger, A., Li, X., & Loutre, M. (1999). Modelling northern hemisphere ice volume over the last 3Ma. *Quaternary Science Reviews*, 18, 1–11.

Blunier, T. et al. (1998). Asynchrony of Antarctic and Greenland climate change during the last glacial period. *Nature*, 394, 739–743.

Blunier, T., & Brook, E. J. (2001). Timing of millennial-scale climate change in Antarctica and Greenland during the last glacial period. *Science*, 291, 109–112.

Blunier, T., Schwander, J., Chappellaz, J., Parrenin, F., & Barnola, J. M. (2004). What was the surface temperature in central Antarctica during the last glacial maximum? *Earth and Planetary Science Letters*, 218(3–4), 379–388. http://doi.org/10.1016/s0012-821x(03)00672-1

Bond, G. et al. (1993). Correlations between climate records from North Atlantic sediments and Greenland ice. *Nature*, 365, 143±147.

Bradley, R. S., & Jonest, P. D. (1993). “Little Ice Age” summer temperature variations: Their nature and relevance to recent global warming trends. *The Holocene*, 3(4), 367–376. http://doi.org/10.1177/095968369300300409

Broecker, W. S. (1998). Paleocean circulation during the last deglaciation: A bipolar seesaw? *Paleoceanography*, 13(2), 119–121.

Broecker, W. S., Bond, G., Klas, M., Bonani, G., & Wolfli, W. (1990). A salt oscillator in the Glacial Atlantic? *Paleoceanograpy*, 5(4), 469–477.

Broecker, W. S., & Denton, G. H. (1990). The role of ocean-atmosphere reorganizations in glacial cycles. *Quaternary Science Reviews*, 9, 305–341.

Bromwich, D. H. (1990). Estimates of Antarctic precipitation. *Nature*, 343(6259), 627–629. http://doi.org/10.1038/343627a0

Brook, E. J., & Buizert, C. (2018). Antarctic and global climate history viewed from ice cores. *Nature*, 558(7709), 200–208. http://doi.org/10.1038/s41586-018-0172-5

Bryan, F., & Oort, A. (1984). Seasonal variation of the global water balance based on aerological data. *Journal of Geophysical Research*, 89(D7), 11717. http://doi.org/10.1029/jd089id07p11717

Caballero, R., & Hanley, J. (2012). Midlatitude eddies, storm-track diffusivity, and poleward moisture transport in warm climates. *Journal of the Atmospheric Sciences*, 69(11), 3237–3250. http://doi.org/10.1175/jas-d-12-035.1

Carter, A., Riley, T. R., Hillenbrand, C.-D., & Rittner, M. (2017). Widespread Antarctic glaciation during the Late Eocene. *Earth and Planetary Science Letters*, 58(2017), 49–57.

Chappell, J. (2002). Sea level changes forced ice breakouts in the Last Glacial cycle: New results from coral terraces. *Quaternary Science Reviews*, 21, 1229–1240.

Chappellaz, J., Blunier, T., Kints, S., Dällenbach, A., Barnola, J.-M., Schwander, J., Raynaud, D., & Stauffer, B. (1997). Changes in the atmospheric CH4 gradient between Greenland and Antarctica during the Holocene. *Journal of Geophysical Research* 102(D13), 15987–15999.

Chappellaz, J., Blunier, T., Raynaud, D., Barnola, J. M., Schwander, J., & Stauffer, B. (1993). Synchronous changes in atmospheric CH4 and Greenland climate between 40 and 8 kyr BP. *Nature*, 366, 443–445.

Charles, C. D., Lynch-Stieglitz, J., Ninnemann, U. S., & Fairbanks, R. G. (1996). Climate connections between the hemisphere revealed by deep-sea sediment core/ice core correlations. *Earth and Planetary Science Letters*, 142, 19–27.

Chen, S., Wang, Y., Cheng, H., Edwards, R. L., Wang, X., Kong, X., & Liu, D. (2016). Strong coupling of Asian Monsoon and Antarctic climates on sub-orbital timescales. *Scientific Reports*, 6(1), 32995. http://doi.org/10.1038/srep32995

Clark, P. U., Archer, D., Pollard, D., Blum, J. D., Rial, J. A., Brovkin, V., . . . Roy, M. (2006). The middle Pleistocene transition: characteristics, mechanisms, and implications for long-term changes in atmospheric pCO2. *Quaternary Science Reviews*, 25(23–24), 3150–3184. http://doi.org/10.1016/j.quascirev.2006.07.008

Clement, A. C., Cane, M. A., & Seager, R. (2001). An orbitally driven tropical source for abrupt climate change. *Journal of Climate*, 14(11), 2369–2375.

Connolley, W. M., & King, J. C. (1993). Atmospheric water-vapour transport to Antarctica inferred from radiosonde data. *Quarterly Journal of the Royal Meteorological Society*, 119, 325–342.

Cuffey, K. M., Clow, G. D., Steig, E. J., Buizertd, C., Fudge, T. J., Koutnikc, M., Waddington, E. D., Alleye, R. B., & Jeffrey, P. (2016). Deglacial temperature history of West Antarctica. *Proceedings of National Academy of Sciences USA*, 113, 14249–14254.

Das, S., & Alley, R. B. (2005). Characterization and formation of melt layers in polar snow: Observations and experiments from West Antarctica. *Journal of Glaciology*, 51(173), 307–313. http://doi.org/10.3189/172756505781829395

Delmonte, B., Andersson, P. S., Hansson, M., Schöberg, H., Petit, J. R., Basile-Doelsch, I., & Maggi, V. (2008). Aeolian dust in East Antarctica (EPICA-Dome C and Vostok): Provenance during glacial ages over the last 800 kyr. *Geophysical Research Letters*, 35(7), n/a—n/a. http://doi.org/10.1029/2008gl033382

Delmonte, B., Petit, J. R., & Maggi, V. (2002). Glacial to Holocene implications of the new 27000-year dust record from the EPICA Dome C (East Antarctica) ice core. *Climate Dynamics*, 18, 647–660.

Dome Fuji Project Members. (2017). State dependence of climatic instability over the past 720,000 years from Antarctic ice cores and climate modelling. *Science Advances*, 3, e1600446.

EPICA Community Members. (2004). Eight glacial cycles from an Antarctic ice core. *Nature*, 429, 623–628.

EPICA Community Members. (2006). One-to-one coupling of glacial climate variability in Greenland and Antarctica. *Nature*, 444(7116), 195–198. http://doi.org/10.1038/nature05301

Etheridge, D. M., Steele, L. P., Langenfeld, R. L., Francey, R. J., Barnola, J. M., & Morgan, V. I. (1996). Natural and anthropogenic changes in atmospheric CO2 over the last 1000 years from air in Antarctic ice and firn. *Journal of Geophysical Research*, 101, 4115–4128.

Fireman, E. (1986). Uranium-series dating of Allan Hills ice. *Journal of Geophysical Research-Solid Earth and Planets*, 91, D539–D544.

Friedlingstein, P., Andrew, R. M., Rogelj, J., Peters, G. P., Canadell, J. G., Knutti, R., . . . Le Quéré, C. (2014). Persistent growth of CO2 emissions and implications for reaching climate targets. *Nature Geoscience*, 7(10), 709–715. http://doi.org/10.1038/ngeo2248

Galeotti, S., DeConto, R., Naish, T., Stocchi, P., Florindo, F., Pagani, M., . . . Zachos, J. C. (2016). Antarctic Ice Sheet variability across the Eocene-Oligocene boundary climate transition. *Science*, 352(6281), 76–80.

Gettelman, A., Walden, V. P., Miloshevich, L. M., Roth, W. L., & Halter, B. (2006). Relative humidity over Antarctica from radiosondes, satellites, and a general circulation model. *Journal of Geophysical Research*, 111(D9). http://doi.org/10.1029/2005jd006636

Gildor, H. (2003). When Earth's freezer door is left ajar. *Eos*, 84(23).

Giovinetto, M. B., & Bentley, C. R. (1985). Surface balance in ice drainage systems of Antarctica. *Antarctica Journal of the United States of America*, 20 (4), 6–13.

Goodwin, I. D. (1990). Snow accumulation and surface topography in the katabatic zone of Eastern Wilkes Land, Antarctica. *Antarctic Science*, 2, 235–242. https://doi.org/10.1017/S0954102090000323

Grove, J. M. (1988). *The Little Ice Age*. Methuen, New York, 498 pp.

Guo, Z. T., Berger, A., Yin, Q. Z., & Qin, L. (2009). Strong asymmetry of hemispheric climates during MIS-13 inferred from correlating China loess and Antarctica ice records. *Climate of the Past*, 5, 21–31. www.clim-past.net/5/21/2009/

Han, X., Wei, F., & Chen, X. (2017). Influence of the anomalous patterns of the Mascarene and Australian highs on precipitation during the prerainy season in South China. *Advances in Meteorology,* 2017, 1–12. https://doi.org/10.1155/2017/640802

Hays, J., Imbrie, J., & Shackleton, N. (1976). Variations in the Earth's orbit: Pacemaker of the ice ages. *Science*, 194, 1121–1132.

Huybers, P. (2011). Combined obliquity and precession pacing of late Pleistocene deglaciations. *Nature*, 480, 229–232. https://doi.org/10.1038/nature10626

Huybers, P., & Wunsch, C. (2005). Obliquity pacing of the late Pleistocene glacial terminations. *Nature*, 434, 491–494.

Imbrie, J., & Imbrie, J. Z. (1980). Modelling the climatic response to orbital variations. *Science*, 207(4434), 943–953. http://doi.org/10.1126/science.207.4434.943

Imbrie, J., et al. (1993). On the structure and origin of major glaciation cycles. 2. The 100,000-year cycle. *Paleoceanography*, 8, 699–735.

Imbrie, J., Hays, J. D., Martinson, D. G., McIntyre, A., Mix, A. C., Morley, J. J., Pisias, N. G., Prell, W. L., & Shackleton, N. J. (1984). The orbital theory of Pleistocene climate: Support from a revised chronology of the marine 5 180 records. In A. L. Berger et al. (Eds.), *Milankovich and Climate*, Part 1, pp. 269–305. Reidel Publishing Company.

Jacobs, S. S., Helmer, H., Doake, C. S. M., Jenkins, A., & Frolich, R. M. (1992). Melting of the ice shelves and the mass balance of Antarctica. *Journal of Glaciology*, 38(130), 275–387.

Jouzel, J., Alley, R. B., Cuffey, K. M., Dansgaard, W., Grootes P., Hoffmann, G., Johnsen, S. J., Koster R. D., Peel, D., Shuman, C. A., Stievenard, M., Stuiver, M., & White, J. (1997). Validity of the temperature reconstruction from water isotopes in ice cores. *Journal of Geophysical Research*, 102(C12), 26,471–26,487.

Jouzel, J., Masson-Delmotte, V., Cattani, O., Dreyfus, G., Falourd, S., Hoffmann, G., . . . Wolff, E. W. (2007). Orbital and millennial antarctic climate variability over the Past 800,000 years. *Science*, 317(5839), 793–796. http://doi.org/10.1126/science.1141038

Jouzel, J., Merlivat, L., & Lorius, C. (1982). Deuterium excess in an East Antarctic ice core suggests higher relative humidity at the oceanic surface during the Last Glacial Maximum. *Nature (London)*, 299, 688–691.

Kawamura, K. et al. (2017). Dome Fuji Ice Core Project Members: State dependence of climatic instability over the past 720,000 years from Antarctic ice cores and climate modelling. *Science Advances*, 3(2). https://doi.org/10.1126/sciadv.1600446 10.1126/sciadv.1600446

Kawashima, K., & Yamada, T. (1997). Experimental studies on the transformation from firn to ice in the wet-snow zone of temperate glaciers. *Annals of Glaciology*, 24, 181–185. http://doi.org/10.3189/s0260305500012143

Köhler, P., & Fischer, H. (2006). Simulating low-frequency changes in atmospheric CO2during the last 740 000 years. *Climate of the Past*, 2, 57–78.

Konijnendijk, T. Y. M., Ziegler, M., & Lourens, L. J. (2015). On the timing and forcing mechanisms of late Pleistocene glacial terminations: Insights from a new high-resolution benthic stable oxygen isotope record of the eastern Mediterranean. *Quaternary Science Reviews*, 129, 308–320. http://doi.org/10.1016/j.quascirev.2015.10.005

Kumar, V., Tiwari, M., Prakash, P., Mohan, R., & Thamban, M. (2021). SST changes in the Indian sector of the Southern Ocean and their teleconnection with the Indian monsoon during the last glacial period. *Paleoceanography and Paleoclimatology*, 36, e2020PA004139. https://doi. org/10.1029/2020PA004139

Kump, L. R., Brantley, S. L., & Arthur, M. A. (2000). Chemical weathering, atmosphericCO2, and climate. *Annual Review of Earth and Planetary Sciences*, 28, 611–667.

Kutbach, J. (1981). Monsoon climate of the early Holocene: Climate experiment with the earth's orbital parameters for 9000 years ago. *Science*, 214, 59–61.

Lang, N., & Wolff, E. W. (2011). Interglacial and glacial variability from the last 800 ka in marine, ice and terrestrial archives. *Climate of the Past*, 7, 361–380. http://doi.org/10.5194/cp-7-361-2011

Langway, C. C. Jr., & Shoji, H. (1990). Past temperature record from the analysis of melt features in the Dye 3, Greenland, Ice Core. *Annals of Glaciology*, 14, 343–344.

Laskar, J., Robutel, P., Joutel, F., Gastineau, M., Correia, A., & Levrard, B. (2004). A long-term numerical solution for the insolation quantities of the Earth. *Astronomy & Astrophysics*, 428, 261–285. http://doi.org/10.1051/0004-6361:20041335

Lisiecki, L. E., & Raymo, M. E. (2005). A pliocene-pleistocene stack of 57 globally distributed benthic – 18O records. *Paleoceanography, 20, PA1003, Data archived at the World Data Center for Paleoclimatology*, Boulder, Colorado, USA.

Liu, H. (1998). Phase modulation effect of the Rubincam insolation variations. *Theoretical and Applied Climatology*, 61, 217–229.

Lorius, C. (1990). Environmental records from polar ice cores. *Philosophical Transactions of the Royal Society* A, 330, 459–462, 459.

Loulergue, L., Schilt, A., Spahni, R., Masson-Delmotte, V., Blunier, T., Lemieux, B., Barnola, J.-M., Raynaud, D., Stocker, T. F., & Chappellaz, J. (2008). Orbital and millennial-scale features of atmospheric CH_4 over the past 800,000 years. *Nature*, 453, 383–386. http://doi.org 10.1038/nature06950

Lüthi, D., Le Floch, M., Bereiter, B., Blunier, T., Barnola, J.-M., Siegenthaler, U., Raynaud, D., Jouzel, J., Fischer, H., Kawamura, K., & Stocker, T. F. (2008). High-resolution carbon dioxide concentration record 650,000–800,000 years before present. *Nature*, 453(7193), 379–382. http://doi.org/10.1038/nature06949

Markle, B. R., & Steig E. J. (2021). Improving temperature reconstructions from ice-core water-isotope records. *The Climate of the Past: Discussions EGU*, 1–70. https://doi.org/10.5194/cp-2021-37

Masson-Delmotte, V., et al. (2010a). EPICA Dome C record of glacial and interglacial intensities. *Quaternary Science Reviews*, 29(1–2), 113–128.

Masuda, K. (1990). Atmospheric heat and water budgets of polar regions: Analysis of FGGE data. *Proc. NIPRSytnp. Polar Meteorology and Glaciology*, 3, 79–88.

McConnell, J. R. et al. (2014). Antarctic-wide array of high-resolution ice core records reveals pervasive lead pollution began in 1889 and persists today. *Scientific Reports*, 4, 5848.

McManus, J., Oppo, D., Cullen, J., & Healey, S. (2003). Marine isotope stage 11 (MIS 11): Analog for holocene and future climate? *Geophysical Monograph Series*, 69–85. http://doi.org/10.1029/137gm06

Milankovitch, M. (1941). *Kanon der Erdbestrahlung und Seine Anwendung auf das Eiszeitenproblem*. Royal Serbian Academy, Belgrade.

Miller, G. H., et al. (2012), Abrupt onset of the Little Ice Age triggered by volcanism and sustained by sea-ice/ocean feedbacks. *Geophysical Research Letters*, 39, L02708. http://doi.org/10.1029/2011GL050168

Miller, K. G., Fairbanks, R. G., & Mountain, G. S. (1987). Tertiary oxygen isotope synthesis, sea-level history and continental margin erosion. *Paleoceanography*, 2, 1–19.

Miller, K. G., Wright, J. D., & Fairbanks, R. G. (1991). Unlocking the Ice House: Oligocene-Miocene oxygen isotopes, eustasy, and margin erosion. *Journal of Geophysical Research: Solid Earth*, 96(B4), 6829–6848. http://doi.org/10.1029/90jb02015

Mulvaney, R., Abram, N. J., Hindmarsh, R. C. A., Arrowsmith, C., Fleet, L., Triest, J., . . . Foord, S. (2012). Recent antarctic peninsula warming relative to holocene climate and ice-shelf history. *Nature*, 489(7414), 141–144. http://doi.org/10.1038/nature11391

Neftel, A., Oeschger, H., Staffelbach, T., & Stauffer, B. (1988). CO_2 record in the Byrd ice core 50,000–5,000 years bp. *Nature*, 331, 609–611. https://doi.org/10.1038/331609a0

Oerter, H., Drücker, C., Kipfstuhl, S., & Wilhelms, F. (2009). Kohnen station—the drilling camp for the EPICA deep ice corein dronning maud land. *Polarforschung*, 78(1–2), 1–23, 2008 (erschienen 2009).

Owens, M. J., Lockwood, M., Hawkins, E., Usoskin, I., Jones, G. S., Barnard, L., . . . Fasullo, J. (2017). The Maunder minimum and the Little Ice Age: An update from recent reconstructions and climate simulations. *Journal of Space Weather and Space Climate*, 7, A33. http://doi.org/10.1051/swsc/201703

Paillard, D. (1998). The timing of Pleistocene glaciations from a simple multiple-state climate model. *Nature*, 391, 378–381.

Parrenin, F., & Paillard, D. (2012). Terminations VI and VIII (~ 530 and ~ 720 kyr BP) tell us the importance of obliquity and precession in the triggering of deglaciations. *Climate of the Past Discussions*, 8(4), 3143–3157. http://doi.org/10.5194/cpd-8-3143-2012

Pedro, J. B., Bostock, H. C., Bitz, C. M., He, F., Vandergoes, M. J., Steig, E. J., Chase, B. M., Krause, C. E., Rasmussen, S. O., Markle, B. R., & Cortese, G. (2016a). The spatial extent and dynamics of the Antarctic Cold Reversal. *Nature Geoscience*, 9(Jan), 51–55.

Pedro, J. B., Jochum, M., Buizert, C., He, F., Barker, S., & Rasmussen, S. O. (2018). Beyond the bipolar seesaw: Toward a process understanding of interhemispheric coupling. *Quaternary Science Reviews*, 192, 27–46. http://doi.org/10.1016/j.quascirev.2018.05.005

Pedro, J. B., Martin, E. J., Steig, M., Jochum, W., & Park, S. O. (2016b). Rasmussen Southern Ocean deep convection as a driver of Antarctic warming events. *Geophysical Research Letters*, 43(Mar), 2192–2199.

Petit, J. R. et al. (1999). Climate and atmospheric history of the past 420,000 years from the Vostok ice core, Antarctica. *Nature*, 399, 429–436.

Prabhu, A., Mahajan, P., Khaladkar, R., & Bawiskar, S. (2009). Connection between Antarctic sea-ice extent and Indian summer monsoon rainfall. *International Journal of Remote Sensing*, 30(13), 3485–3494. http://doi.org/10.1080/01431160802562248

Railsback, L. B., Gibbard, P. L., Head, M. J., Voarintsoa, N. R. G., & Toucanne, S. (2015). An optimized scheme of lettered marine isotope substages for the last 1.0 million years, and the climatostratigraphic nature of isotope stages and substages. *Quaternary Science Reviews*, 111, 94–106. http://doi.org/10.1016/j.quascirev.2015.01.012

Raynaud, D., Barnola, J. M., Souchez, R., Lorrain, R., Petit, J. R., Duval, P., & Lipenkov, V. I. (2005). The record for marine isotopic stage 11. *Nature*, 436, 39–40.

Raynaud, D., Jouzel, J., Barnola, J.-M., Chappellaz, J., Jouzel, R. J., & Lorius, C. (1993). The ice record of greenhouse gases. *Science*, 259, 926–934.

Raynaud, D., Lipenkov, V., Lemieux-Dudon, B., Loutre, M.-F., & Lhomme, N. (2007). The local insolation signature of air content in Antarctic ice: A new step toward an absolute dating of ice records. *Earth and Planetary Science Letters*, 261, 337–349.

Rossignol-Strick, M., Paterne, M., Bassinot, F. C., Emeis, K., & de Lange, G. J. (1998). An unusual mid-Pleistocene monsoon period over Africa and Asia. *Nature*, 392, 269–272.

Rubino, M., Etheridge, D. M., Trudinger, C. M., Allison, C. E., Battle, M. O., Langenfelds, R. L., . . . Francey, R. J. (2013). A revised 1000 year atmospheric δ13C-CO2 record from Law Dome and South Pole, Antarctica. *Journal of Geophysical Research: Atmospheres*, 118(15), 8482–8499. http://doi.org/10.1002/jgrd.50668

Salamatin, A. N. (2000). Paleoclimatic reconstructions based on borehole temperature measurements in ice sheets. Possibilities and limitations. In T. Hondoh (Ed.), *Physics of Ice Core Records*, pp. 243–281. Hokkaido University Press.

Saltzman, B. (2002). *Dynamical Paleoclimatology: Generalized Theory of Global Climate Change*. Academic Press.

Schuepbach, S. et al. (2013). High-resolution mineral dust and sea ice proxy records from the Talos Dome ice core. *Climate of the Past*, 9, 2789–2807.

Schytt, V. (1958). Norwegian—British—Swedish antarctic expedition 1949–1952. *Scientific Results 4*, Glaciology II, Norsk Polarinstitut, Oslo, Norway.

Seroussi, H., Ivins, E. R., Wiens, D. A., & Bondzio, J. (2017). Influence of a West Antarctic mantle plume on ice sheet basal conditions. *Journal of Geophysical Research: Solid Earth*, 122, 7127–7155. http://doi.org/10.1002/2017JB014423

Shakun, J. D., Clark, P. U., He, F., Marcott, S. A., Mix, A. C., Liu, Z., Otto-Bliesner, B., Schmittner, A., & Bard, E. (2012). Global warming preceded by increasing carbon dioxide concentrations during the last deglaciation. *Nature*, 484(7392), 49–54. http://doi.org/10.1038/nature10915

Shepherd, A., Gilbert, L., Muir, A. S., Konrad, H., McMillan, M., Slater, T., Briggs, K. H., Sundal, A. V., Hogg, A. V., & Engdahl, M. (2019). Trends in Antarctic ice sheet elevation and mass. *Geophysical Research Letters*, 46, 8174–8183. http://doi.org/10.1029/2019gl082182

Siddall, M., Rohling, E. J., Almogi-Labin, A., Heml eben, C., Meischner, D., Schmelzer, I., & Smeed, D. A. (2003). Sea-level fluctuations during the last glacial cycle. *Nature*, 423, 853–858. http://doi.org/10.1038/nature01690

Siddall, M., Rohling, E. J., Thompson, W. G., & Waelbroeck, C. (2008). Marine isotope stage 3 sea-level fluctuations: Data synthesis and new outlook. *Reviews of Geophysics*, 46(4). http://doi.org/10.1029/2007rg000226

Siegfried, M. R., & Fricker, H. A. (2021). Illuminating active subglacial lake processes with ICESat-2 laser altimetry. *Geophysical Research Letters*, 48, e2020GL091089. https://doi.org/10.1029/2020GL091089

Sorge, E. (1933). Scientific results of the Wegener Expedition to Greenland. *Geographical Journal*, 81, 333–334.

Sowers, T., & Bender, M. (1995). Climate records covering the last deglaciation. *Science*, 269(5221), 210–214. http://doi.org/10.1126/science.269.5221.210

Stocker, T. F., & Johnsen, S. J. (2003). A minimum thermodynamic model of the bipolar seesaw. *Paleoceanography*, 18, art. no. 1087.

Stocker, T. F., Wright, D. G., & Mysak, L. A. (1992). A zonally averaged, coupled ocean-atmosphere model for paleoclimate studies. *Journal of Climate*, 5, 773±797.

Suganuma, Y., Haneda, Y., Kameo, K., Kubota, Y., Hayashi, H., Itaki, T., . . . Okada, M. (2018). Paleoclimatic and paleoceanographic records through Marine Isotope Stage 19 at the Chiba composite section, central Japan: A key reference for the Early—Middle Pleistocene Subseries boundary. *Quaternary Science Reviews*, 191, 406–430. http://doi.org/10.1016/j.quascirev.2018.04.022

Sverdrup, H. U. 1935. The temperature of the firn on Isachsen's Plateau and general conclusions regarding the temperature of theglaciers on West-Spitzbergen, in Scientific results of the-Norwegian—Swedish Spitzbergen Expedition, 1934. *Geografisker Annaler*, 17, 78–80.

Swithinbank, C. (1957). Norwegian—British—Swedish Antarctic Expedition 1949–1952. In *Scientific Results*, vol. 3. Glaciology I. Norsk Polarinstitut.

Thomas, E. R., Gacitúa, G., Pedro, J. B., King, A. C. F., Markle, B., Potocki, M., & Moser, D. E. (2021). Physical properties of shallow ice cores from Antarctic and sub-Antarctic islands. *The Cryosphere*, 15, 1173–1186. https://doi.org/10.5194/tc-15-1173-2021.

Tzedakis, P. C., Hooghiemstra, H., & Palike, H. (2006). The last 1.35million years at Tenaghi Philippon: Revised chronostratigraphy and long-term vegetation trends. *Quaternary Science Reviews*, 25, 3416–3430.

Tziperman, E., Raymo, M., Huybers, P., &Wunsch, C. (2006). Consequences of pacing the Pleistocene 100 kyr ice ages by nonlinear phase locking to Milankovitch forcing. *Paleoceanography*, 21, PA4206.

van Wessem, J. M., Ligtenberg, S. R. M., Reijmer, C. H., van de Berg, W. J., van den Broeke, M. R., Barrand, N. E., Thomas, E. R., Turner, J., Wuite, J., Scambos, T. A., & van Meijgaard, E. (2016). The modelled surface mass balance of the Antarctic Peninsula at 5.5 km horizontal resolution. *The Cryosphere*, 10, 271–285. https://doi.org/10.5194/tc-10-271-2016

Vavrus, S. J., He, F., Kutzbach, J. E., Ruddiman, W. F., & Tzedakis, P. C. (2018). Glacial inception in marine isotope stage 19: An orbital analog for a natural holocene climate. *Scientific Reports*, 8(1). http://doi.org/10.1038/s41598-018-28419-5

Waelbroeck, C., Jouzel, J., Labeyrie, L. et al. (1995). A comparison of the Vostok ice deuterium record and series from Southern Ocean core MD 88–770 over the last two glacial-interglacial cycles. *Climate Dynamics*, 12, 113–123. https://doi.org/10.1007/BF00223724

Waelbroeck, C., Jouzel, J., Labeyrie, L., Lourius, C., Labracherie, M., Stievenard, M., & Barkov, N. (1995). Comparing the Vostok ice deuterium record and series from Southern Ocean core MD 88–770 over the last two glacial-interglacial cycles. *Climate Dynamics*, 12, 113–123.

Waelbroeck, C., Labeyrie, L., Michel, E., Duplessy, J. C., McManus, J. F., Lambeck, K., Balbon, E., & Labracherie, M. (2002). Sea-level and deep water temperature changes derived from benthic foraminifera isotopic records. *Quaternary Science Reviews*, 21(1–3), 0–305. http://doi.org/10.1016/s0277-3791(01)00101-9

Wang, P., Tian, J., Cheng, X., Liu, C., & Xu, J. (2004). Major Pleistocene stages in a carbon perspective: The South China Sea record and its global comparison. *Paleoceanography*, 19, PA4005. http://doi.org/10.1029/2003PA000991

Watanabe, O. *et al.* (2003). Homogeneous climate variability across East Antarctica over the past three glacial cycles. *Nature*, 422, 509–512.

White, J. W. C., & Steig, E. J. (1998). Timing is everything in a game of two hemispheres. *Nature*, 394(6695), 717–718.

Wolff, E., Fischer, H., & Röthlisberger, R. (2009). Glacial terminations as southern warmings without northern control. *Nature Geoscience*, 2, 206–209.

Wunsch, C. (2003). Greenland-Antarctic phase relations and millennial time-scale climate fluctuations in the Greenland ice-cores. *Quaternary Science Reviews*, 22, 1631–1646.

Yan, Y., Bender, M. L., Brook, E. J., Clifford, H. M., Kemeny, P. C., Kurbatov, A. V., . . . Higgins, J. A. (2019). Two-million-year-old snapshots of atmospheric gases from Antarctic ice. *Nature*, 574(7780), 663–666. http://doi.org/10.1038/s41586-019-1692-3.

Yin, Q. Z., & Berger, A. (2010). Insolation and CO2 contribution to the interglacial climate before and after the mid-brunhes event. *Nature Geoscience*, 3(4), 243–246.

Yin, Q. Z., Berger, A., & Crucifix, M. (2009). Individual and combined effects of ice sheets and precession on MIS-13 climate. *Climate of the Past*, 5, 229–243.

Yin, Q. Z., & Guo, Z. T. (2008). Strong summer monsoon during the cool MIS-13. *Climate of the Past*, 4, 29–34.

Zachos, J. C., Quinn, T. M., & Salamy, K. A. (1996). High-resolution (104years) deep-sea foraminiferal stable isotope records of the Eocene-Oligocene climate transition. *Paleoceanography*, 11(3), 251–266. http://doi.org/10.1029/96pa00571

Ziegler, M., Lourens, L. J., Tuenterand, E., & Reichart, G.-J. (2010). High Arabian Sea productivity conditions during MIS 13—odd monsoon event or intensified overturning circulation at the end of the Mid-Pleistocene transition? *Climate of the Past*, 6, 63–76. www.clim-past.net/6/63/2010/

4 Glacial Geomorphology around Schirmacher Oasis, Antarctica

Jitendra Kumar Pattanaik and
Waseem Ahmad Baba
Department of Geology, School of Environment and Earth Sciences, Central University of Punjab, Bathinda, India

CONTENTS

4.1 INTRODUCTION

Several ice-free areas in Antarctica are located in the marginal or coastal regions where the continental ice sheet meets the ice shelf. These areas provide an opportunity to study geology and another related discipline as the landscape is exposed. Schirmacher oasis marks the northern boundary of the central Dronning Maud land of east Antarctica. It is surrounded by an ice shelf in the north and a continental ice sheet on the southern side (Figure 4.1). Schirmacher oasis is an ice-free area having numerous freshwater lakes. The northern part is characterized by a steep escarpment of rocks while the southern part is overridden by East Antarctica ice sheet (EAIS). The present-day deglaciated landscape of the Schirmacher Oasis is the consequence of long-term climate change (Goudie, 2004) and sub-glacial processes. This oasis consists of low-lying hills capped by thin glacial deposits. The maximum elevation of Schirmacher oasis rises up to 212 m (AMSL) in the central region. The northern region is very steep and may be the result of tectonic activities (Ravikant

DOI: 10.1201/9781003284413-4

and Kundu, 1998). The pattern of landscape in Schirmacher oasis is shaped by the glacial and periglacial processes. Periglacial features include block fields, deflation surface, moraines, caverns pits etc. that are present throughout the oasis. More than 100 lakes are present in the oasis which have been divided based on their geomorphic characteristics (Bormann and Fritzsche 1995; Ravindra 2001). A large amount of debris, remnant lakes and valleys are a part of this landscape (Bormann and Fritzsche, 1995). Based on glacial and topographical features Schirmacher oasis has been divided into four units representing from north to south shelf area, polar ice sheet, the mainland of Schirmacher and lakes. The widespread polar ice comprises close-packed ice impregnated with debris of various sizes cobble to silt size sediments. The mainland of Schirmacher oasis constitutes hills, which are up to 212 m in height, valleys, plains, lakes, and depressions. Because conditions of topography are different with different sediment, accumulation lakes in the mainland of Schirmacher can be easily distinguished as separate sedimentary units (Srivastava et al., 2010).

Indian research stations "Maitri" is located in Schirmacher oasis, which was established in the year 1989. The first Indian Antarctica expedition in 1981–82 open up a new avenue for Indian scientists to conduct various polar research including geology, tectonics, geomorphology, landscape, paleoclimate etc. (Asthana and Chaturvedi, 1998; Sinha and Chatterjee, 2000a; Ravindra, 2001; Bera, 2004, 2006), Scientist from the survey of India mapped the Schirmacher Oasis. In this chapter glacial-geomorphological features, lakes along with the prevailing climate of Schirmacher oasis are discussed. One of the authors was a summer scientific member in the 36th Indian Scientific Expedition to Antarctica and during his stay at Maitri station, he has conducted fieldwork in and around Schirmacher Oasis. Some of the field observations during austral summer are being discussed in this chapter.

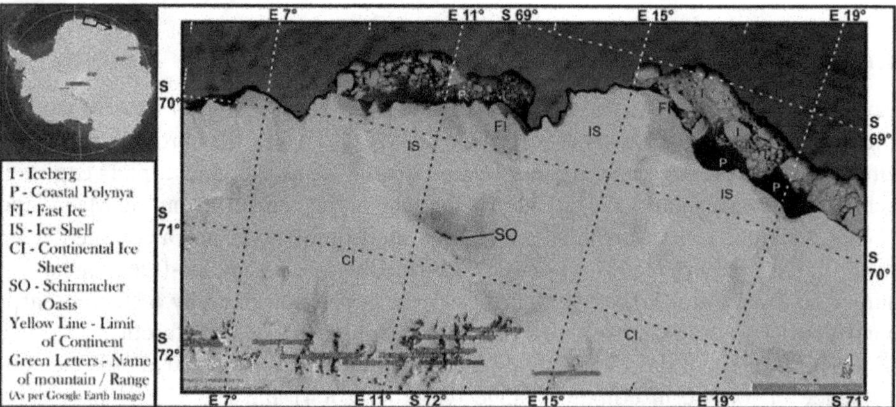

FIGURE 4.1 Google Earth image of the coastal region of Dronning Maud Land, East Antarctica. Various coastal features like iceberg, polynya, fast ice, ice shelf, and continental margins are shown here along with Schirmacher Oasis.

4.1.1 Climate

Schirmacher oasis experiences a continental type climate and is influenced by the Antarctic high with frosty weather with katabatic winds. Most parts of Oasis are ice-free due to which Schirmacher observes a dry climate. This part of Antarctica is witnessing low albedo, low precipitation and high wind velocity (Shumskiy, 1957). Winters are mild with strong winds and snowstorms or blizzards (Phartiyal et al., 2011). The temperature in Oasis ranges from +7.4 °C to –34.5 °C and with an average temperature of –10.2 °C. During winter atmospheric pressure is greatest with the least absolute humidity (Phartiyal et al., 2011). During springtime, pressure decreases with an increase in temperature and humidity causing evaporation and the start of thaw. In summertime temperature and pressure of the atmosphere is highest; winds are weak and very less snowfall; ice melts fast and the meltwater transfers to the shelf. In the autumn season, temperature and humidity decreases with a slight decrease in atmospheric pressure in comparison with summer where the annual mean temperature is –10.5 °C from 1960 to 1980 recorded at Novolazharevskaya, a Russian station. January is the warmest month with an average temperature of –0.9 °C. The maximum temperature was 0.7° C in 1971 with a minimum of –2.5° C in 1966. The velocity of the wind is 10 m/s and the direction of the wind is the east-southeast. The average precipitation is 250–300 mm with a relative humidity of 50%. The climatic influence extends up to 1000 sq. km covering the ice shelf which is outside the influence of katabatic wind (Phartiyal, 2019).

The present-day climate prevailing in Antarctica and the Southern Ocean is due to the interplay of the ocean, ice sheet, atmosphere, and sea ice and their response to climate forcings (Mayewski et al., 2009). For oceanic and atmospheric circulations heat sink located in Antarctica is crucial (Bargagli, 2006). Schirmacher Oasis, which is located at the margin of the continent, encounters atmospheric swings related to the katabatic winds which flow from interior of the continent. The Indian and the Russian research station located in the oasis do monitor the surface-based radiosonde and meteorological parameters (van Lipzig et al., 2004). The surface air temperature of Schirmacher oasis varies seasonally with the maximum being felt in summer and a secondary peak in winter. The peak in summer is due to the warming of the rocky or icy surface (Gajananda et al., 2007). The secondary peak is due to the induction of warm moist air, caused by cyclones that move along the coast (Naithani et al., 1995). The average monthly temperature at Maitri from 1985–2016 has been shown in Figure 4.2. The maximum and minimum temperatures recorded were +4.22 °C and –33.69 °C, respectively. The temperature at Dakshin Gangotri (the first Indian research Station established in Antarctica) has always been lower than at Maitri (Naithani and Dutta 1995). Apart from this, there are also some instances when surface temperature increases due to the blocking of polar high, this is generally observed in the winter season (Kejna, 2003).

The inversion of surface temperature over Antarctica is due to radiative heat loss from the surface of the ice, more so in the polar winter. Philpot and Zillman (1970) established that temperature inversion in Antarctica was 25 °C and decreased to 5 °C near the coast. The upper air temperature affects the Schirmacher oasis. The inversions were recorded from April to October and showed that the inversion layer was

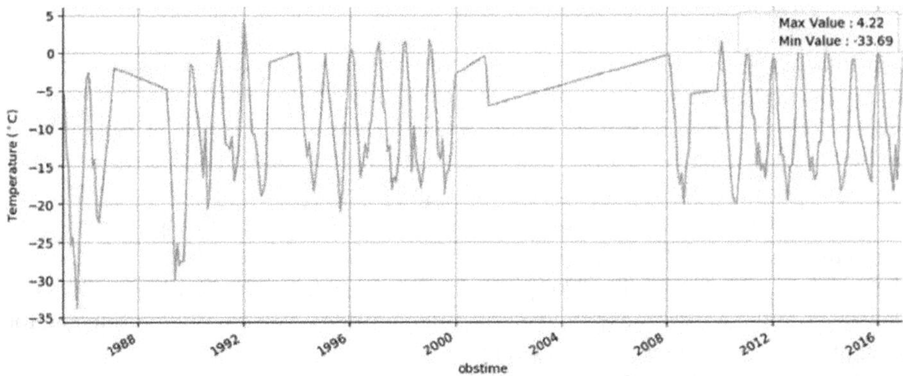

FIGURE 4.2 Long-term monthly average temperature from 1985–2016 at Maitri research Station, Antarctica.

Source: Indian Meteorological Department (IMD).

absent in summer due to the presence of solar radiation all day and night. Tropopause in summer was warmer compared to winter (15–20 °C) (Hosalikar and Machnurkar, 2006) and the variation is dynamic from day to day as weather systems evolve and move. The average cloudiness in Maitri is 4.4 octa. The mean annual 63 clear day sky and 101 overcast days have been observed at Maitri. It has been observed spring is the optimum season for cloud formation. The reason could be the evaporation from the frozen ocean. High clouds are more frequent than other types of clouds in summer, and it has been also seen that cloud cover is lower in summer than in other seasons (Koppar, 1991, 1995).

Precipitation is mainly caused by the moist air above the low level of inversion. The lifting of the warm air by the orography in association with synoptic-scale systems causes more precipitation in coastal areas of Antarctica. Fronts related to cyclones are responsible for snowfall over Schirmacher. Heavy snowfall is related to warm fronts transported southwards. Precipitation at Maitri is mostly in the form of snow and there are 88 days of snowfall approximately. The average maximum snowfall recorded is 43.3 mm in august during 1990–2010. At Maitri station, the accumulation of snow is not dependent on the number of days but on the transport of ice or snow from the southern latitude. Rainfall is very rare and generally occurs when the temperature rises several degrees. In some cases, rain has been observed at Maitri during the summer months in 1996. During this period the temperature was high (12.2 °C) (Rasal and Mohar 2006). The temperature rise accompanied by absorption of latent heat in the lower atmosphere resulted in precipitation at Maitri. It has been seen that Maitri has had less precipitation over the last decade based on the snow anomaly. The trends in temperature indicated cooling over Maitri during 1990–2015. Similar trend was also recorded from Russian station for same period. Speed of winds have significantly decreased indicating cooler surface temperatures. A decrease in average sea level pressure over Maitri during this period indicates SAM (southern annual mode) moving to a high index state in recent decades (Smith and Polvani, 2016; Soni et al., 2017).

4.2 GEOLOGY OF THE AREA

The Schirmacher oasis geology is mapped on a scale of 1:25000 by Antarktiki, 1969; Sengupta, 1986 and the area is also partly mapped on the 1:10000 scale by Wand in 1983. Schirmacher oasis consists of Mesoproterozoic to Neoproterozoic rocks and the Precambrian basement of east Antarctica is exposed in an area of 30 sq. km (Ravich and Kamenev, 1972). The recently published map of GSI showed that Schirmacher consists of granulites, amphibolite and garnet-gneiss as the main rock types. The Schirmacher oasis has been repeatedly metamorphosed. In the early stage, events indicated ductile deformation followed by brittle deformation. Ductile deformation leads to the formation of mesoscopic and macroscopic structures. Granulites of Proterozoic have been superimposed by amphibolite facies (D'Souza et al., 1994; Ravikant and Kundu, 1998). The dominant lithology is the amphibolite facies, quartz garnet gneiss sandwiched with retrograde mafic granulite or amphibolite. The area is intersected by dykes of lamprophyre, basalt etc. (Sengupta, 1986).

Basaltic dykes trend NW-SE and invariably dip vertical to NE, E, and ESE. The thickness varies from a few centimeters to several meters. The basalts are fresh and crop out from a small distance. Vesicular rocks with amygdalas of zeolite, quartz, calcite etc. depict a late stage of hydrothermal alteration, but these are rare. The basalts are phenocrysts or aphyric. Xenoliths can also be found in some cases. The pyroxene and olivine phenocrysts are fresh with a diameter of 10 mm. The age of the basalts is Upper Carboniferous and Jurassic determined by the K-Ar method. The two dykes in the area of different periods are interesting. The dykes and spread of Mesozoic tholeiitic lavas are related to the Gondwanaland breakup (Cox, 1978; Scrutton, 1973; Smith and Hallam, 1970). Antarctica and South Africa formed a continuous landmass before the Jurassic (Smith and Hallam, 1970; Dietz and Holden, 1970). The drifting took place along the Indian ridge in the early Jurassic (Barron et al., 1978). Some basaltic dykes however represent old magmatic events. These may be regarded as the Paleozoic breakup of Gondwanaland (Scrutton, 1973).

4.3 GLACIOMORPHIC LANDFORMS

Numerous U-shaped valleys are present in the Schirmacher Oasis and have criss-crossed each other at different stages of maturity. The more mature valleys are marked by their wide floor occupied by patterned ground or by lakes. These valleys have been formed due to the erosional activity of glaciers. Hanging valleys can also be seen that drain into the main valley. The trend shown by the valleys is mostly ENE which indicates the favorable movement of the glacier. Mature valleys have vertical and steep walls. The hills have flattops or are sub-rounded trending in the EW direction. The surfaces of cliffs are glacially polished. These cliffs are high having elevations of 50 m or higher, thus forming important geomorphic features. Pebbles, coarse sands, and erratic boulders on hill tops suggest a glacial regime before the ice disappeared from the region. The retreat of the ice sheets in a phased manner is indicated by dry glacial valleys and hanging fossil cirques. Polishing and glacial striations are found throughout the oasis and adjoining areas extending up to 5 m to 600 m AMSL. Most of the striations dip towards north to north-east. The northern

TABLE 4.1
Different Types of Glaciomorphic Landforms Found in the Schirmacher Oasis

Erosional landform	Depositional landform	Reference
Roche Moutonnées	Terraces	Shrivastava et al. (2019), Bardin (1971), Gaur et al. (2006)
Striated surface	Patterned ground	Jayapaul and Dharwadkar (2006)
Cliffs/Escarpments	Moraines	Ravindra (2001)
Hills, Valleys	Erratics	Asthana (2017), Gaur et al. (2006)
Block field	Lakes	Asthana et al. (2013)
Glacial cirque	Paleostrand line	Shrivastava et al. (2019)
Frost shattering	Outwash plain	Asthana (2017)
Honeycomb structure	Kettle lakes	Rea et al. (1996); Whalley et al. (1997)

margin of the Schirmacher oasis shows a modification of *roche moutonnées*. Various glaciomorphic landforms are present in this oasis (Table 4.1) and some of the features are modified subsequently due to weathering and sub-glacial processes.

Erratics are a regular feature that can be seen as boulders and perched blocks are placed over the bedrock by the retreat of glaciers (Figure 4.3a). The common erratics in Schirmacher oasis are granite and ultrabasic rocks. The boulders, which are angular to sub-angular in shape, have locally originated. Those rounded in shape have travelled a significant distance. Glacial erratics are important in depicting the flow of the past glacier, its timing, and the direction of retreat. Erratics vary from large boulders to small pebbles and these erratics provide opportunities for exposure dating using cosmogenic radionuclides. Striations marks are observed on the bedrock of the Schirmacher Oasis (Figure 4.3c). Striations indicate the direction of the movement of ice over the bedrock. The movement is essentially N-NE. Striations are rough against the movement of direction and smooth towards the direction of movement (Shrivastava et al., 2019).

Escarpments/cliffs are the walls that stand out of valleys are escarpments. The surface of the cliff shows glacial polish. Most of the north of Schirmacher occurs as cliffs that are 50–100 meters in height. Thus forming an important geomorphic feature (Gaur et al., 2006). Fjord and glacial troughs are linear and deep erosional features indicating the glacial erosion effects where the flow of ice is limited by topography. They are formed by the action of quarrying and ablation as these processes need to form steep troughs. Thus fjords and glacial troughs are deep trenches, incised and excavated by long term erosion in bedrock. These deep landforms are occupied by the sea after glacier retreats. The spectacular scenery of fjords occurs in Schirmacher oasis. The large epishelf lakes occupy the glacial troughs that are connected to the sea from the north beneath the ice shelf.

Block fields are angular masses of rock rubble found as streams of rock. These are derived from mantle weathering as residual deposits, which have been left after the transportation of fine materials (Rea et al., 1996; Whalley et al., 1997). They

FIGURE 4.3 (a) Erratic boulder (b) pressure ridges (c) striation marks (d) landlocked lake. *Note:* Photographs were taken during the 36th Indian Scientific Expedition to Antarctica.

are commonly observed in low lying areas of Schirmacher and along the vicinity of lake margins. Some of these block fields are also demarcated as the path of fossil glaciers in the map of Schirmacher oasis by Ravindra (2001) and Gajananda et al. (2007) (Figure 4.4). Felsenmeer or block fields is a striking feature in the present periglacial environment. Felsenmeer in German means sea of rocks formed by frost riving. Profiles of Felsenmeer consist of boulders that overlie a matrix of fines at a depth below the surface (Ballantyne, 1998; Shrivastava et al., 2019). The prominent geomorphic features occurring in the area are *roche moutonnées* (Figure 4.4). This is formed by the glacial passage over the underlying rock (Benn and Evans, 2014). The hillocks, which are glacially eroded, are gentle on the stoss side and steep on the lee side. Due to the physical weathering, the *Roche Moutonnees* are in the modification stage. The clasts are subrounded to subangular in shape with a size range between pebble and boulder.

The patterned ground of Schirmacher oases is a conspicuous landform because of the repeated thawing and freezing of valley floors. These are typical polygon in shape that is formed on the flat floor of valleys. The polygons are initially irregular and then become regular after attending mature stage of landform. The polygons are texturally unsorted but some coarse-grained ones are surrounded by the fine-grained boulders. The polygon size varies depending on the height and level of patterned

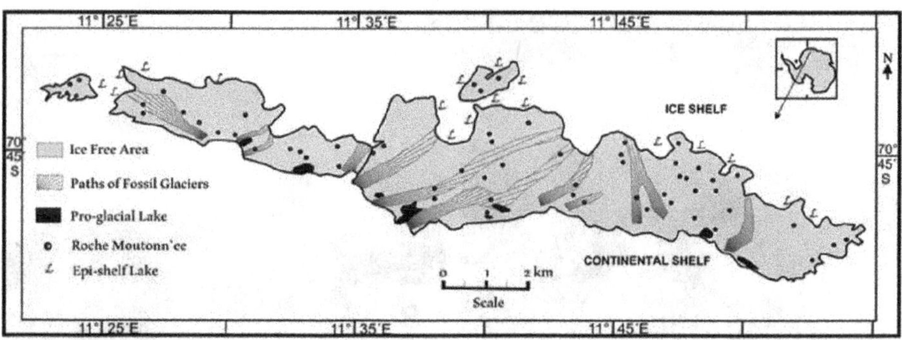

FIGURE 4.4 Map shows locations of *roche moutonnées* and paths of fossil glaciers of the Schirmacher Oasis. *Source:* Modified after after Ravindra, 2001; Beg et al., 2000; Gajananda et al., 2007.

ground. Moraines are present almost everywhere in the area. These are a reflection of glaciation and deglaciation that occurred in the past. On the sides of the valleys occur the lateral moraines while the terminal moraines are periodic (Gaur et al., 2006). Moraines that occur as deposits of ablation are stratified to a certain degree these are known as ablation till or stratified deposits. Piles of recessional and terminal moraines, which are formed due to glacial debris deposition, are also common in the Schirmacher oasis. Recessional moraines indicate the deglaciation stages. Push moraines, which are formed on the ice shelf, have also been encountered. These are formed due to glaciation, during the process unstratified sediment is pushed by the ice mass in a curvilinear or liner ridges. They can be identified by their upward movement from the horizontal position (Asthana, 2017). A well-preserved end moraine trending east-west close to the margin of polar ice consists of clasts that range in size from boulder to granules with minor amounts of the matrix. Bigger clasts are sub-angular to sub-rounded in shape. There are some polished and well-rounded boulders also. This shows the transportation in a glacial environment (Gaur et al., 2006).

Paleostrand line: The northern boundaries of the Schirmacher forming the grounding line for shelf ice which are undulatory are formed due to the action of waves. A pale strandline or shoreline is the demarcation that is 25–30 m above the present mean sea level. Deposition of sand beach occurs in between the spaces of boulders and on terraces on the lee side. The distribution of better sorted and granular sediments on the surface is represented by the winnowing action of waves. The sediments, which are sorted, are clean without clay and silt and fine-grained. Terraces are like benches that are cut in bedrock followed by the deposition of moraines. They are the result of meltwater streams that flow between the valley side and ice. Approximately six glacial episodes of shrinkage have been found from the lowest to the highest point. Shelf being marked as the lowest point. Presently the vertical exposure of each terrace ranges from 1.5–3 m. Terrace deposits don't show any preferred orientation of clasts or any internal stratification. The different types of depositional landforms play a vital role in interpreting the retreat of a glacier. Factors like weathering rate, lichen growth, wind polishing, the position of moraines etc. have been used put to determine the moraine chronology (Bardin, 1971). In Schirmacher oasis terminal moraines are widely exposed which are parallel to continental ice.

4.3.1 Lakes

Schirmacher oasis has approximately 118 lakes and ponds which are permanent there (Bormann and Fritzsche, 1995; Richter and Bormann, 1995). Schirmacher region has also coastal oases, which can be very helpful to gather paleoenvironmental information. The Schirmacher oasis consists mostly of lacustrine sediments. Lakes of Schirmacher oases were previously divided based on morphology, hydrothermal and thermal characteristics (Richter and Bormann, 1995). Later these were simplified and classified based on the polar ice sheet proximity (Ravindra et al., 2001).

a) Epishelf lakes are connected to shelf ice from the northern side. The number of epishelf lakes is 15, which are less than landlocked and proglacial lakes. The lakes show the existence of pressure ridges (Figure 4.3b) along their contact having cracks in the ice. Water coming out through these cracks like founts is seen at the time of high tide in the ocean, which flows like a river over these epishelf lakes. This indicates a subshelf link towards the ocean and a tidal origin of lakes (Swain et al., 2019).
b) Proglacial lakes are attached to the polar ice on the southern side. These lakes vary in size, meltwater from polar ice sheets fills up these proglacial lakes at the time of austral summer.
c) Land locked lakes are surrounded by masses of rock, which are not connected to polar ice sheets or ice shelves. These lakes have originated due to terminal moraine dumping. The areas, which are low lying, acted as locales for accumulating water, which can be seen now as landlocked lakes (Figure 4.3d).

The proglacial and epishelf lakes are mostly covered by ice throughout the year while the periglacial melt in austral summer due to positive conditions in temperature (Ravindra et al., 2002). These lakes constitute 3.4 km^2 of Schirmacher. Lake Priyadarshini, which is close to the Indian research station Maitri, is the largest fresh water lake in this region occupying an area of 0.75 sq. km. This lake was described as a proglacial by (Priddle and Heywood 1980; Verlecar et al., 1996).

The oasis is dominated by a number of freshwater intermontane lakes that occupy valley floors. The lakes differ in shape, depth, and highly unsorted bottom sediments. During winter these lakes freeze and thaw during summer. These lakes get water from polar front ice through the channels of melt water. Some sand pockets can be seen at higher levels, which indicates the recession of these lakes with time (Gaur et al., 2006). Adjacent to the lake vegetation such as mosses and lichens are found (Rai et al., 2011; Verlecar et al., 1996). Organic carbon in the lake is low which is contributed by protozoa, turbellaria, etc. (Rai et al., 2011).

Proglacial lakes are formed at the margin of the ice sheet by the polar ice recession. Landlocked lakes have been surrounded by the landmass occupying the earlier paths of the glacier. The epishelf lakes are formed in between the ice shelf and Schirmacher

Oasis. They receive meltwater from the eastern and western parts (Ravindra, 2001). At the time of austral summer, meltwater passes through lakes, which are landlocked from proglacial lakes to the Epishelf lakes. This helps in forming a sedimentary record which can provide information about sedimentation nature (Bennike and Björck, 2000; Cook and McConnell, 2001). Many channels of meltwater fall in the lakes of Epishelf, yet they maintain the level indicating that they might be joined to the ocean underlying. This information gap was filled by the Indian Antarctic expedition at the time of austral summer.

Schirmacher Oasis and its adjoining areas are typical examples of polar and periglacial environments. Deglaciation in Schirmacher took place during late Quaternary because of vertical thinning and horizontal recession. In Schirmacher oasis, preserved striation mark on bed rock indicate the movement of ice/glacier in the north and northeastern direction. These have been observed from the highest points of Schirmacher. The next was NE direction, indicating ice movement controlled by the valleys. This shows the present position after the final recession (Shrivastava et al., 2019).

4.3.2 Paleoclimatic Reconstruction of Schirmacher Using Lake Records

Lakes of Antarctica are sensitive to changes in the environment; their sediments provide information about past changes in climate and biogeochemistry (Smeltzer and Swain, 1985; Warrier et al., 2017). Scientists from India have studied actively ice-free regions around Maitri station. The first attempt to study sedimentary records was mainly focused on the Schirmacher Oasis, to depict past changes and lacustrine sedimentology and transport of palynodebris (Sinha and Chatterjee, 2000b; Sinha et al., 2000; Bera, 2004; Sharma et al., 2007). The quartz grains from lake sediments were studied to know the glacio-fluvial and aeolian deposition of sediments (Asthana et al., 2009; Shrivastava et al., 2009). Since the last decade, more improved techniques and systematic studies were undertaken to depict the changes in past climate and environment. Various proxy and analytical techniques like sedimentology, biogenic silica, geochemistry, and paleontological parameters were studied. Phartiyal et al. (2011) reconstructed the paleoclimate of the Schirmacher oasis from 13 ka to the present. The results showed dramatic changes in the size of lakes, possibly due to the retreat of glaciers feeding them. A study conducted by the Geological Survey of India indicated the characteristics of sediments in epishelf lakes of Schirmacher depend on the composition of source rock, transport of sediment, weathering and depositional processes. Lake sediments helped to infer the transporting conditions and or various sources of provenance (Shrivastava et al., 2012). Granulometric, field observations and morphological studies on quartz grains showed comparable results, except for the strength and extent of glacio-fluvial processes.

The present-day landlocked lakes were glacial lakes and the whole Schirmacher was dominated by glaciers during 13–12.5 cal ka (Phartiyal et al., 2011). During the beginning of the Holocene glaciation at 11.5 ka, five large proglacial lakes were formed due to the retreating glaciers. Drying of landlocked lakes was formed due to the withdrawal of glaciers, less precipitation, reduced snow accumulation sublimation of the lakes etc. on the contrary epishelf and proglacial lakes have a consistent source of water from the ice shelf and continental ice sheet. Multi proxy data from land-locked lakes in the central part of Schirmacher has provided a better constraint for paleoclimatic evolution in the region (Srivastava et al., 2018). This revealed various phases of cooler and warmer intervals, highlighted by fluctuations in different statistical and sedimentological parameters. Physical weathering has mostly controlled the sediment and composition of the clay fraction. Clay mineralogy depicts a shift in weathering and climate around 42 ka. This shows the onset of warming in the area before LGM. This warm spell was not strong to alter the clay chemistry. Proxy records show short-term changes in climate at the time of Late Quaternary. Paleoclimatic history of Lake Priyadarshani from pollen spores indicates that the region was under a cold and dry climate from 10 to 9 ka BP followed by a prolong warm period and a moist type of climate from 9 to 2.4 ka BP. From 2.4 to 1 ka onwards cold and dry conditions prevailed in Schirmacher oasis (Sharma et al., 2007).

4.3.3 Effect of Thermal Conductivity of Rocks on Ice Sheets

Thermal conductivity of rocks plays a significant role in the ice sheet recession in the form of glacier snout and polar ice sheets. The different rock types exposed at Schirmacher have varying thermal conductivity ranging from 2.77 ± 0.18$Wm^{-1}k^{-1}$ in lamprophyres to 6.1 ±0.0.37$Wm^{-1}k^{-1}$ in metapelites. Because of uneven thermal conductivities, their reaction varies on ice mass in particular at austral summer, when the temperature is higher than 0 °C. This variation in thermal conductivity can be seen in mean annual recession for the Schirmacher and Dakshin Gangotri glaciers. However, both areas have the same meteorological parameters. However, the topographic and physiographic variations cannot be ruled out. The area close to the Schirmacher has shown more ablation than the other areas of the ice shelf. The polar ice sheets show that rocky exposure plays a vital role under the values in thermal conductivity. It has been observed that rocks with better thermal conductivity values have a higher recession and vice versa. These facts indicate that recession in Schirmacher and surrounding areas is influenced by the thermal conductivity of rocks (Swain, 2019).

4.3.4 Summary

Schirmacher Oasis is an ice-free area of Donning Maud Land, East Antarctica. This area is sandwiched between a continental ice sheet in the south and an ice shelf in the north. The ice-free region provides opportunities to study lithology, paleoclimate, glacial process, and glacial geomorphology of Antarctica. The landscape and the glacial-geomorphological features found in the Schirmacher oasis help to understand the glacial process active in the region. Striations on bedrocks and erratics indicate the glacier's movements and deglaciation phases of the Oasis. A considerable amount of sediments found in the oasis as till, pattern ground, moraines, block fields and outwash plains witnessed the past glacial processes. Retreat and advancement of polar ice have resulted in the formation of various landforms such as valleys, glacial troughs, cliffs, moraines and *roche moutonnées*. Sediment archives from the lakes provide paleoclimatic information about the oasis. Permanent Indian research station Maitri and Russian research station Novolazarevskaya provide a platform for the researchers for conducting various research works in the Schirmacher oasis.

ACKNOWLEDGMENT

JKP is thankful to the NCPOR and MOES for providing opportunities to be part of the summer scientific member in the 36th Indian Scientific Expedition to Antarctica. Authors are thankful to the Central University of Punjab, Bathinda for providing the administrative and infrastructural facilities.

REFERENCES

Asthana, R. (2017). *Geomorphology of the polar periglacial environments at Schirmacher Oasis and Larsemann Hills East Antarctica with special reference to the sedimentation characteristics*. PhD Thesis submitted to Banaras Hindu University, Varanasi.

Asthana, R., and Chaturvedi, A. (1998). The grain-size behaviour and morphoscopy of supraglacial sediments, south of Schirmacher Oasis, E. Antarctica. *Journal of the Geological Society of India*, 52(5), 557–568.

Asthana, R., Shrivastava, P. K., Beg, M. J., Swain, A. K., Dharwadkar, A., Roy, S. K., and Srivastava, H. B. (2013). Sedimentary processes in two different polar periglacial environments: Examples from Schirmacher Oasis and Larsemann Hills, East Antarctica. *Geological Society, London, Special Publications*, 381(1), 411–427.

Asthana, R., Shrivastava, P. K., Beg, M. J., Shome, S., and Kachroo, K. (2009). Surface microtextures of quartz grains from glaciolacustrine sediment core from Priyadarshini Lake, Schirmacher Oasis, East Antarctica as revealed under scanning electron microscope. *Indian Journal of Geosciences*, 63, 205–214.

Ballantyne, C. K. (1998). Age and significance of mountain-top detritus. *Permafrost and Periglacial Processes*, 9(4), 327–345.

Bardin, V. J. (1971). Moraines of Antarctica. In: R. J. Adie (Ed.), *Antarctic geology and geophysics*. Oslo Universitetes forlaget, pp. 663–667.

Bargagli, R. (2006). *Antarctic ecosystems: environmental contamination, climate change, and human impact* (Vol. 175). Springer Science & Business Media.

Barron, E. J., Harrison, C. G. A., and Hay, W. W. (1978). A revised reconstruction of the Southern Continents. *EOS*, 59, 436–449.

Beg M. J., Prasad A. V. K., Chaturvedi, A. (2000). Interim Report on Glaciological Studies in the Austral Summer of 19th Indian Antarctic Expedition. In: *Scientific Report of Nineteenth Indian Expedition to Antarctica, Tech. Pub. No.17, Department of Ocean Development*, Govt. of India, New Delhi, pp. 121–126.

Benn, D., and Evans, D. J. (2014). *Glaciers and glaciation*. Routledge.

Bennike, O., and Björck, S. (2000). Lake sediment coring in South Greenland in 1999. *Geology of Greenland Survey Bulletin*, 186, 60–64.

Bera, S. K. (2004). Late Holocene palaeo-winds and climatic changes in Eastern Antarctica as indicated by long-distance transported pollen—spores and local microbiota in polar lake core sediments. *Current Science*, 86(11), 1485–1488.

Bera, S. K. (2006). *Reconstruction of Palaeoclimate from Holocene Sequences of Lake Sediment Schirmacher Oasis, East Antarctica*. Nineteenth Indian Expedition to Antarctica, Scientific Report, 2004. Department of Ocean Development, Technical Publication No. 17, pp. 33–40.

Bormann, P., and Fritzsche, D. (eds) (1995). *The Schihnacher Oasis, Queen Maud Land, East Antarctica, and its surroundings*. Justus Perthes Verlag Gotha.

Cook, S. J., and Mcconnell, J. W. (2001). Lake sediment geochemical methods in the Canadian Shield, Cordillera and Appalachia. *Geological Society, London, Special Publications*, 185, 125–149.

Cox, K. G. (1978). Flood basalts, subduction and the break-up of Gondwanaland. *Nature*, 274(5666), 47–49.

D'Souza, M. J., Beg, M. J., Ravindra, R., and Chaturvedi, A. (1994). Occurrence of alkaline lamprophyre dyke from Schirmacher range, Dronning Maud Land, East Antarctica. *Technical Publications*, 6, 161–172.

Dietz, R. S., and Holden, J. C. (1970). Reconstruction of Pangaea: Breakup and dispersion of continents, Permian to present. *Journal of Geophysical Research*, 75(26), 4939–4956.

Gajananda, K., Dutta, H. N., and Lagun, V. E. (2007). An episode of coastal advection fog over East Antarctica. *Current Science*, 654–659.

Gaur, M. P., Dharwadkar, A., and Asthana, R. (2006). *Glaciomorphic Land Forms in Parts of Schirmarcher Oasis, Central Dronning Maud Land, East Antarctica*. Eighteenth Indian Expedition to Antarctica, Scientific Report, 2002. Department of Ocean Development. Technical Publication No. 16, pp. 81–93.

Goudie, A. S. (Ed.), 2004. *Encyclopedia of Geomorphology* (Vol. 1). Routledge Ltd., p. 578.

Hosalikar, K. S., and Machnurkar, P. N. (2006). *Report on Meteorological and Climatological Studies at Maitri, 25 Antarctica*. Fourteenth Indian Expedition to Antarctica, Scientific Report, 1998. Department of Ocean Development. Technical Publication No. 12, pp. 25–46.

Jayapaul, D., and Dharwadkar, A. (2006). *Glaciomorphological Studies in Parts of Schirmacher Oasis, Central Dronning Maud Land, East Antarctica*. Twentieth Indian Expedition to Antarctica, Scientific Report, 2006. Ministry of Earth Sciences. Technical Publication No. 18, pp. 131–138.

Kejna, M. (2003). Trends of air temperature of the Antarctic during the period 1958–2000. *Polish Polar Research*, 99–126.

Koppar, A. L. (1991). Cloud observation at Antarctica. *Mausam*, 42(2), 307–308.

Mayewski, P. A., Meredith, M. P., Summerhayes, C. P., Turner, J., Worby, A., Barrett, P. J., .and Bromwich, D. (2009). State of the Antarctic and Southern Ocean climate system. *Reviews of Geophysics*, 47(1).

Naithani, J., and Dutta, H. N. (1995). Acoustic sounder measurements of the planetary boundary layer at Maitri, Antarctica. *Boundary-Layer Meteorology*, 76(1–2), 199–207.

Phartiyal, B. (2019). *Climatic History and Landscape Evolution of Schirmacher Oasis, East Antarctica during Late Quaternary: A Multi-Proxy Study*. Twenty-Fifth Indian Antarctic Expedition 2006–2007. Ministry of Earth Sciences. Technical Publication No. 23, pp. 73–105.

Phartiyal, B., Sharma, A., and Bera, S. K. (2011). Glacial lakes and geomorphological evolution of Schirmacher Oasis, East Antarctica, during late quaternary. *Quaternary International*, 235(1–2), 128–136.

Philpot, H. R., and Zillman, J. W. (1970). The surface temperature inversion over the Antarctic continent. *Journal of Geophysical Research*, 75(21), 4161–4169.

Priddle, J., and Heywood, R. B. (1980). Evolution of Antarctic lake ecosystems. *Biological Journal of the Linnean Society*, 14(1), 51–66.

Rai, H., Khare, R., Nayaka, S., Upreti, D. K., and Gupta, R. K. (2011). Lichen synusiae in East Antarctica (Schirmacher Oasis and Larsemann Hills): substratum and morphological preferences. *Czech Polar Reports*, 1(2), 65–77.

Rasal, A. S., and Mohar, D. R. (2006). *Meteorological Studies Carried Out during the 15th Indian Scientific Expedition to Antarctica, 1996*. Fifteenth Indian Expedition to Antarctica, Scientific Report, 1999. Department of Ocean Development. Technical Publication No. I3, pp. 39-62.

Ravich, M. G., and Kamenev, E. N. (1972). *Kristallicheskii Fundament Antarkticheskoi Platformy (Crystalline Basement of the Antarctic Platform)*. Gidrometeoizdat.

Ravikant, V., and Kundu, A. (1998). Reaction textures of retrograde pressure-temperature deformation paths from granulites of Schirmacher Hills, East Antarctica. *Journal of the Geological Society of India*, 51(3), 305–314.

Ravindra, R. (2001). Geomorphology of Schirrmacher Oasis, East Antarctica. *Geological Survey of India, Special Publication*, 53, 379–390.

Ravindra, R., Chaturvedi, A., and Beg, M. J. (2002). Melt water lakes of Schirmacher Oasis-their genetic aspects and classification. *Advances in Marine and Antarctic Sciences*, 301–313.

Rea, B. R., Whalley, W. B., Rainey, M. M., and Gordon, J. E. (1996). Blockfields, old or new? Evidence and implications from some plateaus in northern Norway. *Geomorphology*, 15(2), 109–121.

Bormann, P., and Fritzsche, D. (1995). *The Schirmacher Oasis, Queen Maud Land, East Antarctica, and its surroundings: 64 tables*. Peterm Geogr Mitt Erg-h 289, Justus Perthes Verlag Gotha, 448 p.

Richter, W., Bormann, P., 1995. Hydrology. In: Bormann, P., Fritzsche, D. (Eds.), *The Schirmacher Oasis, Queen Maud Land, East Antarctica, and Its Surroundings*. Justus Perthes Verlag Gotha, pp. 259–319 (Chapter 8).

Scrutton, R. A. (1973). The age relationship of igneous activity and continental break-up. *Geological Magazine*, 110(3), 227–234.

sediment Schirmacher Oasis, East Antarctica. (XIX IAE: 1999–2000). DOD publication

Sengupta, Sudipta. (1986). *Geology of Schirmacher Range (Dakshin Gangotri), East Antarctica*. Scientific Report of Third Indian Expedition, Technical Report No. 3, Department of Ocean Development, New Delhi, pp. 187–217.

Sharma, C., Chauhan, M. S., and Sinha, R. (2007). Studies on Holocene climatic changes from Priyadarshini Lake sediments, East Antarctica: the palynological evidence. *Journal-Geological Society of India*, 69(1), 92.

Shrivastava, P. K., Roy, S. K., Srivastava, H. B., and Dharwadkar, A. (2019). Estimation of Paleo-ice Sheet Thickness and Evolution of Landforms in Schirmacher Oasis and Adjoining Area, cDML, East Antarctica. *Journal of the Geological Society of India*, 93(6), 638–644.

Shrivastava, P. K., Asthana, R., Beg, M. J., and Singh, J. (2009). Climatic fluctuation imprinted in quartz grains of lake sediments from Schirmacher Oasis and Larsemann Hills area, East Antarctica. *Indian Journal of Geosciences*, 63, 87–96.

Shrivastava, P. K., Asthana, R., Roy, S. K., Swain, A. K., and Dharwadkar, A. (2012). Provenance and depositional environment of epi-shelf lake sediment from Schirmacher Oasis, East Antarctica, vis-à-vis scanning electron microscopy of quartz grain, size distribution and chemical parameters. *Polar Science*, 6(2), 165–182.

Shumskiy, P. A. (1957). Glaciological and geomorphological reconnaissance in the Antarctic in 1956. *Journal of Glaciology*, 3(21), 54–61.

Smith K. L., and Polvani, L. M. (2016). Spatial patterns of recent Antarctic surface temperature trends and the importance of natural variability: Lessons from multiple reconstructions and the CMIP5 models. *Climate Dynamics*, 48(7), 2653–2670.

Sinha, R., and Chatterjee, A. (2000a). Mineralogy of lacustrine sediments in the Schirmacher range area, eastern Antarctica. *Journal-Geological Society of India*, 56(1), 39–46.

Sinha, R., and Chatterjee, A. (2000b). Lacustrine sedimentology in the Schirmacher Range Area, East Antarctica. Journal of Geological Society of India, v.56, pp. 39–45.

Sinha, R., Sharma, C., and Chauhan, M. S. (2000). Sedimentological and pollen studies of Lake Priyadarshini, Eastern Antarctica. *Paleobotanist*, 49, 1–8.

Smeltzer, E. R. I. C., and Swain, E. B. (1985). Answering lake management questions with paleolimnology. *Lake Reservory Management*, 1, 268–274.

Smith, A. G., and Hallam, A. (1970). The fit of the southern continents. *Nature*, 225(5228), 139–144.

Soni, V. K., Sateesh, M., Das, A. K., and Peshin, S. K. (2017, June). Progress in meteorological studies around Indian stations in Antarctica. *Proc Indian Natl Sci Acad*, 83, 461–467.

Srivastava, A. K., Ingle, P. S., and Khare, N. (2010). Textural characteristics, distribution pattern and provenance of heavy minerals in glacial sediments of Schirmacher Oasis, East Antarctica. *Journal of the Geological Society of India*, 75(2), 393–402.

Srivastava, H. B., Shrivastava, P. K., Roy, S. K., Beg, M. J., Asthana, R., Govil, P., and Verma, K. (2018). Transition in late quaternary paleoclimate in Schirmacher region, East Antarctica as revealed from Lake sediments. *Journal of the Geological Society of India*, 91(6), 651–663.

Swain, A. K. (2019). Influence of thermal conductivity of rocks on Polar ice sheet recession near Schirmacher Oasis, East Antarctica. *Journal of the Geological Society of India*, 93(4), 455–465.

Swain, A. K., and Chaturvedi, A. (2019). *Bathymetric Profiling of Lakes of Schirmacher Oasis, East Antarctica Twenty seventh Indian Antarctic expedition 2007–2009.* Ministry of Earth Sciences. Technical publication No. 25, pp. 45–72

Van Lipzig, N. P. M., Turner, J., Colwell, S. R., and van Den Broeke, M. R. (2004). The near-surface wind field over the Antarctic continent. *International Journal of Climatology: A Journal of the Royal Meteorological Society*, 24(15), 1973–1982.

Verlecar, X. N., Dhargalkar, V. K., and Matondkar, S. G. P. (1996). *Ecobiological studies of the freshwater lakes at Schirmacher Oasis, Antarctica. Department of Ocean Development, India. Twelfth Indian expedition to Antarctica, scientific report, 1996.* Department of Ocean Development technical publication No.10, pp. 233–257

Wand, W., 1983. Geological observations in the Schirmacher Oasis, Queen Maud Land, East Antarctica—Preliminary Accounts, Geodat. *Geophys.* Veroff. NKGG., Reihe I, Heft 9, pp. 85–89 (in Russian).

Warrier, A. K., Mahesh, B. S., and Mohan, R. (2017). Lake Sediment Studies in Ice-Free Regions of East Antarctica—An Indian Perspective. *Proceedings of the Indian National Science Academy*, 83(2), 289–297.

Whalley, W. B., Rea, B. R., Rainey, M. M., and McAlister, J. J. (1997). Rock weathering in blockfields: some preliminary data from mountain plateaus in North Norway. *Geological Society, London, Special Publications*, 120(1), 133–145.

5 Significance of Foraminiferal Studies from the Southern High Latitudes in Assessing Global Climate Change

An Overview

Subodh Kumar Chaturvedi[1] *and Neloy Khare*[2]

[1] Institute of Hydrocarbon, Energy and Georesources, ONGC Centre for Advanced Studies, University of Lucknow, India

[2] Ministry of Earth Sciences, Government of India, New Delhi, India

CONTENTS

DOI: 10.1201/9781003284413-5

5.1 INTRODUCTION

The southern high latitudes comprising of the Southern Ocean and Antarctica play an important role in global climatic change. The ocean surrounding Antarctica is the locale for the origin of the world's deep and intermediate waters, which constitute a major conduit for the exchange of heat and dissolved gases such as carbon dioxide and oxygen between the atmosphere and the oceans. Antarctic Bottom Water and Antarctic Intermediate Water produced in the Southern Ocean near the Antarctic continent are major components of the global thermohaline circulation. In the Southern Ocean, the world's largest surface current, the Antarctic Circumpolar Current, controls the exchange of salt, heat, and nutrients between the Pacific, Atlantic and Indian Oceans. Recently, the Southern Ocean has been recognized as the site of absorption from the atmosphere of much of the carbon dioxide emitted by human activities, which gives this ocean another prominent role in the global warming story. What happens in the south, affects the entire earth. The teleconnection between the southern high latitudes and the Indian monsoon system is currently a major topic of ongoing research. Recognizing that the Southern Ocean plays an important role in controlling the Earth's climate, considerable efforts are being made to understand the changes in the physicochemical and biological structure of the Southern Ocean that have taken place throughout the geologic past.

Like the Southern Ocean, the Antarctica continent too plays a key role in the global climate change. Among the factors that are critical to understand the Earth's climatic history is the changing extent of the ice. The ice extent regulates the Earth's albedo and thus can significantly enhance or reduce climate change. Ice is also important in that the volume of land ice controls the global sea level. Any increase or decrease in the ice extent will lower or raise the sea level, respectively. Antarctica contains the largest ice sheets on Earth, and so has had a major effect on sea level through time. Most of Antarctica is covered with ice, except in the peripheral regions, parts of which may remain exposed throughout the year or during the austral summer. Some of these peripheral regions contain lakes fed by the melting glaciers or ice sheets, or filled with seawater. Such lakes can be used to study the changing ice extent and sea-level history of the Antarctica, and have provided extensive records of the changes in the ice extent over the geologic past.

Several proxies are used to reconstruct the past. One such frequently used proxy in marine sediments is characteristics of foraminifera. Foraminifera are exclusively marine microorganisms and very sensitive to environmental/climatic change. That explains their extensive use as proxies for temperature, salinity, productivity, circulation, sedimentation, and several other processes in paleoclimatic/paleoceanographic reconstruction. A large number of foraminiferal studies have been carried out to understand the influence of various water masses on foraminifera and the geologic history of the southern high latitudes. In this paper, we provide an overview of the salient findings of the foraminiferal studies carried out on samples collected from the southern high latitudes. The results expand on earlier work by Kennet (1980), who compiled the Cenozoic paleoclimatic and paleoceanographic history of the Southern Ocean, and Ehrmann (1994), who synthesized the Cenozoic climatic changes in the Antarctic region, based on the studies carried out mainly from the Atlantic and

Indian sectors of the Southern Ocean. The present work is thus the first complete synthesis of foraminiferal studies from the southern high latitudes.

5.2 THE SOUTHERN OCEAN

Oceanographers regard the Southern Ocean as everything south of the sub-Antarctic Front, which defines the northern edge of the Antarctic Circumpolar Current at around 45 °S. Geopolitically, the International Hydrographic Organization (IHO) 2000 agreed that the Southern Ocean should be formally recognized as the ocean south of 60 °S, which coincides with the northern limit of the area of the Antarctic Treaty (Pandey et al., 2006). Here we are concerned only with the oceanographer's definition (Figure 5.1). From north to south, the fronts and zones of the Southern

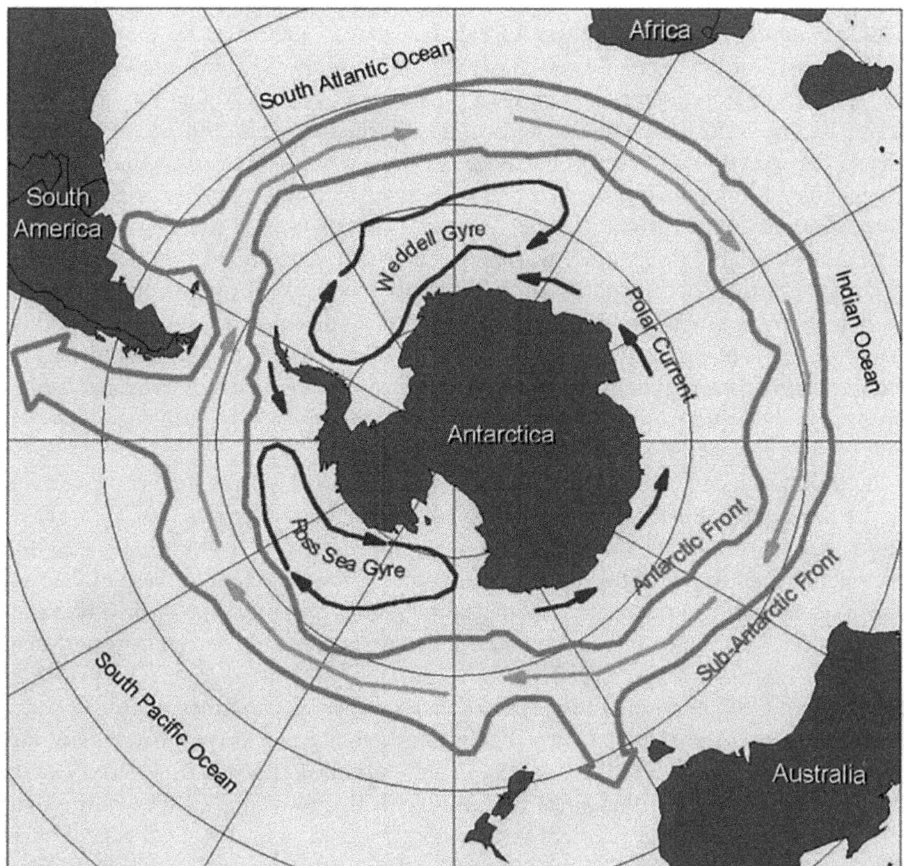

FIGURE 5.1 The area considered for the present review of the foraminiferal studies from the southern high latitudes. Various hydrographic features of the region are also marked on the map.

Ocean are the sub-Antarctic Front (SAF) at around 45 °S, the Polar Frontal Zone (PFZ), and the Polar Front (PF) at around 48 ºS and the Antarctic Zone (AZ) (Whiteworth, 1980). The fronts coincide with current jets that transport most of the water in the Antarctic Circumpolar Current (Nowlin and Clifford, 1982). Between the fronts lie zones of relatively uniform water mass properties. The Southern Ocean is influenced by many important processes taking place in and around as previously summarized by many workers (Rintoul et al., 2002; Pollard et al., 2002; Pandey et al., 2006) and highlighted here.

The pronounced meridional gradient in surface properties separates waters of the Southern Ocean from the warmer and saltier waters of the subtropical circulations. Deacon (1933, 1937) called this hydrographic boundary, the Subtropical Convergence, a term replaced by the Subtropical Front (STF) in the recent years (Clifford, 1983; Hofman, 1985). The transition from warm, light subtropical water in the north to cold, dense Antarctic water in the south, occurs in a step-like manner rather than as a gradual change across the breadth of the Southern Ocean (Deacon, 1933). The detailed mapping of the Southern Ocean fronts was attempted by many workers (Belkin and Gordon, 1996; Orsi et al., 1995). South of the STF, the eastward flow of the Antarctic Circumpolar Current (ACC) extends unbroken around the globe. It is driven by the world's mightiest westerly winds (Trenberth et al., 1990), found approximately between 45° S and 55° S. Because of the land mass distribution, the ACC is a unique global link that connects all major oceans. Deep water masses originating far to the north cross the current and well up towards the surface, so entering the subpolar regime where they mix laterally with Antarctic shelf waters (Orsi, 1995). In contrast to the long-established demarcation between the subtropical regime and the ACC, there is yet no well-defined boundary between this current and the subpolar regime (Deacon, 1937) recognized a westward flow adjacent to Antarctica, which is driven by the prevailing easterly winds found to the south of about 65 °S. Isotherms and isohalines shoal at the transition between westerly and easterly winds, inducing clockwise flow under geostrophic balance (Deacon, 1937; Deacon, 1982).

The proximity of Subantarctic Surface Water (SASW) to the much warmer and saltier Subtropical Surface Waters (STSW) at the northern edge of the ACC produces large property gradients, the original indicators of the STF. Deacon (1937) gave the first global description of this upper-water front. He noted that regardless of the season or ocean basin, surface temperature (salinity) changes as large as 4–5 °C (0.5 psu) marking the location of the STF; its northern side is generally warmer (saltier) than 11.5 °C (34.9 psu) (Deacon, 1982).

Deacon (1982) examined temperature and salinity distributions at 100 m water depth and observed that the front lies within a band across which temperatures increase northward from 10 °C to 12 °C and salinities from 34.6 psu to 35.0 psu at 100 m. He suggested that since the salinity field shows smaller seasonality than temperatures, the haline gradient is the more reliable indicator. One exception is the southeast Pacific sector (Orsi, 1995). The sea surface expression of most of the frontal systems is a sharp southward temperature drop with consequent changes in planktic biotic assemblages (Barker and Thoma, 2004).

The highly barotropic Antarctic Circumpolar Current (ACC) reaches the ocean floor and can mix North Atlantic Deep Water (NADW) and deep waters from the

Indian and Pacific Oceans. The mixture of these deep waters, the Circumpolar Deep Water (CDW), then spreads back into other ocean basins in one form or another. The CDW wells up to the surface, south of the ACC as a response to Ekman pumping, forced by the westerly winds that push surface waters to the north. This water spreads onto the continental shelves, mixes with continental shelf water, cools, becomes dense, and then sinks to form the Antarctic Bottom Water (AABW) especially in and around the Weddell and Ross Seas. Antarctic Intermediate Water (AAIW) forms north of the Polar Front and sinks north below subtropical waters at the Subantarctic Front. The northward advection of AABW, CDW and AAIW from the Southern Ocean to other ocean basins has a profound effect on the global ocean circulation and thus on global climate (Matsumoto et al., 2001). Geochemical and climatic signals of the Southern Ocean processes are transmitted throughout the world's oceans. The process of upwelling in the Southern Ocean stimulates considerable primary productivity, amounting to nearly a third of the global oceanic total. The sea-ice distribution around Antarctica and the nutrient structure within the cold Antarctic Circumpolar Current control the biogenic sedimentary provinces of the Southern Ocean. About two-thirds of the silica flux into the global ocean is removed by siliceous microorganisms in the Southern Ocean. The Southern Ocean is also the region of major wind stress, as a result of which the mixed layer is rather deep, especially during the winter, which in turn affects the foraminifera.

5.3 ANTARCTICA

Antarctica covers some 13,209,000 sq. km, an area that varies due to changing ice shelves. It is the coldest and windiest spot on the planet, with the lowest temperature ever recorded on Earth (–89.2 °C) and winds commonly measured at up to 200 km per hour. It is also the highest continent, though its highest Point (the Vinson Massif) only reaches 16,066 ft. (4,897 m). There are numerous small- to large-sized water bodies (lakes) in the Antarctic coastal area. Most were formed after the last deglaciation, and contain the climatic history of the Holocene period. Older sediments have been obtained from some lakes, some being as old as 40 kyr and others in exceptional cases dating back to isotopic stage 5e. Antarctic lake sediments have revealed significant regional climatic changes showing that different parts of Antarctica responded in different ways to the global climatic changes. Lake sediment records have been used to document changes in the regional temperature over the last several thousand years, sea-level variations, and isostatic rebound since the last deglaciation (Roberts et al., 2001; Piskova et al., 2019).

5.4 FORAMINIFERAL STUDIES

Given the potential of foraminiferal proxies to understand environmental conditions, a brief analysis of the major previous studies carried out in the southern high latitudes (south of 30 °S) is presented here. Those papers with only a part of their study area falling south of 30 °S latitude have also been considered. This work is the first comprehensive compilation and discussion of foraminiferal studies carried out so far from southern high latitudes.

5.4.1 Foraminiferal Studies from Antarctic Lakes

Although many foraminiferal studies have been carried out from the Southern Ocean waters adjacent to Antarctica, only a few studies have discussed foraminiferal distributions in the Antarctic lake sediments and their paleoclimatic significance (Cornelius and Gooday, 2004) (Table 5.1). Pushkin et al. (1991) provided a preliminary checklist of various fauna including foraminifera from the Fish Tail Bay in the Bunger Oasis of East Antarctica. Later, Cromer et al. (2005) used various proxies, including changes in foraminiferal characteristics, to infer productivity changes during the Holocene in Ace Lake in the Vestfold Hills of East Antarctica. Subsequently, Cromer et al. (2006) reconstructed the glacial history of Antarctica for the last 130,000 years based on the sediment characteristics in a core collected from the Larsemann Hills region. Smith et al. (2007) reconstructed the history of ice extent of the George VI Ice Shelf between 9600 cal yr BP and 7730 cal yr BP, based on micropaleontological (diatoms/foraminifera) evidence along with data on stable isotopes ($\delta^{18}O$, $\delta^{13}C$), geochemistry (total organic carbon (TOC), total nitrogen (TN), C/N ratios) and grain-size from cores from Moutonnee Lake. The limited foraminiferal studies from the Antarctic lakes are because of the sparse recovery of foraminifera from these environments, most of which are fresh water.

5.4.2 Foraminiferal Studies from Land-Based Sections in Antarctica

Foraminifera has also been recovered from land-based sections in Antarctica and used to infer past climatic/oceanographic changes (Table 5.2). Webb et al. (1984, 1996) used foraminifera from the trans Antarctic mountains to infer depositional environment and changes in the ice extent as well as in sea-level history. A series of glacial advancements and retreats during the Pliocene, as well as sea-level transgressions

TABLE 5.1
Foraminiferal Studies from the Antarctica Lake Sediments

Sl. no	Authors	Year	Region	Remarks
1.	Pushkin et al.	1991	Fish Tail Bay in the Bunger Oasis	Provided a preliminary checklist of various fauna including foraminifera.
2.	Cromer et al.	2005	Ace Lake, Vestfold Hills, Antarctica	Evolution of microfossil content over the Holocene; its relationship with productivity changes.
3.	Smith et al.	In press	Moutonnee Lake	Inferred the absence of George VI Ice Shelf between 9600 cal yr BP and 7730 cal yr BP, based on micropaleontological (diatoms/foraminifera), stable isotopic ($\delta^{18}O$, $\delta^{13}C$), geochemical (total organic carbon (TOC), total nitrogen (TN), C/N ratios), and grain-size analyses.

TABLE 5.2
Foraminiferal Studies from the Exposed Sections in Antarctica

Sl. no	Authors	Year	Region	Remarks
1.	Webb et al.	1996	Middle Beardmore Glacier-Queen Alexandra Range, Transantarctic Mountains	Inferred depositional environment of Sirius Group Formation based on foraminifera.
2.	Webb et al.	1984	Pliocene Sirius Formation in the Reedy, Beardmore, and Ferrar glacier areas of the Transantarctic Mountains	Reconstructed transgressing-regressing sea/ice extent during Mesozoic and Cenozoic times.

and regressions, were inferred based on the foraminiferal characteristics of the Sirius Group. However, such studies are very limited as there are few rock exposures on the continent, and most of the rocks exposed are igneous or metamorphic.

5.5 FORAMINIFERAL STUDIES FROM THE SOUTHERN OCEAN

Numerous foraminiferal studies have been carried out on the surface and subsurface sediments of the Southern Ocean to understand ecology, productivity, and water mass characteristics and reconstruct past changes in this region (Table 5.3). The major foraminiferal studies carried out so far, from the Southern Ocean are discussed below.

5.5.1 Distribution and Ecology

Since the pioneering work of Chapman (1916) and Heron-Allen and Earland (1932), many researchers have documented the surface distribution of foraminifera, and their relationship with the physicochemical parameters of the Southern Ocean (Table 5.3). The foraminiferal distribution near the Antarctic Peninsula is mainly controlled by the extent of the ice sheet today and in the past (Temnikow and Lipps, 1975), whereas the abundance and diversity of foraminifera in the region further north are controlled by the carbonate compensation depth and productivity (Parker and Berger, 1971; Anderson, 1975b; Ward et al., 1987; Mackensen and Douglas, 1989). Theyer (1971) divided the foraminifera from the Pacific Antarctic region into four distinct foraminiferal assemblages. In contrast, Boltovskoy et al. (1996) identified five distinct assemblages of planktic foraminifera from the Atlantic sector of the Southern Ocean. The foraminiferal assemblages include Subtropical (north of 31 °S, 80% warm water individuals); Warm-Transitional (34°–37 °S, 35% warm water individuals); Transitional (37°–49 °S, 99% cold water individuals); Subantarctic (49–55 °S. 100% cold water individuals); and Antarctic (south of 56 °S, 100% cold water individuals), and roughly correspond with the various oceanographic fronts in the region. Sediment traps have been deployed to understand the temporal changes in

TABLE 5.3
Foraminiferal Studies from the Waters Adjacent to Antarctica up to 30 °S Latitude

Sl. no	Authors	Year	Region	Remarks
1.	Maria et al.	2022	Global Ocean	The ecosystem model of biogeographical patterns of non-spinose species of foraminifera exhibits maximum biomass concentrations in mid- to high-latitude waters and upwelling areas. In response to ocean warming and associated changes in primary production and ecological dynamics, the global average loss in nonspinose foraminifera biomass is projected to be between 8% and 11% by 2050 and between 14% and 18% by 2100.
2.	Petrizzo et al.	2022	SE Indian Ocean (IODP Site U1513)	$\delta^{18}O$ values of planktonic foraminifera suggest a decline in surface water paleotemperatures of 4°C within the Santonian.
3.	David et al.	2021	The Southern Ocean, Atlantic sector	A new sea ice proxy based on the $^{18}O/^{16}O$ ratio of foraminifera from the Atlantic sector that yields an estimate of winter sea ice extent consistent with modern observations
4.	Joseph et al.	2021	Southern Ocean	The ecological and geochemical data indicate that habitat shifts were dictated by a northward migration of food supply (primary production) into the sub-Antarctic Zone and poorly oxygenated seawater at depth during this Antarctic cooling interval.
5.	Brian et al.	2020	The Southern Ocean, southern South Atlantic and the southern Indian Ocean	The geographic distribution of *Antarcticella pauciloculata* (Jenkins) and *Zeauvigerina waiparaensis* (Jenkins), as well as *Eoglobigerina maudrisensis* n. sp. from just above the K/Pg in the southern South Atlantic and the southern Indian Ocean, helps define the extent of the Austral Biogeographic Province and provides evidence for marine communication via marine seaways across.
6.	Sandi et al.	2020	The Southern Ocean, south of Africa	The nitrogen isotope data suggest that, under modern conditions, the late-summer ammonium recycling signal outweighs that of wintertime decomposition on the annually integrated ^{15}N of sinking.
7.	Andrew et al.	2019	The Southern Ocean, south of Tasmania	Based on boron isotopes and carbon isotopes in planktonic foraminifera, it has been inferred that the variations in upwelling intensity and the distribution of Southern Ocean water masses played a key role in regulating atmospheric CO_2 during the last glacial-interglacial cycle.

Sl. no	Authors	Year	Region	Remarks
8.	Anna et al.	2019	Southern Ocean	A sediment trap study from the West Antarctic Peninsula reported a distinct seasonal production having the highest shell fluxes during the warmest and most productive months of the year
9.	Thomas et al.	2019	Pacific Southern Ocean	Multiproxy data suggest that there are different processes of deep- and bottom-water formation around Antarctica.
10.	Thomas et al.	2018	The Southern Ocean, South of Australia	Results for nine transects from Australia to Antarctica in 2008–2015 showed low levels of PIC compared to Northern Hemisphere polar waters. Coccolithophores slightly exceeded the biomass of diatoms in Subantarctic waters, but their abundance decreased more than 30-fold poleward, while diatom abundances increased.
11.	Meir et al.	2017	Atlantic and the Southern Ocean (Atlantic sector)	A compilation of benthic $\delta^{18}O$ suggests that while the shallow proto-ACC supplied the energy for deep ocean convection in the Southern Hemisphere, the onset of the interhemispheric northern circulation cell was due to the significant Eocene-Oligocene Transition (EOT) intensification of deepwater formation in the North Atlantic driven by the Nordic anti-estuarine circulation.
12.	Rebecca et al.	2017	Cosgrove Ice Shelf, Antarctica	The glacial history and paleoceanographic conditions reflect that the ice shelf persisted until 2.3 cal. kyr BP, when TOC and diatom abundance increased as the bay opened and coastal areas deglaciated. The retreat of the Cosgrove Ice Shelf may be due to the influence of Circumpolar Deep Water and possibly to internal glacial dynamics.
13.	Woelders et al.	2017	Southern Ocean, South Atlantic	The environmental perturbations during the latest Maastrichtian warming event were less severe than those following the K-Pg boundary impact.
14.	Jacob et al.	2016	Southern Ocean, South Atlantic	No evidence for the greater incursion of Antarctic Intermediate Water into the South Atlantic during either the Younger Dryas or Heinrich Stadial 1 were observed
15.	Julie et al.	2016	The Southern Ocean, the oceanic regions around Crozet and Kerguelen Islands	The low Living Population of Foraminifera standing stocks in late austral summer in the Southern Ocean contrasted with the presence of high densities of heavily silicified diatoms. Further, the polar/subpolar foraminiferal assemblages are characterized by high abundances of G. uvula in the iron-enriched waters surrounding the French Subantarctic Islands

(Continued)

TABLE 5.3
(Continued)

Sl. no	Authors	Year	Region	Remarks
16.	Mazumder et al.	2016	Indian Sector of Southern Ocean	The Q-mode cluster analysis of different parameters of *Neogloboquadrina pachyderma* suggests that sea surface temperature; sea surface salinity and nutrients are the main ecological factors controlling the morphological characteristics.
17.	Wei et al.	2016	Southern Ocean, South Atlantic	Down-core sedimentary Nd isotope records suggest continuous deep-water production in the North Atlantic and export to the South Atlantic and the Southern Ocean.
18.	Petersen and Schrag	2015	Southern Ocean	The long-term change in $\delta^{18}O$ seen in planktonic foraminifera is predominantly due to changes in ice volume.
19.	Mohan et al.	2015	Western Indian Ocean Sector of Southern Ocean	The planktic foraminiferal isotopes show the presence of heavier isotopes in the surface sediment foraminifera as compared to plankton tows, thus confirming secondary calcification.
20.	Tapia et al.	2015	Southern Ocean, South Pacific	A 200 kyr record of paired Mg/Ca ratios and stable oxygen isotope from the surface and deep-dwelling planktonic foraminifera, from the South Pacific Gyre, suggest the variability in Southern Ocean Intermediate Waters is probably driven by the changes in the intensity of the Southern Westerly Winds.
21.	Thomas et al.	2015	The Southern Ocean, southwest Pacific	Carbon and oxygen isotope records from epibenthic foraminifera of sediment cores reveal pronounced freshwater input by melting sea ice into the glacial Antarctic Intermediate Water (AAIW) significantly hampered the downward expansion of southwest Pacific AAIW.
22.	Benedetto et al.	2014	Southern Ocean, Southwest Pacific	The isotopic records from nearby Hawke Bay, east of the North Island of New Zealand, exhibited several changes in thermocline depth indicating switches between distal subtropical and proximal subantarctic influences during the early deglaciation ending only after the Antarctic Cold Reversal.
23.	Franck et al.	2014	Southern Ocean	The eDNA diversity of Foraminifera in 42 deep-sea sediment samples collected across different scales in the Southern Ocean was analyzed. The majority of the operational taxonomic units are represented by a few sequences comprising several well-known deep-sea morphospecies.

Sl. no	Authors	Year	Region	Remarks
24.	Friederike et al.	2014	Atlantic Sector of the Southern Ocean	Transparent exopolymer particles likely played an important role by aggregating these pico- and nanoplankton small cells and making them more readily available to mesozooplankton grazers.
25.	Mazumder et al.	2014	Indian Sector of Southern Ocean	The study suggests that the ecological parameters are the governing factors for the morphological characteristics of planktonic foraminiferal species *G. bulloides*.
26.	Naik et al.	2014	Western Indian Ocean Sector of Southern Ocean	The temporal variation in planktic foraminiferal abundance, stable isotopic ratio ($\delta^{18}O$) and trace elemental ratio (Mg/Ca) of *Globigerina bulloides* implied that over the last glacial-interglacial cycle, the hydrography of the southwestern Indian Ocean was driven by strengthened westerlies, ARC as well as a migrating subtropical front.
27.	Timothy et al.	2014	Maud Rise in the Southern Ocean	The Earth System Climate Model shows that stratification increased and the nutricline strengthened at the onset of the Paleocene-Eocene thermal maximum.
28.	Lower et al.	2013	Indian Ocean sector of Southern Ocean	The data collectively indicate water conditions that are significantly changed from those that prevailed in the Coorong Lagoon for most of the Holocene.
29.	Pena et al.	2013	Pacific Sector of Southern Ocean	A comparison of Nd isotopes in foraminifera shells for the last 30kyr indicates a latitudinal shift and/or a change in the convection depth of intermediate waters in the Southern Ocean prior to the onset of the deglaciation.
30.	Ziegler et al.	2013	Atlantic Sector of the Southern Ocean	The $\delta^{13}C$ values of intermediate- and bottom-dwelling foraminifera suggest a relationship between the efficiency of the biological pump and fertilization by dust-borne iron.
31.	Khare and Chaturvedi	2012	Indian sector of Southern Ocean	The study of total planktonic foraminiferal assemblage, relative abundance and the oxygen isotopic values of the indicator planktic species *Globigerina bulloides* have traced the signatures of salinity linked variations of different water masses along the N-S transect in the Indian Sector of the Southern Ocean.
32.	Khare and Chaturvedi	2012	Indian sector of Southern Ocean	The characteristic patterns of the values of oxygen and carbon isotopes are associated with various frontal systems and/or water masses along the N-S transect in the Indian Sector of the Southern Ocean.

(*Continued*)

TABLE 5.3 (Continued)

Sl. no	Authors	Year	Region	Remarks
33.	Khare et al.	2012	Indian sector of Southern Ocean	Nonexistence of any relationship between the coiling direction and reproductive modes have been observed in planktonic foraminifera *Neogloboquadrina pachyderma, Globigerinita glutinata* along the N-S transect in the Indian Sector of the Southern Ocean.
34.	Abrantes et al.	2011	Atlantic Sector of the Southern Ocean	Multidecadal variability and sea surface temperatures (SSTs) appear to be responding to long-term solar insolation variability. AD 1850 marks a major shift in the phytoplankton community. These changes are interpreted as a response to a reduction in the summer and/or annual upwelling and more frequent fall-winter upwelling-like events.
35.	Atsushi et al.	2011	Pacific Sector of Southern Ocean	Identification of a total of five reliable chemostratigraphic datums, based on the fundamental structural changes in the $^{87}Sr/^{86}Sr$ curve and paired simultaneous $\delta^{13}C$ and $\delta^{18}O$ events were made.
36.	Caniupán et al.	2011	Pacific Sector of Southern Ocean	Differences in the new SST record reflect regional cooling related to the proximal location of the southern Patagonian Ice Sheet and related meltwater supply at least during the LGM consistent with the fact that no longer an SST cooling trend is observed.
37.	Huber et al.	2011	Atlantic Sector of the Southern Ocean	The dramatic changes in planktic foraminiferal assemblages across the Aptian/Albian boundary interval (AABI) suggest major changes in carbonate chemistry, vertical stratification, or productivity in the surface mixed layer occurred during the last 1 myr of the Aptian.
38.	Jan et al.	2011	Southern Ocean	Novel lineages of Southern Ocean deep-sea foraminifera revealed by environmental DNA sequencing
39.	Tiwari et al.	2011	Indian sector of the Southern Ocean	The planktic foraminifera secretes their shells in isotopic equilibrium with seawater and the planktic foraminifera from the core top sediments yield values akin to that obtained from plankton net samples.
40.	Abouchami et al.	2010	Atlantic Sector of the Southern Ocean	The study demonstrates unambiguously a change in Cd isotope fractionation along with changes in nutrients inventories and water mass distribution across the PF, definitively establishing Cd isotopes as water mass tracer and primary productivity proxy.

Sl. no	Authors	Year	Region	Remarks
41.	Anderson et al.	2010	Southern Ocean	Bottom-dwelling foraminifera records the buildup of ^{14}C-depleted CO_2 in the deep ocean during the last glacial period, more than 21,000 years ago, and the shells of surface dwellers record the transfer of ^{14}C-depleted CO_2 to the surface between 617,000 and 21,000 years ago.
42.	Elderfield et al.	2010	Pacific Sector of Southern Ocean	Mg/Ca record for *Uvigerina* spp. generated for the Southern Ocean over the past 440,000 years from Ocean Drilling Program Site 1123 shows variability that correlates with climate oscillations.
43.	Khare et al.	2010	Indian Sector of Southern Ocean	The coiling direction and absolute abundance of *Neogloboquadrina pachyderma* can be related to ambient temperature and salinity values.
44.	Scher et al.	2010	Atlantic Sector of the Southern Ocean	The lack of significant Nd variability at Walvis Ridge combined with large excursions at Southern Ocean sites argues for a southern source of non-radiogenic Nd to the Southern most likely resulting from an increase in the weathering flux from Antarctica during ice sheet development.
45.	Takahashi et al.	2010	Indian Sector of Southern Ocean	A Continuous Plankton Recorder (CPR) is suggested to be one of the ideal methods to monitor organisms that are indicators of environmental change.
46.	Bergami et al.	2009	Pacific Sector of Southern Ocean	Results document that diversity of planktonic foraminifera, a number of specimens and variations in test morphology are related to regional differences in water properties (temperature, salinity, and DCM depth).
47.	Khare et al.	2009a	Indian Ocean sector of Southern Ocean	The oxygen isotopic analysis of *Globigerina bulloides* from 23 surface sediment samples collected across the north-south transect of the Indian sector of the Southern Ocean reveals an increasing trend in the $\delta^{18}O$ value towards higher latitude. Such an increase in $\delta^{18}O$ values is inversely related to the temperature changes along the transect.
48.	Khare et al.	2009b	Indian Ocean sector of Southern Ocean	The coiling directions *Neogloboquadrina pachyderma* co-vary with temperature and salinity, the abundance of sinistrally coiled forms increasing towards higher latitudes
49.	Khare et al.	2009c	Indian Sector of Southern Ocean	The average number of chambers in *Neogloboquadrina pachyderma* was correlated with average temperature and salinity.

(*Continued*)

TABLE 5.3 (Continued)

Sl. no	Authors	Year	Region	Remarks
50.	Kim et al.	2009	Indian Sector of Southern Ocean	Paleoclimate records are primarily influenced by local conditions and less subjected to long-distance lateral transport than other organic proxies in the Southern Ocean.
51.	Mazumder et al.	2009	Indian Sector of Southern Ocean	The study shows that the ecological parameters do not have any major role in the morphological variations of planktic foraminiferal species *G. glutinata*. This signifies the cosmopolitanism of this species.
52.	Moy et al.	2009	Southern Ocean	The study shows that the modern shell weights are 30–35% lower than those from the sediments of the underlying Holocene age. The loss in shell weight is consistent with reduced calcification today induced by ocean acidification.
53.	Roberts et al.	2008	Southern Ocean	With CO_2 continuing to enter the ocean detectable reduction in calcification of pteropods and other marine calcifiers could result in changes to the distribution of species and the structure of Southern Ocean ecosystems.
54.	Billups et al.	2008	Atlantic Sector of the Southern Ocean	Reduced nutrient utilization during warmer intervals would be consistent with reduced carbon isotope fractionation during air-sea gas exchange during relatively warmer conditions.
55.	Diz et al.	2008	Indian Ocean sector of Southern Ocean	The study suggests that the $\delta^{13}C$ and DIC of *Epistominella exigua* potentially more faithfully record the amplitude of ambient bottom water changes than *Fontbotia wuellerstorfi*, notably in settings such as the Southern Ocean that experienced substantial changes through time in the organic carbon supply to the seafloor.
56.	Khare et al.	2008	Indian Sector of Southern Ocean	The results of micropaleontological studies carried out on the samples collected during the Pilot Expedition to the Southern Ocean (PESO) were collated.
57.	Passchier and Krissek	2008	Ross Sea, Antarctica	Results reveal a strong coupling between East Antarctic continental temperatures and ice volume, with deep-sea temperatures and concentrations of atmospheric carbon over long timescales.
58.	Hayward et al.	2007	Campbell and Bounty Plateaux, New Zealand	Investigated the factors affecting the benthic foraminiferal distribution.

Sl. no	Authors	Year	Region	Remarks
59.	Khare and Chaturvedi	2007	Indian Sector of Southern Ocean	The relationship between salinity, nutrient content and the planktonic foraminiferal population was established in the south-western Indian Ocean.
60.	Khare et al.	2007	Indian Sector of Southern Ocean	The study indicates that up to ~31 °S latitude, salinity may influence the oxygen isotopic composition of the foraminiferal tests. However, beyond 31 °S (further south) latitude, salinity does not appear to influence the *G. bulloides* oxygen isotopic composition significantly.
61.	Robinson et al.	2007	Chile Margin, Southeast Pacific	Inferred denitrification changes over the last 70 kyr, based on benthic foraminiferal oxygen isotopic and bulk sediment nitrogen isotopic analysis; the linear relationship between Antarctic climate changes and southeast Pacific denitrification changes.
62.	Saraswat et al.	2007	Indian Sector of Southern Ocean	The planktic foraminiferal abundance in general increases as the sea water becomes more alkaline in the Indian Ocean sector of Southern Ocean.
63.	Scott et al	2007	Southern Ocean	Assessed the evolution and morphological changes in the *Globorotalia puncticulata* throughout the Pliocene and Quaternary, as a result of changing hydrography.
64.	Cook et al.	2006	Chatham Rise	Benthic foraminiferal Cd/Ca and Mg/Ca analysis-based changes in nutrient availability in Antarctic Intermediate waters during the last 150 kyr.
65.	Friedrich and Meier	2006	Weddell Sea, Antarctic Ocean	Oxygen isotopic composition of calcareous dinoflagellates is comparable to shallow dwelling planktic foraminifera, whereas carbon isotope shows a constant offset; thus potential application of dinoflagellate $\delta^{18}O$ for paleoclimatic reconstruction.
66.	Hunt and Hosie	2006a	Southern Ocean south of Australia	Effect of ice extent and seawater temperature on the plankton community including foraminifera.
67.	Hunt and Hosie	2006b	The Southern Ocean south of Australia	Seasonal changes in the plankton community including foraminifera from continuous plankton recorder.
68.	Khare and Chaturvedi	2006a	Indian Sector of Southern Ocean	The study shows that nutrients play a significant role in determining the size of the foraminiferal population in the Indian sector of Southern Ocean.

(Continued)

TABLE 5.3
(Continued)

Sl. no	Authors	Year	Region	Remarks
69.	Khare and Chaturvedi	2006b	Indian Sector of Southern Ocean	The study shows an apparent relationship between the mechanical stress on foraminiferal tests caused due to the strength of prevailing ocean currents.
70.	Pawlowski et al.	2006	Polar Oceans	Inferred genetically almost identical morphotypes of *Epistominella exigua, Cibicides wuellerstorfi* and *Oridorsalis umbonatus* from both the north and south polar oceans.
71.	Pudsey et al.	2006	Prince Gustav Channel, Larsen Inlet	Applied benthic foraminiferal assemblage to reconstruct Holocene ice-extent history of Larsen area, northeast Antarctic Peninsula; major ice-shelf break-up episode during mid-Holocene.
72.	Salvi et al.	2006	Pennell Trough, Ross Sea	Reconstructed the Ross ice-shelf movement history for the Late Pleistocene and Holocene based on X-rays and volume magnetic susceptibility (VMS), clay mineralogy and foraminiferal studies.
73.	Suhr and Pond	2006	Arthur Harbor, Anvers Island, Antarctica	Inferred selective feeding in benthic foraminiferal species *Cassidulina crassa* based on fatty acid biomarker analyses.
74.	Webb and Strong	2006	Margins of Victoria Land Basin	Reconstructed Upper Oligocene-Lower Miocene paleoceanographic history based on benthic foraminiferal distribution.
75.	Domack et al.	2005	Larsen ice shelf	Evidence of continuous thinning of Larsen B ice shelf, based on oxygen isotopic composition of planktic foraminifera.
76.	Gallagher et al.	2005	Otway Basin, southeastern Australia	Reported dissolution and extinction of several benthic foraminifers as a result of Late Cretaceous, Ocean Anoxic Event.
77.	Hillenbrand et al.	2005	Western Bellingshausen Sea, West Antarctica	Inferred last glacial ice-extent history based on various sedimentological proxies supported by foraminiferal abundance.
78.	Hunt and Hosie	2005	South of Australia	Marked differences were noticed in continuous zooplankton recorder collected zooplankton assemblages in different frontal zones.
79.	Northcote and Neil	2005	Campbell Plateau, New Zealand	Effect of productivity and water mass characteristics on planktic foraminiferal distribution, and dissimilarities in foraminiferal assemblage in sediment trap and surface sediments.

Sl. no	Authors	Year	Region	Remarks
80.	Pawlowski et al.	2005	Ross Ice Shelf	Reported 14 Allogromiid foraminifera from sediment samples and carried out phylogenetic analysis on partial small subunit rDNA sequences of the same.
81.	Robert et al.	2005	Southwest Africa	Late Miocene to early Pliocene evolution of continental margin off southwest Africa based on clay mineralogy, coarse siliciclastics and benthic foraminifer accumulation rates.
82.	Sicre et al.	2005	Southern Indian Ocean	Reported inconsistency in sea-surface-temperature estimated from alkenone analysis and foraminiferal assemblage and isotopic ratio; attributed it to advection of detrital alkenone in Agulhas Current.
83.	Williams et al.	2005	World Ocean	Assessed the applicability of the foraminiferal oxygen isotopic ratio to estimate Late Miocene seawater temperature.
84.	Bianchi and Gersonde	2004	Atlantic Sector of Southern Ocean	Reconstructed the Last glacial to Holocene paleoclimatic history of the region and correlated it with the global climatic variations.
85.	Compton et al.	2004	Western shelf of South Africa	Applied strontium isotopic ($^{87}Sr/^{86}Sr$) composition of biogenic (including foraminifera) and abiogenic phases to determine phosphogenic episodes during Late Cenozoic.
86.	Cornelius and Gooday	2004	Western Weddell Sea	Reported 158 live foraminiferal species from multicorer samples and correlated the distribution with depth and phytodetritus.
87.	Cortese et al.	2004	Subantarctic Atlantic Ocean	Applied benthic foraminiferal oxygen isotopic ratio along with radiolarian SST to determine salinity and thus changes in Agulhas current over the last 450 kyr.
88.	Gooday and Pawlowski	2004	Weddell Sea	Reported a new genus and species, viz. *Conqueria laevis*
89.	Gooday et al.	2004	Weddell Sea	Molecular study of a new monothalamus foraminifera.
90.	Hayward et al.	2004a	Bounty Trough, New Zealand	Statistical analysis of benthic foraminifera reflects productivity, bottom water oxygenation and circulation changes.
91.	Hayward et al.	2004b	Southwest Pacific, east of New Zealand	Changes in benthic foraminiferal distribution are caused by surface water productivity and bottom water changes.
92.	Holbourn et al.	2004	Wombat Plateau and the Great Australian Bight	Benthic foraminiferal isotopic analysis to interpret paleoceanographic conditions.
93.	Murray and Pudsey	2004	Larsen Ice Shelf, Weddell Sea	Agglutinated benthic foraminifera dominate the dead assemblage and are more useful for paleoecological interpretations.

(*Continued*)

TABLE 5.3 (Continued)

Sl. no	Authors	Year	Region	Remarks
94.	Pilskaln et al.	2004	Prydz Bay, E Antarctica	Influence of sea-ice and productivity on biogeochemical flux in sediment traps.
95.	Sabbatini et al.	2004c	Polar front region (Pacific sector)	Southernmost record of live and dead benthic foraminiferal community including both calcareous and agglutinated specimens, in the Pacific sector.
96.	Schumacher and Lazarus	2004	Southern high-latitudes	Productivity in the Southern Ocean increased at Eocene/Oligocene boundary.
97.	Volbers and Henrich	2004	South Atlantic	Inferred changes in lysocline during the last glacial based on changes in the preservation of *Globigerina bulloides*.
98.	Andres et al.	2003	Great Australian Bight	Inferred presence of two rapid cold reversals (between 13.1 and 11.1 kyr BP), based on foraminiferal oxygen isotopic and sedimentary iron content.
99.	Billups and Schrag	2003	The subtropical Indian Ocean and the Weddell Sea	Concluded that the foraminiferal Mg/Ca ratio is helpful to reconstructing Cenozoic ice-volume record despite the possible recycling of Mg and Ca on such long time scales.
100.	Channell et al.	2003	Agulhas Ridge in the Subantarctic South Atlantic	Applied foraminiferal $^{87}Sr/^{86}Sr$, $\delta^{18}O$ and $\delta^{13}C$ ratios to corroborate Eocene to Miocene magnetostratigraphy and chemostratigraphy.
101.	Clarke and Johnston	2003	Antarctica	Documented the marine fauna from the Antarctica region.
102.	Diekmann et al.	2003	Southeast Atlantic sector of the Southern Ocean	Reconstructed the middle Miocene to Pleistocene paleoceanographic history of the region.
103.	Gallagher et al.	2003	Southeastern Australia	Applied foraminiferal census count as supporting evidence to reconstruct Pliocene climatic history.
104.	Gray et al.	2003	Deception Island, Antarctica	Recorded the changing abundance of benthic foraminiferal species since the last volcanic eruption.
105.	Hillenbrand et al.	2003	Bellingshausen and Amundsen seas	Biogeochemical parameters, including foraminifera from the surface sediments, show dominant control of productivity in the faunal distribution and abundance.
106.	Lovell and Trego	2003	Port Foster, Deception Island (South Shetland Islands, Antarctica)	Assessed the effect of volcanic eruption on the benthic community, including foraminifera and reported significant post-volcanic recovery.

Sl. no	Authors	Year	Region	Remarks
107.	Morigi et al.	2003	Southern Ocean (Pacific sector)	Based on foraminiferal oxygen isotopes inferred that climatic changes in this sector of the Southern Ocean are synchronous with northern hemisphere climatic changes.
108.	Suhr et al.	2003	Western Antarctic Peninsula shelf	Evidence for selective feeding in benthic foraminifera based on the biochemical analysis of benthic foraminifera.
109.	Theissen et al.	2003	Prydz Bay, E Antarctica	Reconstructed Pleistocene ice-sheet history of the region based on benthic and planktic foraminiferal oxygen isotopes.
110.	Villa et al.	2003	Pacific continental margin of the Antarctic Peninsula	Developed biostratigraphy and paleoecology on the basis of calcareous nannofossils, diatoms, planktic and benthic foraminifera and inferred marked differences in microfossil assemblage over glacial-interglacial scales.
111.	Billups	2002	Sub-Antarctic South Atlantic	Inferred changes in deep-water circulation and Southern Ocean climate, through Late Miocene early Pliocene based on benthic foraminiferal isotopic analysis.
112.	Billups and Schrag	2002	Southern Ocean	Explored the applicability of Mg/Ca and $\delta^{18}O$ analysis for pre-Pleistocene paleoclimatic changes.
113.	Chen et al.	2002	Southeast Atlantic	Changes in abundance of certain species are more frequent than glacial cycles; foraminiferal transfer function based on sea surface temperature precedes, glacial ice-volume changes.
114.	Cortese and Abelmann	2002	ODP Site 1089 (Southern Ocean, Atlantic Sector)	Suggested a lead of Southern Ocean paleotemperature changes compared to the global ice-volume changes, based on radiolarian census and benthic foraminiferal isotopic analysis.
115.	D'Hondt and Arthur	2002	Southern Ocean	Inferred presence of at least three deepwater masses in the late Maastrichtian ocean based on stable isotopic analysis of benthic foraminifera.
116.	Diekmann and Kuhn	2002	Southeastern South Atlantic	Inferred presence of 130 kyr cyclicity in Polar front movement, between 1200 ka and 650 ka, in the study area.
117.	Duplessy et al.	2002	World Oceans	Reviewed the literature to reconstruct the deepwater characteristics during the last glacial maximum, from the benthic foraminiferal isotopic analysis.
118.	Galeotti et al.	2002	Subantarctic Zone south of South Africa	Planktic foraminiferal assemblage allows the development of geochronology for a Middle Eocene-Early Pliocene sequence.

(*Continued*)

TABLE 5.3 (Continued)

Sl. no	Authors	Year	Region	Remarks
119.	Hayward et al.	2002	East of New Zealand, Southwest Pacific Ocean	Foraminiferal distribution in surface sediments is affected by various factors associated with depth, including, bottom water oxygen, organic matter flux, and seasonality in ecological parameters.
120.	Hebbeln et al.	2002	Southern Peru-Chile Current, SE Pacific	Reported highest relative productivity during the last glacial maximum out of the last 33 000 yrs, based on planktic foraminiferal abundance and attributed it to the northward movement of Antarctic Circumpolar Current.
121.	Howe et al.	2002	Northern Scotia Sea, South Atlantic Ocean	Used foraminiferal abundance to support dinoflagellate based paleoceanographic reconstruction.
122.	Kanfoush et al.	2002	South Atlantic	Inferred low influence of ice sheets at the beginning of interglacial periods as evident from low ice-rafted debris and high foraminiferal abundance.
123.	Lucchi et al.	2002	Antarctic Peninsula, Pacific margin	Inferred change in sedimentary processes over the glacial-interglacial periods based on different parameters including foraminiferal content.
124.	Majewski	2002	Kerguelen Plateau	Inferred paleoecological changes based on foraminiferal census count, isotopic analysis, morphology, and coiling direction of *Globorotalia praescitula* and *G. zealandica*.
125.	Marchant and Guzman	2002	Chilean waters of the southeastern Pacific Ocean and the Antarctic	Provided a checklist along with the biogeographic details of the planktic foraminifera.
126.	Mazaud et al.	2002	Southern Indian Ocean	Reported differences in alkenone derived and foraminiferal transfer function based on sea surface temperature and attributed it to water column stratification.
127.	Melis et al.	2002	Ross Sea (Coulman Island and Cape Adare areas)	Applied foraminiferal preservation and census count to infer ice-extent history during Late Quaternary.
128.	Murphy et al.	2002	Meteor Rise, subantarctic South Atlantic	Inferred presence of sea-level change, Antarctic ice-sheet instability, ice-rafting associated with glacial-interglacial changes, since late Pliocene.
129.	Pawlowski et al.	2002	Explorers Cove, McMurdo Sound, Antarctica	Identified 49 species of allogromiid foraminifera grouped in 28 genera, out of 27 morphotypes, based on molecular phylogenetic analysis.

Sl. no	Authors	Year	Region	Remarks
130.	Robert et al.	2002	Kerguelen Plateau and Maud Rise (Southern Ocean)	The used accumulation rate of benthic foraminifera along with other proxies to reconstruct the Late Eocene-Oligocene ice extent and environmental history of the Antarctic region.
131.	Schrag et al.	2002	North and South Atlantic Ocean	The deep-water temperature during the last glacial maximum was within 1°C of the freezing point of seawater.
132.	Volbers and Henrich	2002	Eastern South Atlantic	Inferred massive dissolution of biogenic calcite well above hydrographic lysocline, based on dissolution stages of *Globigerina bulloides*.
133.	Waelbroeck et al.	2002	Southern Indian Ocean	Inferred sea-level change over glacial-interglacial time-scale, from the benthic foraminiferal isotopic changes.
134.	Yoon et al.	2002	Western Antarctic	Reconstructed changes in productivity and ice extent over the last 15000 yr, based on foraminiferal abundance and other proxies.
135.	Brathauer et al.	2001	Subantarctic Atlantic Ocean	Developed chronology by calibrating *Cycladophora davisiana* abundance with oxygen isotopic records.
136.	Vargas et al.	2001	South Pacific	Identified four different species of *G. truncatulinoides* based on molecular, morphological, and ecological evidence.
137.	Gallagher et al.	2001	Gippsland Basin, southeastern Australia	Inferred water column structure during Miocene, based on foraminiferal census count.
138.	Igarashi et al.	2001	Lutzow-Holm Bay, East Antarctica	Inferred carbonate dissolution level and water circulation changes during the Holocene, based arenaceous, calcareous benthic, and planktonic foraminifera.
139.	Klass	2001	Atlantic sector of the Southern Ocean	Reported influence of productivity on the distribution of planktic foraminifera based on plankton net samples.
140.	Mackensen et al.	2001	Subantarctic eastern Atlantic	Inferred reduced formation of North Atlantic Deep Water and bottom water in the southwestern Weddell Sea, compensated by increased bottom water formation at the northern rim of Weddell sea.
141.	Matsumoto et al.	2001	Atlantic and Pacific sectors of the Southern Ocean	$\delta^{18}O$ analysis of *Neogloboquadrina pachyderma* and *Cibicidoides* shows no apparent change in Southern Ocean hydrography over the last glacial maximum and Holocene.
142.	Petrizzo	2001	Kerguelen Plateau	Biostratigraphic zonation of Cretaceous sediments based on the foraminifera.

(*Continued*)

TABLE 5.3
(Continued)

Sl. no	Authors	Year	Region	Remarks
143.	Quilty	2001	Mac. Robertson Shelf, East Antarctica	Reported reworked foraminifera of Paleogene age in Cenozoic sediments.
144.	Schumacher	2001	Southern South Atlantic Ocean and the northern Angola-Basin	Described depth habitat of live benthic foraminifera based on multi-core surface sediment samples.
145.	Sikes et al.	2001	Subtropical Front (STF) east of New Zealand	Attributed the discrepancies reported in alkenone and foraminiferal assemblage-based sea-surface-temperature records to seasonality in alkenone and foraminiferal flux.
146.	Cann et al.	2000	River Murray, southeastern South Australia	Foraminiferal distribution and sea-level history are based on the relative abundance of various foraminiferal species.
147.	Elderfield and Rickaby	2000	Subantarctic and Southern Ocean	Cd/Ca analysis of planktic foraminifera shows comparable phosphate intake in the Subantarctic whereas much-reduced utilization in the polar Southern Ocean during the last glacial maximum.
148.	Hebbeln et al.	2000	Peru-Chile margin, southeast Pacific	Influence of ecological parameters in the surface distribution and isotopic composition of foraminifera.
149.	Hodell et al.	2000	Polar Frontal Zone, South Atlantic sector	Inferred oceanographic conditions during Marine Isotopic Stage 11, based on oxygen and carbon isotopic analysis of *Neogloboquadrina pachyderma*.
150.	Li et al.	2000	Maxwell Bay, King George Island, Antarctica	Inferred changes in lysocline, carbonate compensation depth and productivity over glacial-interglacial period based on changing abundance of benthic foraminifera.
151.	Licht	2000	Ross Sea, Antarctica	Dated the ice retreat and advance during the Late Quaternary based on foraminiferal ^{14}C dates.
152.	Nelson et al.	2000	Southwest Pacific off eastern New Zealand	Inferred changes in the position of the Subtropical Convergence Zone, during the last glacial, based on planktic foraminiferal species, transfer function data and stable oxygen and carbon isotopic analysis.
153.	Oppo and Horowitz	2000	Western South Atlantic	Applied benthic foraminiferal Cd/Ca and $\delta^{13}C$ analysis to infer glacial deep-water changes.
154.	Perry	2000	Ross Sea, Antarctica	Identified three foraminiferal assemblages in the Ross Sea: Pleistocene carbonate-bank fauna, Recent-calcareous fauna, and Holocene agglutinated fauna and correlated their distribution with the regional oceanography and ice-shelf retreat.

Sl. no	Authors	Year	Region	Remarks
155.	Petrizzo	2000	Exmouth Plateau, NW Australia	Presented planktic foraminiferal based zonation for the Upper Cretaceous.
156.	Pudsey	2000	West of the Antarctic Peninsula	Changing sedimentation rate and affecting factors during last three glacial cycles, based on multiproxy including foraminiferal abundance.
157.				
158.	Yoon et al.	2000	Maxwell and Admiralty Bays, King George Island, Shetland Islands	Reconstructed paleoceanographic conditions in the Bay and deglaciation of Antarctica during the last 6000 yr based on benthic foraminiferal isotopic analysis.
159.	Barbieri et al.	1999	Ross Sea, Antarctica	Inferred role of bacterial colonies in post-depositional preservation of benthic foraminifera.
160.	Domack et al.	1999	Ross Sea, Antarctica	Inferred remobilization of sediments based on the dating of foraminiferal calcite.
161.	Martinez et al.	1999	Southeastern Indian Ocean	Statistical analysis of foraminiferal distribution and oxygen isotopic ratio indicates changes in thermocline depth, sea-surface temperature, upwelling, etc. during the last glacial maximum.
162.	Rickaby and Elderfield	1999	Subantarctic Southern Ocean	Cd/Ca analysis of *Globigerina bulloides* shows the insignificant change in phosphate utilization over glacial-interglacial time-scales
163.	Allen et al.	1998	Antarctica	Applied fractal dimension analysis to understand test construction and grain-size selection in agglutinated foraminifera.
164.	Gersonde et al.	1998	Gunnerus Ridge and Kainan Maru Seamount (Indian sector)	Inferred sedimentation rates during Cenozoic based on multiproxy including foraminiferal abundance.
165.	Gilbert et al.	1998	Northwest Weddell Sea	Inferred cyclic changes in bottom water activity, surface productivity and carbonate dissolution depths based on sedimentological, foraminiferal and 3.5 kHz sub-bottom profiles.
166.	Niebler and Gersonde	1998	South Atlantic Ocean	Developed a planktic foraminiferal abundance-based transfer function
167.	Oslick	1998	Southern Indian Ocean	Refined the knowledge of depth habitat of certain planktic foraminiferal species and applied foraminiferal $\delta^{18}O$ analysis to infer past water column structure.
168.	Pichon et al.	1998	West Indian Subtropical, Subantarctic and Polar Front Zones	Applied the microfossil presence to infer the age of sediments.

(*Continued*)

TABLE 5.3
(Continued)

Sl. no	Authors	Year	Region	Remarks
169.	Saidova	1998	Southern Ocean	Identified and mapped forty benthic foraminiferal communities.
170.	Vorobeva	1998	Galindez Island, Argentine Islands (Antarctic Peninsula)	Documented distribution of various benthic microorganisms including foraminifera, and correlated it with water depth and type of substrate.
171.	Harloff and Mackensen	1997	The Scotia Sea and Argentine Basin, southwestern South Atlantic Ocean	Influence of ecological parameters on the distribution of foraminifera recovered from surface sediment samples.
172.	Kucera et al.	1997	Walvis Ridge and Rio Grande Rise, South Atlantic Ocean	Inferred variation in lysocline during the Cretaceous, based on changes in the degree of fragmentation of planktic foraminifera.
173.	McGowran et al.	1997	South Australia	Regenerated the changes in Leeuwin current since the late middle Eocene, based on foraminiferal assemblages.
174.	Okada and Wells	1997	South Australia	Applied planktic foraminiferal isotopic analysis to develop chronology.
175.	Rathburn and De Deckker	1997	Coral Sea, Australia and Prydz Bay, Antarctica	Potential application of Mg/Ca and Sr/Ca ratio of selected benthic foraminiferal species for paleotemperature estimation over –2 to 6 °C range.
176.	Rathburn et al.	1997	Prydz Bay region, Antarctica	Ice-extent and productivity history over the last ~8000 yr based on foraminiferal abundance and isotopes.
177.	Wells and Okada	1997	South Chatham Rise, southeastern New Zealand	Applied planktic foraminiferal isotopic analysis to develop a chronology for the core, while the foraminiferal transfer function to infer the paleoceanographic history of the southern margin of the subtropical convergence zone.
178.	Boltovskoy et al.	1996	Southwestern Atlantic	Described planktic foraminiferal distribution from plankton samples, with reference to the ecological parameters.
179.	Chen	1996	Bransfield Strait, West Antarctica	Reported 21 foraminiferal species from a late Quaternary core, and clubbed them in three groups comprising calcareous, siliceous, and mixed tests; discussed ecological implications of each assemblage.
180.	Diester-Haass	1996	Kerguelen Plateau, Southern Indian Ocean	Late Eocene-Oligocene productivity and water-mass changes based on multiproxy including benthic foraminiferal abundance.

Sl. no	Authors	Year	Region	Remarks
181.	Gooday et al.	1996	Explorers Cove, Antarctica	Surface distribution of benthic foraminifera.
182.	Nomura	1995	Kerguelen Plateau, southeast Indian Ocean	Reconstructed deep-water circulation changes based on quantitative analysis of benthic foraminifera.
183.	Bowser et al.	1995	McMurdo Sound, Antarctica	Questioned the taxonomic status of large agglutinated foraminifera recovered from surface sediments.
184.	Chang and Yoon	1995	Marian Cove, South Shetland Islands, West Antarctica	Documented depth distribution of foraminifera from surface sediments.
185.	Diester-Haass	1995	Maud Rise, Antarctica	Inferred paleo productivity during middle Eocene to the early Oligocene based on microfossil accumulation rates; 430 kyr cyclicity in productivity.
186.	Jennings et al.	1995	Western Ross Sea	Reconstructed the ice extent history for the last 30,000 yr based on benthic foraminiferal abundance.
187.	Mackensen et al.	1995	South Atlantic Ocean	Reported effect of various environmental parameters on the benthic foraminiferal distribution.
188.	Mead and Hodell	1995	Maud Rise, Antarctica	Reported a significant increase in marine $^{87}Sr/^{86}Sr$ towards the Eocene/Oligocene boundary, based on foraminiferal $^{87}Sr/^{86}Sr$ analysis, and attributed the increase to a combination of parameters including hydrothermal activity, Himalayan uplift, and glaciations.
189.	Niebler	1995	Southern South Atlantic Ocean	Developed foraminiferal transfer function, and evaluated the applicability of foraminiferal isotopic analysis for paleoclimatic reconstruction.
190.	Widmark	1995	World Oceans	Inferred benthic foraminiferal distribution throughout the world oceans during the Cretaceous.
191.	Flower and Kennett	1994	Southern High Latitudes	Reconstructed changes in the East Antarctic Ice Sheet (EIAS) during the Miocene; major growth of the EIAS during 14.8–12.9 Ma.
192.	Gooday et al.	1994	Explorers Cove, Antarctica	The surface distribution of benthic foraminifera resembles deep-sea foraminifera.
193.	Gore et al.	1994	Vestfold Hills, East Antarctica	The reported presence of older reworked foraminifera in surface sediments.
194.	Ishman and Domack	1994	Bellingshausen Sea	Reported two foraminiferal assemblages from surface sediments, viz. *Bulimina aculeata* and *Fursenkoina* spp. Assemblages; distribution is controlled by water masses and not by sediment type or productivity.

(*Continued*)

TABLE 5.3 (Continued)

Sl. no	Authors	Year	Region	Remarks
195.	Jones and Pudsey	1994	Western Antarctic Ocean	Surface distribution of benthic foraminifera.
196.	Melles et al.	1994	Weddell Sea	Documented effect of physicochemical factors on various sedimentological processes, based on different parameters including foraminiferal abundance and isotopic analysis.
197.	Bernhard	1993	McMurdo Sound, Antarctica	Laboratory culture of foraminifera to understand, and respond to anoxic conditions and hydrogen sulfide.
198.	Boersma and Huber	1993	Atlantic and Indian Ocean sectors of the Southern Ocean	Studied water-mass characteristics based on biserial planktic foraminifera.
199.	Broecker	1993	Antarctic Ocean	Reviewed the discrepancy between benthic foraminiferal carbon isotopic and Cd/Ca based paleocirculation estimates.
200.	Cann and Clarke	1993	Esperance, Western Australia, and Spencer Gulf, South Australia	Attributed the abundant presence of megascopic *Marginopora vertebralis* to warm Leeuwin Current.
201.	Keller	1993	Kerguelen Plateau, Broken Ridge and Maud Rise, Weddell Sea	Applied foraminiferal abundance as a biostrtigraphic tool.
202.	Lu and Keller	1993	The Antarctic Indian Ocean	Inferred three major changes in the planktic foraminiferal composition, during the Paleocene and Eocene, based on isotopic and abundance changes.
203.	Mackensen et al.	1993	Eastern South Atlantic Polar Front region	Influence of productivity and water-circulation of the foraminiferal distribution.
204.	MacLeod	1993	Kerguelen Plateau, Australian Antarctic Basin	Inferred evolution of triserial and biserial planktic foraminifera during Maastrichtian-Danian, based on morphologic and isotopic analysis.
205.	Shemesh et al.	1993	Southern Ocean	Inferred lower than Holocene productivity during last glacial maximum based on foraminiferal isotopic analysis.
206.	Bernhard and Bowser	1992	Antarctic Ocean	Observed differential response of benthic foraminiferal species to bacterial biofilm and role of pseudopodia in food collection, under laboratory culture.
207.	Charles and Fairbanks	1992	Southern Atlantic	Large and rapid variation in North Atlantic deepwater formation, based on benthic foraminiferal isotopic analysis.

Sl. no	Authors	Year	Region	Remarks
208.	Clarke and Crame	1992	Southern Ocean	Foraminiferal distribution and its application for paleoclimatic reconstruction.
209.	Healy-Williams	1992	Southern Indian Ocean	Different morphotypes of *Neogloboquadrina pachyderma* are isotopically distinct and in disequilibrium with the ambient seawater isotopic composition.
210.	Hovan and Rea	1992	South Indian Ocean	Decreased strength of zonal winds during the Paleocene/Eocene boundary, based on benthic foraminiferal oxygen isotopic analysis.
211.	Huber	1992a	Southern high-latitudes	Identified changes in Antarctic biogeography and land-sea adjustments during Campanian-Maastrichtian time based on planktic foraminiferal abundance changes.
212.	Huber	1992b	Circum-Antarctic region	Presented foraminiferal abundance-based zonation for Upper Cretaceous sediments from the circum-Antarctic region.
213.	Kurihara and Kennett	1992	Southwest Pacific	Inferred that the foraminiferal assemblage changed very marginally in response to the changing Antarctic ice-sheet extent during Neogene.
214.	Mackensen and Ehrmann	1992	Maud Rise and Kerguelen Plateau	Reconstructed Middle Eocene through Early Oligocene climate history and Paleoceanography, based on different parameters including foraminiferal isotopic analysis.
215.	Shemesh et al.	1992	South Atlantic	Calibration of biogenic silica-based paleotemperature equation with foraminiferal oxygen isotope-based paleotemperature equation.
216.	Zachos et al.	1992	Kerguelen Plateau, southern Indian Ocean	Inferred Antarctic ice sheet expansion during the Oligocene, based on benthic foraminiferal isotopic analysis.
217.	Garrison et al.	1991	Weddell Sea	Minor presence of foraminifera in the water samples.
218.	Gowing and Garrison	1991	Weddell/Scotia Seas	Reported the abundance of large sarcodinid protozooplankton, including foraminifera and correlated it with environmental factors.
219.	Hodell and Warnke	1991	Subantarctic South Atlantic	Inferred warming conditions during early Pliocene, cooling during mid-Pliocene and distinct cooling during late Pliocene, based on foraminiferal isotopic analysis.
220.	Scherer	1991	Ice Stream B, West Antarctica	Reconstructed the Cenozoic ice-sheet history based on microfossil content.
221.	Mackensen et al.	1990	Weddell Sea	Identified distinct assemblages from the surface sediment samples and correlated their distribution to ecological conditions.

(Continued)

TABLE 5.3 (Continued)

Sl. no	Authors	Year	Region	Remarks
222.	Schroder	1990	Prydz Bay (Antarctica) and Lancaster Sound (Canadian Arctic)	Compared the distribution of agglutinated benthic foraminifera from the north and south polar regions.
223.	Domack et al.	1989	Wilkes Land continental shelf and slope, East Antarctica	Applied tandem accelerator mass-spectrometer (TAMS) for ^{14}C dating to infer glacial-interglacial ice-sheet history.
224.	Harwood et al.	1989	Ross Ice Shelf	Inferred the age and origin of sediments based on microfossil presence and abundance.
225.	Mackensen and Douglas	1989	The Weddell Sea and the California continental borderland	Documented foraminiferal distribution from short cores and suggested the changes in foraminiferal assemblages in the top several centimeters to changing geo-chemical conditions.
226.	Mackensen et al.	1989	Eastern Weddell Sea	Inferred increased surface water productivity in the Southern Ocean during interglacial periods, based on benthic and planktic foraminiferal isotopic analysis.
227.	Stott and Kennett	1989	Weddell Sea, Antarctica	Carbon isotopic analysis of foraminiferal shows that Antarctica was not a major source of deep-water during Paleocene and that Cretaceous/Tertiary extinction lead productivity changes were not as intense as reported earlier.
228.	Zhu	1989	Southern Ocean	Foraminifera as food for Antarctic krill
229.	Barker et al.	1988	Weddell Sea	Described the sedimentologic characteristics of the cores.
230.	Domack	1988	East Antarctic continental shelf	Inferred possible correlation between relict and modern glaciomarine sediments.
231.	Gostin et al.	1988	South Australia	Benthic and planktic foraminifera from algal mats.
232.	Zheng and Wu	1988	Northwestern Antarctic	Identified 40 genera and 140 species of benthic foraminifera and correlated their distribution with physicochemical characteristics of the ambient seawater.
233.	Alexander and DeLaca	1987	Antarctica	Studied mode of attachment and feeding mechanism of epifaunal *Cibicides refulgens* on the valves of the free-swimming Antarctic scallop, *Adamussium colbecki*.
234.	Barrera	1987	Southern Ocean	Inferred Maastrichtian-Campanian seawater temperature based on benthic foraminiferal oxygen isotopic analysis.

Sl. no	Authors	Year	Region	Remarks
235.	Gersonde and Wefer	1987	Drake Passage, Bransfield Strait and west of South Orkney Islands (Powell Basin)	Foraminifera constitutes a minor part of the material trapped in the sediment traps.
236.	Li et al.	1987	Western sea, Antarctic Peninsula	150 foraminiferal species were reported and differentiated into assemblages; the effect of physicochemical parameters on foraminiferal distribution was discussed.
237.	Mead and Kennett	1987	Southwest Atlantic	Identified three faunal assemblages from 25 core-top samples and attributed them to different water masses.
238.	Pharr and Williams	1987	Southern Indian Ocean	Studied the evolutionary changes in the *Globorotalia truncatulinoides* based on allometric shape changes during the growth.
239.	Ward et al.	1987	McMurdo Sound	Identified three main benthic foraminiferal assemblages and suggested that the carbonate compensation depth mainly controls the foraminiferal distribution.
240.	Bowser et al.	1986	Antarctica	Found extracellular matrix of thin branching fibers, associated with *A. rara*'s pseudopodia by using electron microscopy.
241.	Krasheninnikov and Basov	1986	Southwest Atlantic	Reconstructed Mesozoic and Cenozoic paleoceanographic history based on microfossil abundance.
242.	Labeyrie et al.	1986	Southern Ocean	Marked changes in the hydrological structure of the Southern Ocean during the last 60 kyr based on oxygen isotopic composition of planktic and benthic foraminifera.
243.	Bowser and DeLaca	1985	Antarctica	Studied food transport in foraminiferal pseudopodia by using phase-contrast light microscopy.
244.	Mead	1985	Polar Front region of the southwest Atlantic.	Described 27 species of calcareous benthic foraminifera from 62 trigger-core tops.
245.	Howard and Prell	1984	The Subantarctic Indian Ocean	Interpreted the foraminiferal abundance-based SST lag over radiolarian SST during stage 6–5e transition as a varying response to the changes in surface and subsurface water.
246.	Labeyrie et al.	1984a	Southern Ocean	Diatom oxygen isotopic composition is a potential and more sensitive proxy as compared to foraminiferal oxygen isotopic composition, for paleotemperature estimation.
247.	Labeyrie et al.	1984b	Antarctic Ocean	Inferred water column structure during last glacial maximum based on diatom and foraminiferal oxygen isotopic ratio.

(*Continued*)

TABLE 5.3
(Continued)

Sl. no	Authors	Year	Region	Remarks
248.	Lindenberg and Auras	1984	Kerguelen plateau	Reported influence of sediments, nutrients, and carbonate compensation depth on surface distribution of foraminifera.
249.	Marsh and Buddemeier	1984	Antarctic Ocean	Inferred a possible link between seawater radionuclide concentration and foraminifera.
250.	Corliss	1983	The southeast Indian Ocean	Documented foraminiferal distribution from core-top samples and reported two foraminiferal assemblages associated with water-mass characteristics.
251.	Corliss	1983	Circum-Polar region	Based on foraminiferal abundance, interpreted glacial-interglacial shifts in Antarctic circum-polar current, probably due to changes in the formation of North Atlantic Deep Water.
252.	Corliss and Thunell	1983	Southeast Indian Ocean sector of the Southern Ocean	Explored the relationship between carbonate dissolution and deep-water circulation based on planktic foraminiferal fragmentation and benthic foraminifera.
253.	Guptha	1983	Off Queen Maud Land	Reported calcareous and arenaceous benthic foraminifera as well as planktic foraminifera from the shelf region.
254.	Leckie and Webb	1983	Ross Sea, Antarctica	Reconstructed late Oligocene-early Miocene history, based on foraminiferal assemblages.
255.	Loutit et al.	1983	Equatorial and southwest Pacific Ocean	Inferred paleoceanographic conditions in the equatorial and southern Pacific Ocean during Miocene based on foraminiferal abundance and isotopic ratio.
256.	Malmgren	1983	Maurice Ewing Bank (eastern Falkland Plateau), South Atlantic Ocean	Prepared dissolution susceptibility order of planktic foraminifera based on various sedimentological and faunal characteristics.
257.	Nancy	1983	Southern Indian Ocean	Less elongate and more conical forms of *Globorotalia truncatulinoides* are reported in a south to north transect.
258.	Curry and Lohmann	1982	Western South Atlantic	Inferred glacial-interglacial changes in North Atlantic Deep Water (NADW) and Antarctic Bottom Water (AABW), based on foraminiferal isotopic analysis.
259.	Peterson and Lohmann	1982	South Atlantic	Significant increase in the amount of Antarctic Bottom water entering into the western south Atlantic at ~700,000 yr BP, based on benthic foraminiferal abundance.
260.	Shchedrina	1982	Indian sector of the Southern Ocean	Documented the distribution of agglutinated foraminifera.

Sl. no	Authors	Year	Region	Remarks
261.	Boltovskoy and Watanabe	1981	The Antarctic sector of the Atlantic Ocean	Distribution of planktic foraminifera based on plankton samples.
262.	DeLaca et al.	1981	McMurdo Sound, Antarctica	Reported that certain foraminiferal species can feed on both particulate and dissolved organic material.
263.	Healy-Williams and Williams	1981	Southern Indian Ocean	The abundant presence of highly conical forms of *Globorotalia truncatulinoides* in tropical waters, whereas more compressed, bi-convex forms in cool water.
264.	Lena	1981	Adelaide, Anvers, and Brabant Islands.	A reported total of 42 genera and 66 foraminiferal species from 32 bottom samples; agglutinated specimens were more abundant.
265.	Lena	1981	South Atlantic Ocean	Planktic foraminiferal distribution from core-top samples.
266.	Loutit	1981	Subantarctic, southwest Pacific	Late Miocene-early Pliocene paleoceanographic and Antarctic ice extent history based on foraminiferal isotopes.
267.	Milam and Anderson	1981	Adelie-George V continental shelf and slope, Antarctica	Documented the benthic foraminiferal distribution.
268.	Boltovskoy and Watanabe	1980	Tierra del Fuego, Islas Georgia del Sur	Identified 14 planktic and 108 benthic foraminiferal taxa, including a few reworked benthic foraminifera of Eocene age, from 30 dredge and piston core samples; demarcated carbonate compensation depth.
269.	Kennett	1980	Southern Ocean	Compiled the Cenozoic changes in the circum-Antarctic circulation based on microfossil abundance changes.
270.	Lena	1980	The northwest sector of the Antarctic Peninsula	Reported 50 species of benthic foraminifera out of which 60% were calcareous. However, agglutinated specimens are more abundant.
271.	Anderson et al.	1979	Ross and Weddell Seas	Reported few displaced benthic foraminifera in the sediments recovered from the shelf region' discussed the origin of different types of sediments.
272.	Boltovskoy and Watanabe	1979	Southwestern Atlantic and its Antarctic sector	Reported thirteen planktic foraminiferal species in 82 plankton samples; correlated the foraminiferal distribution with hydrological properties.
273.	Corliss	1979a	The southeast Indian Ocean	Inferred changes in the relative amount and extent of Antarctic Bottom water and Indian Deep Water over the last 500,000 yr BP.

(*Continued*)

TABLE 5.3
(Continued)

Sl. no	Authors	Year	Region	Remarks
274.	Corliss	1979b	The southeast Indian Ocean	Distribution of recent benthic foraminifera from 64 trigger-core top samples.
275.	Finger	1975	Deception Island	Documented effect of volcanic eruption and other granulometric and physicochemical parameters on benthic foraminiferal community.
276.	Malmgren and Kennett	1978	Sub-Antarctic waters of the southern Indian Ocean	Potential application of mean size variation of *Globigerina bulloides* to infer Late Quaternary paleoclimatic changes.
277.	Melguen et al.	1978	Vema Channel (south Atlantic Ocean)	Inferred changing Antarctic Bottom Water activity during the Late Cenozoic, based on changes in foraminiferal dissolution.
278.	Williams and Keany	1978	Sub-Antarctic Indian Ocean	Inferred paleoceanographic history based on foraminiferal distribution and fragmentation index; reported increased dissolution during glacial periods and intense Antarctic bottom-water activity.
279.	Luz	1977	South Pacific	Inferred sea-surface temperature for the last 150,000 years, based on the foraminiferal transfer function.
280.	Margolis et al.	1977	Circum-Antarctic region	Reviewed the oxygen and carbon isotope analysis of calcareous nannofossils and foraminifera based paleotemperature and paleoceanographic studies.
281.	Scheibnerova	1977	Indian Ocean	Summarized the distribution of benthic foraminifera during the Cretaceous.
282.	Anderson	1975a	Weddell Sea	Documented foraminiferal distribution in the surface sediments.
283.	Anderson	1975b	Weddell Sea	The surface distribution of foraminifera shows that the carbon compensation depth is controlled by the water-mass characteristics and productivity, rather than the water depth.
284.	Temnikow and Lipps	1975	The western coast of the Antarctic Peninsula	Foraminiferal distribution is affected by the amount of sea-ice abrasion.
285.	Melguen and Thiede	1974	Vema channel, Rio Grande rise (Southwest Atlantic Ocean)	Inferred microfossil dissolution depths based on the distribution in surface sediments.
286.	Malmgren and Kennett	1973	South Pacific	Statistical analysis of planktic foraminiferal abundance and coiling direction ratio of *Globigerina pachyderma* reveals six assemblages corresponding with different water masses.

Sl. no	Authors	Year	Region	Remarks
287.	Sissingh	1973	Breid Bay area (Antarctica)	Reported 66 foraminiferal taxa from 6 core and four water samples.
288.	Fillon	1972	Ross Sea	Inferred a link between the Subantarctic and Antarctic South Pacific circulation and bottom water production in the Ross Sea, based on benthic foraminiferal distribution.
289.	Keany and Kennett	1972	Northern Antarctic and Subantarctic waters south of Australia and New Zealand	Reconstructed middle Pliocene to early Pleistocene paleoclimatic history based on foraminiferal assemblages and coiling direction of *Globigerina pachyderma*.
290.	Echols	1971	Scotia Sea	Distribution of foraminifera and the influence of ecological parameters on foraminiferal distribution.
291.	Herb	1971	Drake Passage	Surface distribution of foraminifera and its relationship with physicochemical characteristics.
292.	Huddestun	1971	Southern Ocean	Good correlation between paleoclimatic information obtained from radiolaria, planktic foraminifera and calcareous nannofossils.
293.	Parker and Berger	1971	South Pacific	Documented foraminiferal distribution and explained in terms of ecological parameters, especially foraminiferal lysocline and compensation depth.
294.	Theyer	1971	Pacific-Antarctic	Identified four foraminiferal assemblages.
295.	Belyaeva	1970	Southern Ocean	Discussed foraminiferal distribution about water-mass characteristics.
296.	McKnight	1962	Antarctic coasts	Documented the distribution of foraminifera.
297.	Heron-Allen and Earland	1932	The Falkland Islands and adjacent seas	Reported the foraminiferal distribution.
298.	Chapman	1916	Ross Sea	Distribution of foraminifera.

the foraminiferal population (Gersonde and Wefer, 1987; Pilskaln et al., 2004; Anna et al., 2019).

Researchers have documented the influence of depth, productivity, oxygenation and circulation on foraminiferal diversity and abundance (Mackensen et al., 1995; Boltovskoy et al., 1996; Harloff and Mackensen, 1997; Cornelius and Gooday, 2004; Hayward et al, 2002, 2004a, 2004b). Klass (2001) and Hillenbrand et al. (2003) reported that productivity provided a dominant control on foraminiferal distribution. Seasonal changes in the faunal composition in response to the changes in physicochemical parameters have been reported, based on the continuous plankton tow-sampling during different seasons and across different frontal zones (Hunt and Hosie, 2005, 2006a), as well as from the sediment trap studies (Pilskaln et al., 2004). Northcote and Neil (2005) reported increased foraminiferal flux as a result of the high productivity during the austral spring season. However, the faunal composition as received in the sediment traps was an order of magnitude different than that reported in the core-top sediments collected near the sediment traps. Sabbatini et al. (2004) also noted differences in the live and dead faunal composition and reported the presence of both calcareous and agglutinated foraminifera from as far south as 63 °S latitudes.

Though the diversity of calcareous species was found to be high, comparatively more agglutinated specimens comprise the live foraminiferal population (Lena, 1980, 1981; Mackensen et al., 1995). Taphonomic post-depositional dissolution of calcareous fauna results in the dominance of agglutinated fauna in the dead and fossil assemblages at certain locations (Murray and Pudsey, 2004). However, in general, agglutinated foraminifera do not survive the post-depositional taphonomic processes (Mackensen et al., 1995). As the Southern Ocean is surrounded by and separated from distinct water masses, many endemic foraminiferal species have been reported from this southern region (Gooday et al., 1996). Several new species have also been reported from the region (Gooday and Pawlowski, 2004; Gooday et al., 2004). Nevertheless, the taxonomic status of certain agglutinated foraminiferal species found in southern high-latitude waters has been questioned (Bowser et al., 1995). Gooday (1994) and Gooday et al. (1996) noted that the faunal composition at very shallow sites in Explorers Cove was similar to that found at much deeper depths elsewhere in the world ocean. They attribute this to the conditions at Explorers Cove being similar to those at deeper water settings elsewhere.

Molecular systematic studies have also been carried out to understand the differences between Antarctic foraminifer and those from elsewhere (Pawlowski et al., 2006; Franck et al., 2014), as well as to differentiate the morphotypes of single species (Vargas et al, 2001; Pawlowski et al., 2002, 2005). Pawlowski et al. (2006) noted that a few benthic foraminiferal species recovered from both the southern and northern high-latitudes are genetically similar, thus indicating cross-equator migration of those species. The studies also revealed the capability of selected benthic foraminiferal species to inhabit a wide range of water depths. Oslick (1998) refined knowledge of the depth habitat of planktic foraminiferal species, whose oxygen isotopic composition is frequently used for paleoceanographic/paleoclimatic reconstructions. However, foraminiferal distribution and diversity have still not been explored fully in many parts of the Southern Ocean.

5.5.2 Biostratigraphic Studies

Changes in the seawater properties may result in significant changes in the foraminiferal community, including changes in species composition or the extinction or introduction of species. Besides this, morphological changes have also been reported in various species over the years as a mark of adaptation to changed physicochemical conditions (MacLeod, 1993). Such changes act as markers for establishing the chronology of the sediments, or their biostratigraphic zonation (Majewski, 2002). As a result, foraminifera has proved effective as biostratigraphic tools in the Southern Ocean (Majewski, 2002; Webb and Strong, 2006).

Huber (1992b) established planktic foraminifera-based biostratigraphic zonation for the Upper Cretaceous sediments of the circum-Antarctic region. Petrizzo (2000,

2001) applied foraminiferal abundance and diversity to biostratigraphic zonation of the Late Cretaceous sediments from the southern high latitudes. Kennett (1978) provided a comprehensive review of planktic foraminiferal evolution in the southern high latitudes for the Cenozoic, observing the major changes in planktic foraminiferal diversity, morphology and abundance associated with changes in water mass. Galeotti et al. (2002) used planktic foraminifera for the Middle Eocene-early Pliocene biostratigraphic zonation of cores recovered from the Agulhas ridge. Villa et al. (2003) developed comprehensive Quaternary biostratigraphy based on the faunal content of the cores collected from the Pacific continental margin of Antarctica. Majewski (2003) showed that the changes in the coiling direction of the globorotalids can be used as a biostratigraphic marker for the mid-Miocene in the southern Indian Ocean. Foraminiferal isotopic changes have also been used to constrain the biostratigraphy developed based on another microfauna (Brathauer et al., 2001). Overall, the recovery of long sediment cores as part of the ocean drilling program and deep-sea drilling project has led to the development of comprehensive foraminiferal biostratigraphy for Southern Ocean sediments

5.5.3 Past Circulation Changes

Foraminifera, being extremely sensitive to slight changes in ambient conditions, have widely been used for reconstructing the past changes in the thermohaline circulation in the Southern Ocean. Mackensen et al. (1993) showed that the different water masses in the eastern South Atlantic, Polar Front region are characterized by specific benthic foraminiferal assemblages, thus paving the way for the use of benthic foraminiferal assemblage to trace past water mass changes there. Both changes in foraminiferal assemblage, as well as in the isotopic composition of foraminiferal tests, have been used to reconstruct the changing geographic extent of polar fronts as well regions of formation of deep and intermediate waters. The elemental composition (Cd/Ca) of foraminiferal tests have also been used to infer past circulation changes, but discrepancies have been observed in productivity estimates based on foraminiferal isotopic composition and elemental analysis (Broecker, 1993).

Boersma and Huber (1993) inferred that the surface water masses in the Atlantic and Indian sectors of the Southern Ocean were different during the late Cretaceous-early Tertiary. D'Hondt and Arthur (2002), based on the benthic foraminiferal oxygen isotopic composition, concluded that at least three deep water masses occupied the world ocean including the southern high latitudes during the Mastrichtian. Kurihara and Kennett (1992) concluded that the deep water conditions in the southwest Pacific remained almost the same throughout the Neogene. Volbers and Henrich (2002) developed a novel technique to trace the corrosiveness of seawater based on the relative degree of ultrastructural breakdown of the planktic foraminiferal species Globigerina bulloides, and later on, used the same proxy to infer past circulation changes in the south Atlantic region. A shallower lysocline depth during the last glacial maximum was attributed to the presence of corrosive water mass at the water depth occupied by the North Atlantic deep water at present (Volbers and Henrich, 2004). Holbourn et al. (2004) identified the presence of cooler, well-ventilated water at about 700 m water depth in the Great Australian Bight region over the last 16 Ma, based on changes in the stable oxygen and carbon isotopic composition of foraminifera, and noted that the cooling of the bottom water in this region started much earlier than that in the Indian Ocean. Diekmann and Kuhn (2002) inferred that in the southeastern south Atlantic, the Polar Front started moving towards the north during glacial periods after 1150 ka. Distinct differences in the seawater circulation between glacial and interglacial periods seem to have occurred after 1200 ka in this region and were most prominent at about 650 ka BP. A 130 ka cyclicity was noticed in the frontal movement in this area that took place between 1200–650 ka BP and was considered to roughly correspond to the global 100 ka cycle. Hodell et al. (2000) noted increased input of North Atlantic deep water to the southern high latitudes during isotope stage 11. Gilbert et al. (1998) inferred that the production of Weddell Sea bottom water was considerably reduced during cold glacial periods. Labeyrie et al. (1986) noted that the Polar Front was covered by melt water from the melting ice-bergs during 35–17 ka BP. Peterson and Lohman (1982) concluded that the depth of transition between deep and bottom water in the south Atlantic became notably shallower at about 700 ka, suggesting an increase in the column of deep water. The contribution of Antarctic Bottom Water to the south Atlantic was significantly less before 700 ka when Circum-Polar Deep water contributed the most of the water flowing north in this region. Williams and Keany (1978) showed increased Antarctic bottom water production during glacials. Corliss (1982) suggested that changes in the Antarctic Circumpolar Current during the last 440,000 years were linked to changes in NADW production. A significant reduction in the NADW production during glacials was suggested by Curry and Lohmann (1982). Martinez et al. (1999) inferred that the differences between the Australasian Mediterranean Water (AAMW) and the Indian Central Water (ICW) during the last glacial maximum, were much less than they are at present. Matsumoto et al. (2001) inferred that Southern Ocean hydrography, including the position of oceanic fronts in the Atlantic and Pacific sectors of the Southern Ocean, remained almost unchanged throughout Holocene and glacial times. Oppo and Horowitz (2000) inferred a reduced influence of glacial North Atlantic Intermediate Water on the intermediate water of the world's oceans based on the foraminiferal Cd/Ca and carbon isotopic composition.

As mentioned before, the Southern Ocean is a major location for deep water formation, and changes in the amount and position of deep water are thought to be an important control of climatic variations. Mackensen et al. (2001) concluded that during the last glacial maximum, deep water formed south of the Antarctic Circumpolar Current, whereas, during the Holocene, the location for deep water formation shifted to the southwestern Weddell Sea. Charles and Fairbanks (1992) inferred rapid changes in the North Atlantic Deep Water and a sudden increase in its production following the last deglaciation.

Foraminifera has also been used to trace the evolution and history of regional currents. McGowran et al. (1997) inferred that the Leeuwin current off Australia originated during the late-middle Eocene, much earlier than suggested before. A 100 ka cyclicity was noted in the strength of the Leeuwin Current, in response to glacial-interglacial cooling and warming, roughly corresponding with known orbital cyclicity.

5.5.4 Foraminifera and Paleo-Productivity Reconstruction

Organic matter production is reportedly the most important factor affecting foraminiferal distribution (Gooday and Rathburn, 1999; Zwaan et al., 1999). Changes in organic matter production result in changes in foraminiferal assemblage as well as in the isotopic composition of foraminiferal tests. Changes in foraminiferal abundance, assemblage, and isotopic composition have been used to infer changes in productivity. Stott and Kennett (1989) used carbon isotopic analysis of foraminifera to infer changes in productivity in the Southern Ocean during the Cretaceous-tertiary transition. Rickaby and Elderfield (1999) and Elderfield and Rickaby (2000) showed that the foraminiferal Cd/Ca be used to trace past productivity changes from the Southern Ocean.

Stott and Kennett (1989) concluded that the Southern Ocean experienced a considerable rise in productivity following the Cretaceous-Tertiary extinction. Diester-Haass (1995, 1996) reconstructed the productivity changes in the Indian sector of the Southern Ocean and the Maud Rise region of the Antarctic, for the Eocene and Oligocene. Productivity was low during the late Eocene, and at a maximum during the early Oligocene. This follows a cooling trend. Productivity was low during the latter part of the early Oligocene but again increased during the late Oligocene. She also documented a general trend of increasing productivity towards the late Oligocene, as well as a 400–30 kyr cyclicity in productivity (Diester-Haass, 1995, 1996). Later on, Robert et al. (2002) also identified orbitally controlled 400 ka cyclicity in marine productivity during the late Eocene-Oligocene. Schumacher and Lazarus (2004) also inferred increased productivity during the Eocene-Oligocene transition. Diekmann et al. (2003) reconstructed the productivity changes in the southeast Atlantic sector of the Southern Ocean during the Middle Miocene to Pleistocene. Productivity was comparatively low during the warmer Miocene but increased with cooling towards the Miocene-Pliocene boundary. Region-specific productivity changes are reported since the late Pliocene, probably reflecting the development of different oceanic fronts during this period. Mackensen et al. (1989) used the isotopic composition of foraminifera to infer increased productivity in the Southern Ocean during interglacial periods. This trend is confirmed by Li et al. (2000), who inferred increased productivity in the Maxwell Bay, King George Island, region following the deglaciation; the glacial period was marked by significantly reduced productivity. Shemesh et al. (1993) also noted that the last glacial productivity was significantly lower than the Holocene productivity, which led to their questioning the proposed lowering of glacial atmospheric carbon-di-oxide as a result of increased sequestration of CO_2 in the high latitudes. For the Holocene, Rathburn et al. (1997) inferred increased productivity during 2700–3400 yr BP on the Fram Bank near Prydz Bay, whereas changes were also noted during 6000 and 7000–7500 yr BP.

The trend towards increased productivity during interglacials is opposite to that reported by Hebbeln et al. (2002), who found that the productivity during the past 33 ka in the southern Peru-Chile Current (PCC) off Chile was highest during the last glacial maximum, and decreased significantly during the early to mid-Holocene. However, the coastal currents off western South America are part of the upwelling system of the Humboldt Current, which is thought to have been intensified—giving rise to more productivity—when winds were stronger during glacials.

5.5.5 Ice-Sheet Dynamics

The changes from glacial to interglacial periods were also marked by changes in the extent of ice sheets, which in turn affected other related processes. Though foraminiferal abundance is low to negligible in the regions that are suitable to study past changes in ice-sheet extent, the mere presence or absence of foraminiferal tests has been used to interpret ice-sheet dynamics in past. To develop a modern analogue for the past faunal response to an ice-sheet extent, foraminiferal presence in Recent sediments has been correlated with seasonal ice sheet changes (Hunt and Hosie, 2006a). Zachos et al. (1992) concluded that the Antarctic ice sheet increased in size for a short time during the earliest Oligocene and further that the ice sheet extent was comparable with the Holocene. Flower and Kennett (1994) inferred a significant increase in the East Antarctic ice sheet between 14.8–12.9 Ma. Loutit (1981) also inferred a significant increase in the Antarctic ice-sheet volume during the latest Miocene, while the ice-sheet volume remained comparatively stable during the late middle-early Late Miocene. Murphy et al. (2002) proposed that the rising sea levels during the northern hemisphere interglacials might have led to the destabilization and subsequent breakup of large ice sheets in the southern high latitudes after the Late Pliocene. Hodell and Warnke (1991) inferred increases in sea ice during the late Pliocene period, prior to the widespread increase in the Antarctic ice sheet during the major climatic change at 2.4 Ma. Theissen et al. (2003) reconstructed changes in the East Antarctic ice sheet during the Pleistocene, from Prydz Bay and the adjacent Lambert Glacier—Amery Ice Shelf region. In general, the ice sheet was found to expand, with intermittent warming and reduction in ice volume during the early-middle Pleistocene. Signatures of northern hemispheric ice sheet expansion during the mid-Pleistocene are also noticed at this location. Melis et al. (2002) reconstructed the Late Pleistocene-Holocene ice-sheet history of the Ross Sea region by using the preservation status of the foraminifera. Kanfoush et al. (2002) noted that the interglacial periods in the Weddell Sea region are characterized by a significant reduction in ice-rafted debris suggestive of reduced ice extent. Glacial terminations I and V show a stepwise reduction in ice-rafted debris interspersed with periods of increased debris flow. Scherer (1991) noted changes in the extent of the West Antarctic ice sheet during the Cenozoic.

The contribution of natural versus anthropogenic factors to the large scale melting and break-up of ice sheets has also been explored. Domack et al. (2005) reported that though the Larsen ice shelf broke up recently due to persistent recent warming, it had been thinning since the beginning of the Holocene. This reflects the fact that in general there has been large scale ice sheet retreat after each glacial maximum (Salvi et al., 2006). The history of the ice sheet between the last glacial and the Holocene has mainly been reconstructed from the marine regions adjoining Antarctic Peninsula. The ice sheet extent has been reported to fluctuate even during the supposedly stable Holocene, with large scale break-up during the mid-Holocene (Pudsey et al., 2006).

Different sedimentation regimes during the glacial-interglacial period have also been inferred based on the foraminiferal and sedimentological characteristics (Hillenbrand et al., 2005).

5.5.6 Seawater Temperature Reconstruction

Seawater temperature is one of the most important climatic parameters. Precise estimation of seawater temperature over the geologic past can help not only to understand the response of the region to global climatic variation but also as input to climate prediction models. Responses to global temperature changes are region specific, so to reconstruct seawater temperature change for a given region, region-specific proxies have to be developed. A Southern Ocean specific transfer function for past seawater temperature estimation has been developed based on the quantitative distribution of planktic foraminifera (Niebler, 1995). Cann and Clarke (1993) noted that the presence of a warm Leeuwin Current can be inferred from the presence of megascopic Marginopora vertebralis. Recently, well-established techniques like Mg/Ca analysis of planktic foraminifera, have also been shown to be effective tools for reconstructing sea-surface temperature from the southern high latitudes (Rathburn and De Deckker, 1997; Billups and Schrag, 2003). Additionally, Sr/Ca ratio has also been found to record the seawater temperature between the –2 to 6 °C range. However, large interspecific variability has been noted between the Mg/Ca ratio of the test and seawater temperature (Rathburn and De Deckker, 1997). Friedrich and Meier (2006) used the planktic foraminiferal isotopic composition to assess the suitability of the calcareous dinoflagellate cyst oxygen isotopic composition to reconstruct past climatic conditions. It was noted that while the carbon isotopic composition has some secondary unknown control, dinoflagellate cyst oxygen isotopic composition accurately records the seawater temperature. Sicre et al. (2005) reported differences in the sea-surface temperature estimated from alkenone and foraminiferal assemblages. The divergence between the two signals was attributed to the advection of detrital alkenones. Similar differences noted by Mazaud et al. (2002) were attributed to water column stratification, whereas Sikes et al. (2001) attributed the differences to the seasonality in the alkenone and foraminiferal flux.

Mackensen and Ehrman (1992) inferred a cooling trend beginning at 49.5 Ma, in both the Indian and Atlantic sectors of the Southern Ocean. Billups and Schrag (2002) reconstructed the Southern Ocean seawater temperature during the Oligocene through Pleistocene by using foraminiferal Mg/Ca. Mid Miocene development of the Antarctic ice sheets resulted in about a 3 °C cooling of the Southern Ocean. Late Pliocene cooling was of the order of about 4 °C, and synchronous with widespread northern hemispheric glaciation (Billups and Schrag, 2002). In another study, Williams et al. (2005) reported widespread cooling during the Pliocene, based on the oxygen isotopic composition of the planktic foraminifera. The Pliocene cooling estimates were in contrast to the well-established warming estimated based on the transfer function of the planktic foraminiferal assemblages. The differences were assigned to the secondary calcification or the secretion of the calcite below the mixed layer. Gallagher et al. (2003) reported relatively cooler conditions towards the basal part of the early Pliocene, changing to stable warmer conditions during much of the early Pliocene in the marine basins off southeastern Australia. The widespread fluctuation was noticed during the Middle to late Pliocene, in this region. Hodell and Warnke (1991) inferred a warmer Southern Ocean during the early Pliocene and cooling during the mid-Pliocene.

The seawater temperature fluctuations during the Brunhes (Brunhes, the final normal polarity chron in the Quaternary, preceded by the Matuyama) were comparatively more in number and magnitude than during Matuyama magnetic epoch (approximately 781,000 years ago). The Matuyama was comparatively cooler than the Brunhes (Keany and Kennett, 1972). Hodell et al. (2000) reconstructed the seawater temperature of the southern high latitudes for the last 450 ka and noted that marine isotopic stage 11 was almost as warm as other interglacials, except that the warming period lasted longer. Cortese et al. (2004) reported a warm sea surface temperature anomaly during marine isotopic stage 10 from the South Atlantic Ocean and linked it to east Antarctic moisture source anomalies. Cortese and Abelmann (2002) reported a 7 °C warming (3 °C higher than at present) of the Southern Ocean during Termination II. Their study showed that Southern Ocean warming preceded global ice volume changes. Climatic instability with rapid temperature fluctuations as large as noted during the glacial-interglacial transitions was also noted during marine isotopic stages 3 and 4. Rapid cooling episodes (Younger Dryas and Antarctic Cold Reversal) were noted during both Terminations I and II. Bianchi and Gersonde (2004) reported that the last glacial-interglacial warming history of the Atlantic sector of the Southern Ocean is similar to that reported from other parts of the Southern Ocean as well as Antarctica. However, region-specific cooling during the 9–7 kyr BP was also reported and attributed to the expansion of the Weddell Gyre circulation during this period. Andres et al. (2003), based on the oxygen isotopic composition of planktic foraminifera reported two rapid cooling episodes during the last glacial-interglacial transition that were synchronous with the Antarctic Cold Reversal and the Oceanic Cold Reversal reported in ice cores from Antarctica. Although Southern Ocean climatic changes are in general synchronous with the climatic changes reported from Antarctic ice cores, and different from the climatic changes reported from the northern hemisphere, Morigi et al. (2003) noted that climatic variations in the Pacific sector of the Southern Ocean do appear to be synchronous with the northern hemisphere climatic changes. He proposed that the see-saw pattern seen in the climatic changes between the northern and southern hemispheres was confined to the south of the Polar Front in the southern hemisphere, while the region north of the Polar Front mimicked the northern hemispheric climatic changes. Mazaud et al. (2002) also noted that the southern hemisphere did not warm before the northern hemisphere.

5.5.7 Foraminiferal Response to Abrupt Events

Besides the studies carried out to understand the gradual climatic and oceanographic changes during the geologic past, foraminiferal studies have also been carried out to understand the effect of sudden abrupt events at southern high latitudes. Deception Island has a history of volcanic eruptions. The last major volcanic eruption in this region occurred in 1967–70. It severely affected the faunal community in the nearby regions (Gallardo and Castillo, 1968). Significant recovery was reported in all benthic faunal communities including foraminifera, within a few years after the eruption (Finger, 1975; Lovell and Trego, 2003). However, Gray et al. (2003) reported that the recovery continued only until the mid-seventies.

5.5.8 Other Inferences

In view of perennial and seasonal anoxic conditions at certain locations in the southern high latitudes, Bernhard (1993) conducted laboratory culture experiments to understand the response of southern high latitude foraminifera to anoxic and high hydrogen sulfide conditions. Anoxic conditions do not result in any statistically significant change in the faunal composition. Gersonde et al. (1998) inferred changes in the sedimentation rate in the Indian sector of the Southern Ocean, throughout the Cenozoic. Domack et al. (1999) reported large-scale reworking of bank sediments in the Antarctic, thus indicating the need for careful dating of the sediments before drawing paleoceanographic inferences. Similar inferences were also drawn by Gore et al. (1994) based on studies carried out in the Vestfold Hills region. Barbieri et al. (1999) inferred bacterial decay of partially dissolved tests and post-depositional changes in the foraminiferal assemblage. Igarashi et al. (2001) inferred that the lysocline in the Lutzow-Holm Bay region, during the Holocene was at the present-day water depth of 300–400 m. Robert et al. (2002) inferred the late Eocene-Oligocene erosional history of Antarctica. The late Eocene was marked by chemical weathering, whereas physical weathering prevailed during the Eocene-Oligocene transition. Robinson et al. (2007) inferred a link between the intensity of denitrification in the southeast Pacific and climatic changes in the Antarctic, thus establishing the dominant effect of Antarctica on the oceanic processes in the nearby oceans. The study of Moy et al. (2009) shows that the modern shell weights in modern Southern Ocean planktonic foraminifera are 30–35% lower than those from the sediments of underlying Holocene-age. The loss in shell weight is consistent with reduced calcification today induced by ocean acidification. The ecosystem model of biogeographical patterns of non-spinose species of foraminifera developed by Maria et al. 2022 exhibits maximum biomass concentrations in mid- to high-latitude waters and upwelling areas. In response to ocean warming and associated changes in primary production and ecological dynamics, they noted the global average loss in nonspinose foraminifera biomass to be between 8% and 11% by 2050 and between 14% and 18% by the year 2100.

5.5.8.1 Gray Areas

The future foraminiferal investigations in the Southern Ocean must address the following important issues:

- Determination of the dynamics of the formation and distribution of water masses, currents, and sea ice.
- Investigating the relationship between oceanic and atmospheric circulation systems and the physical basis for biological productivity.
- Assessment of the distribution, sources and sinks of organic carbon.
- Studies on the chemical oceanography of the Southern Ocean.
- Temporal variations in biological productivity.
- Assessment of sustainable living and non-living potentials.
- Paleoclimatic studies using proxy indicators in the sediment cores to be retrieved from the Southern Ocean and evaluation of the impact of the Southern Ocean climate in other parts of the globe.

- Improved understanding of the role of the Southern Ocean in modulating the global oceanic circulation patterns and climate, especially over the Indian ocean.
- Development of operational capabilities to deploy and support investigations in a range of scientific disciplines in the Southern Ocean.

ACKNOWLEDGMENTS

The authors are thankful to the Secretary, Ministry of Earth Sciences, New Delhi, and Vice-Chancellor, University of Lucknow, for their permission to publish the manuscript. Authors are thankful to Dr. Rajeev Saraswat, Principal Scientist at National Institute of Oceanography, Goa, for critically reviewing the manuscript and offering his valuable suggestions.

REFERENCES

Abouchami, W., Galer, S. J., Middag, R., DeBaar, H. J., Andreae, M. O., Feldmann, H., Laan, P., Raczek, I., 2010. *CD isotopes, a proxy for water mass and nutrient uptake in the Southern Ocean*. American Geophysical Union, URL: www.agu.org

Abrantes, F., Rodrigues, T., Montanari, B., Santos, C., Witt, L., Lopes, C., Voelker, A.H.L., 2011. Climate of the last millennium at the southern pole of the North Atlantic Oscillation: an innershelf sediment record of flooding and upwelling. *Climate Research* 48(2–3): 261–280.

Alexander, S.P., DeLaca, T.E., 1987. Feeding adaptations of the foraminiferan cibicides refulgens living epizoically and parasitically on the Antarctic scallop *Adamussium colbecki. Biological Bulletin, Marine Biological Laboratory, Woods Hole* 173: 136–159.

Allen, K., Roberts, S., Murray, J.W., 1998. Fractal grain distribution in agglutinated foraminifera. *Paleobiology* 24: 349–358.

Anderson, J.B., 1975a. Ecology and distribution of foraminifera in the Weddell Sea of Antarctica. *Micropaleontology* 21: 69–96.

Anderson, J.B., 1975b. Factors controlling $CaCO_3$ dissolution in the Weddell Sea from foraminiferal distribution patterns. *Marine Geology* 19: 315–332.

Anderson, J.B., Kurtz, D.D., Weaver, F.M., 1979. Sedimentation on the Antarctic continental slope. In: Doyle L.J., Pilkey O.H. (Eds.), *Geology of continental slopes.* Society of Economic Paleontologists and Mineralogists; Tulsa, OK (USA), Spec. Publ. SEPM, (no. 27), pp. 265–283.

Anderson, R.F., Carr, M.E., 2010. Uncorking the Southern Ocean's vintage CO2. *Science* 328(5982): 1117–1118.

Andres, M.S., Bernasconi, S.M., McKenzie, J.A., Rohl, U., 2003. Southern Ocean deglacial record supports global Younger Dryas. *Earth and Planetary Science Letters* 216: 515–524.

Andrew, D.M., Martin R P., William R H., Jelle, B., Matthew J C., Eva C., Carles, P., Michael K.G., Thomas B.C., 2019. Varied contribution of the Southern Ocean to deglacial atmospheric CO2 rise. *Nature Geoscience* 12(12): 1006–1011.

Anna, M., Katharine, R.H., Jennifer, P., Daniela, N.S., Kirsty, M.E., Victoria, P., Frank, J.C.P., Melanie, J.L., Michael, P.M., Chloe, L. T; Sharon, S., Hugh, D., 2019. Temporal variability in foraminiferal morphology and geochemistry at the West Antarctic Peninsula: a sediment trap study. *Biogeosciences* 16(16): 3267–3282.

Atsushi, A., Boo-Keun, K., Takanori, N., Hiroyuki, T., 2011 January 15. Chemostratigraphic documentation of a complete Miocene intermediate-depth section in the Southern Ocean: Ocean Drilling Program Site 1120, Campbell Plateau off New Zealand. *Marine Geology* 279(1–4): 52–62.

Barbieri, R., D'Onofrio, S., Melis, R., Westall, F., 1999. r-Selected benthic foraminifera with associated bacterial colonies in Upper Pleistocene sediments of the Ross Sea (Antarctica): implications for calcium carbonate preservation. *Palaeogeography, Palaeoclimatology, Palaeoecology* 149: 41–57.

Barker, P.F., Kennett, J.P., Scientific Party, 1988. Weddell Sea palaeoceanography: Preliminary results of ODP Leg 113. *Palaeogeography, Palaeoclimatology, Palaeoecology* 67: 75–102.

Barker, P.F., Thoma, E., 2004. Origin, Signature and paleoclimatic influence of the Antarctic Circumpolar Current. *Earth-Science Review* 66: 143–162.

Barrera, E., 1987. *Isotopic paleotemperatures: 1. Effect of diagenesis. 2. Late Cretaceous temperatures.* Ph. D. Thesis, Case Western Reserve Univ., Cleveland, OH (USA).

Belkin, I.M., Gordon, A.L., 1996. Southern Ocean fronts from the Greenwich meridian to Tasmania. *Journal of Geophysical Research* 101: 3675–3696.

Belyaeva, N.V., 1970. Regularities in the distribution of planktonic foraminifera in the water and sediments of the Southern Ocean. In: Holdgate, M.W. (Ed.), *The South Indian and Pacific Oceans.* Academic Press, New York, pp. 154–161.

Benedetto, S., Elisabeth, L.S., Aurora, C.E., Mea S.C., Kathryn, A.R., 2014. Southwest Pacific subtropics responded to last deglacial warming with changes in shallow water sources. *Paleoceanography* 29(6): 595–611.

Bergami, C., Capotondi, L., Langone, L., Giglio, F., Ravaioli, M., 2009. Distribution of living planktonic foraminifera in the Ross Sea and the Pacific sector of the Southern Ocean (Antarctica). *Marine Micropaleontology* 73 (1–2): 37–48.

Bernhard, J.M., 1993. Experimental and field evidence of Antarctic foraminiferal tolerance to anoxia and hydrogen sulfide. *Marine Micropaleontology* 20: 203–213.

Bernhard, J.M., Bowser, S.S., 1992. Bacterial biofilms as a trophic resource for certain benthic foraminifera. *Marine Ecology Progress Series* 83: 263–272.

Bianchi, C., Gersonde, R., 2004. Climate evolution at the last deglaciation: the role of the Southern Ocean. *Earth and Planetary Science Letters* 228: 407–424

Billups, K., 2002. Late Miocene through early Pliocene deep water circulation and climate change viewed from the Subantarctic South Atlantic. *Palaeogeography, Palaeoclimatology, Palaeoecology* 185: 287–307.

Billups, K., Kelly, C., Pierce, E., 2008. The late Miocene to early Pliocene climate transition in the Southern Ocean. *Palaeogeography, Palaeoclimatology, Palaeoecology* 267(1–2): 31–40.

Billups, K., Schrag, D.P., 2002. Paleotemperatures and ice volume of the past 27 myr revisited with paired Mg/Ca and 18O/16O measurements on benthic foraminifera. *Paleoceanography* 17: 3-1-3-11.

Billups, K., Schrag, D.P., 2003. Application of benthic foraminiferal Mg/Ca ratios to questions of Cenozoic climate change. *Earth and Planetary Science Letters* 209: 181–195.

Boersma, A., Huber, B., 1993. Origins and biogeography of Cretaceous/Tertiary biserial planktonic foraminifera in the Southern Oceans. *Antarctic Journal of the United States* 28: 109.

Boltovskoy, E., Boltovskoy, D., Correa, N., Brandini, F., 1996. Planktic foraminifera from the southwestern Atlantic (30°-60°S): species-specific patterns in the upper 50 m. *Marine Micropaleontology* 28: 53–72

Boltovskoy, E., Watanabe, S., 1979. Foraminifera from 82 surface plankton samples taken during the 'islas orcadas 5/75' oceanographic cruise in the southwestern Atlantic and its Antarctic sector (May-June, 1975). *Rev. Mus. Argent. Cienc. Nat. Bernardino Rivadavia, Inst. Nac. Invest. Cienc. Nat. (Argent.) (Hidrobiol.)* 5: 229–240.

Boltovskoy, E., Watanabe, S., 1980. Foraminiferans from quaternary sediments between tierra del fuego and south georgia islands. *Rev. Mus. Argent. Cienc. Nat. Bernardino Rivadavia Inst. Nac. Invest. Cienc. Nat.* (Argent.) (Geol.) 8: 95–124.

Boltovskoy, E., Watanabe, S., 1981. Live planktonic foraminifera from the Antarctic sector of the Atlantic Ocean. *Rev. Mus. Argent. Cienc. Nat. Bernardino Rivadavia Inst. Nac. Invest. Cienc. Nat.* (Argent.) (Hidrobiol.) 4: 169–104.

Bowser, S.S., Gooday, A.J., Alexander, S.P., Bernhard, J.M., 1995. Larger agglutinated foraminifera of McMurdo Sound, Antarctica: Are *Astrammina rara* and *Notodendrodes antarctikos* allogromiids incognito? *Marine Micropaleontology* 26: 75–88.

Bowser, S.S., DeLaca, T.E., 1985. Rapid intracellular motility and dynamic membrane events in an Antarctic foraminifer. *Cell Biology International Reports* 9: 901–910.

Bowser, S.S., Ted, E., Delaca, Conly, L., Rieder., 1986. Novel extracellular matrix and microtubule cables associated with pseudopodia of Astrammina rara, a carnivorous antarctic foraminifer. *Journal of Ultrastructure and Molecular Structure Research* 94: 149–160.

Brathauer, U., Abelmann, A., Gersonde, R., Niebler, H.S., Futterer, D.K., 2001. Calibration of cycladophora davisiana events versus oxygen isotope stratigraphy in the subantarctic Atlantic Ocean—a stratigraphic tool for carbonate-poor quaternary sediments. *Marine Geology* 175: 167–181.

Brian, T.H., Rose, P.M., Kenneth, G.M., 2020. Planktonic Foraminiferal Endemism at Southern High Latitudes Following the Terminal Cretaceous Extinction. *Journal of Foraminiferal Research* 50 (4): 382–402.

Broecker, W.S., 1993. An oceanographic explanation for the apparent carbon isotope-cadmium discordancy in the glacial Antarctic? *Paleoceanography* 8: 137–139.

Caniupán, M., Lamy, F., Lange, C.B., Kaiser, J., Arz, H., Kilian, R., Urrea, O.B., Aracena, C., Hebbeln, D., Kissel, C., Laj, C., Mollenhauer, G., Tiedemann, R., 2011. Millennial-scale sea surface temperature and Patagonian Ice Sheet changes off southernmost Chile (53 degree S) over the past?60 kyr. *Paleoceanography* 26(3), PA3221. doi:10.1029/2010PA002049.

Cann, J.H., Bourman, R.P., Barnett, E.J., 2000. Holocene foraminifera as indicators of relative estuarine-lagoonal and oceanic influences in estuarine sediments of the river Murray, south Australia. *Quaternary Research* 53: 378–391.

Cann, J.H., Clarke, J.D.A., 1993. The significance of marginopora vertebralis (foraminifera) in surficial sediments at esperance, western Australia, and in last interglacial sediments in the northern Spencer Gulf, south Australia. *Marine Geology* 111: 171–187.

Chang, S.K., Ho, I.Y., 1995. Foraminiferal assemblages from bottom sediments at Marian Cove, South Shetland Islands, West Antarctica. *Marine Micropaleontology* 26: 223–232.

Channell, J.E.T., Galeotti, S., Martin, E.E., Billups, K., Scher, H.D., Stoner, J.S., 2003. Eocene to Miocene magnetostratigraphy, biostratigraphy, and chemostratigraphy at ODP site 1090 (sub-Antarctic South Atlantic). *Bulletin of the Geological Society of America* 115: 607–623.

Chapman, F., 1916. Report on the foraminifera and ostracoda out of marine muds from soundings in the Ross Sea. British Antarctic Expedition 1907–1909, Reports of Scientific Investigations. *Geology* 2: 53–80, 1–6.

Charles, C.D., Fairbanks, R.G., 1992. Evidence from Southern Ocean sediments for the effect of North Atlantic deep-water flux on climate. *Nature* 355: 416–419.

Chen, C., 1996. Preliminary study on late quaternary foraminiferal assemblage and its significance in environment from the Bransfield stratid, west Antarctica. *Antarct.Res./Nanji Yanjiu* 83: 13–19.

Chen, M.-T., Chang, Y.-P., Chang, C.-C., Wang, L.-W., Wang, C.-H, Yu, E.-F., 2002. Late Quaternary sea-surface temperature variations in the southeast Atlantic: a planktic foraminifer faunal record of the past 600000 yr (IMAGES II MD962085). *Marine Geology* 180: 163–181.

Clarke, A., Crame, J.A., 1992. The Southern Ocean benthic fauna and climate change: a historical perspective. *Philosophical Transactions of the Royal Society, London, Series B* 338: 299–309.

Clarke, A., Johnston, N.M., 2003. Antarctic marine diversity. *Oceanography and Marine Biology Annual Review* 48: 47–114.

Clifford, M. A., 1983 *A descriptive study of the zonation of the Antarctic Circumpolar Current and its relation to wind stress and ice cover.* Unpublished MS thesis, Texas A&M University 93 pp.

Compton, J.S., Wigley, R., McMillan, I.K., 2004. Late Cenozoic phosphogenesis on the western shelf of South Africa in the vicinity of the Cape Canyon. *Marine Geology* 206: 19–40.

Cook, M.R., Elderfield, H., Zahn, R., Pahnke, K., 2006. High-resolution trace metal analysis of benthic foraminifera reveals nutrient excursions in Antarctic Intermediate. *Geochimica et Cosmochimica Acta* 70: 109.

Corliss, B.H., 1979a. Quaternary Antarctic Bottom-Water history: Deep-sea benthonic foraminiferal evidence from the southeast Indian Ocean. *Quaternary Research* 12: 271–289.

Corliss, B.H., 1979b. Recent deep-sea benthonic foraminiferal distributions in the southeast Indian Ocean: Inferred bottom-water routes and ecological implications. *Marine Geology* 31: 115–138.

Corliss, B.H., 1982. Linkage of North Atlantic and Southern Ocean deep-water circulation during glacial intervals. *Nature* 298: 458–460.

Corliss, B.H., 1983. Distribution of Holocene deep-sea benthonic foraminifera in the southwest Indian Ocean. Deep-Sea Research Part A. *Oceanographic Research Papers* 30: 95–117.

Corliss, B.H., 1983. Quaternary circulation of the Antarctic circumpolar current. Deep-Sea Research Part A. *Oceanographic Research Papers* 30: 47–61.

Corliss, B.H., Thunell, R.C., 1983. Carbonate sedimentation beneath the Antarctic Circumpolar Current during the late Quaternary. *Marine Geology* 51: 293–326.

Cornelius, N., Gooday, A.J., 2004. Live (stained) deep-sea benthic foraminiferans in the western Weddell Sea: Trends in abundance, diversity and taxonomic composition along a depth transect. *Deep-Sea Research* II 51: 1571–1602.

Cortese, G., Abelmann, A., 2002. Radiolarian-based paleotemperatures during the last 160 kyr at ODP site 1089 (Southern Ocean, Atlantic sector). *Palaeogeography, Palaeoclimatology, Palaeoecology* 182: 259–286.

Cortese, G., Abelmann, Andrea, Gersonde, Rainer, 2004. A glacial warm-water anomaly in the subantarctic Atlantic Ocean, near the Agulhas Retroflection. *Earth and Planetary Science Letters* 222: 767–778.

Cromer, L., Gibson, J.A.E., Kerrie, M., Swadling, David, Ritz, A., 2005. Faunal microfossils: Indicators of Holocene ecological change in a saline Antarctic lake. *Palaeogeography, Palaeoclimatology, Palaeoecology* 221: 83–97.

Cromer, L., Gibson, J.A.E., Swadling, K.M., Hodgson, D.A., 2006. Evidence for a faunal refuge in the Larsemann Hills, East Antarctica, during the Last Glacial Maximum. *Journal of Biogeography* 33(7): 1314–1323.

Curry, W.B., Lohmann, G.P., 1982. Carbon isotopic changes in benthic foraminifera from the western South Atlantic: Reconstruction of glacial abyssal circulation patterns. *Quaternary Research* 18: 218–235.

David, C.L., Zanna, C., Karen, E.K., Earle, A.W., 2021. Tracking Southern Ocean Sea Ice Extent With Winter Water: A New Method Based on the Oxygen Isotopic Signature of Foraminifera. *Paleoceanography and Paleoclimatology* 36(6). http://dx.doi.org/10.1029/2020PA004095

Deacon, G.E.R., 1933. A general account of the hydrology of the Southern Atlantic Ocean. *Discovery Reports* 7: 71–238

Deacon, G.E.R., 1937. The hydrology of the Southern Ocean. Discovery Reports 15: 1–24.

Deacon, G.E.R., 1982. Physical and biological zonation in the Southern Ocean, *Deep-Sea Research* 29: 1–15

DeLaca, T.E., Karl, D.M., Lipps, J.H., 1981. Direct use of dissolved organic carbon by agglutinated benthic foraminifera. *Nature* 289: 287–289.

D'Hondt, S., Arthur, M.A., 2002. Deep water in the late maastrichtian ocean. *Paleoceanography* 17: 8-1-8-11.

Diekmann, B., Faelker, M., Kuhn, G. (2003). Environmental history of the south-eastern south Atlantic since the middle miocene: Evidence from the sedimentological records of ODP sites 1088 and 1092. *Sedimentology* 50: 511–529.

Diekmann, B., Kuhn, G., 2002. Sedimentary record of the mid-Pleistocene climate transition in the southeastern South Atlantic (ODP Site 1090). *Palaeogeography, Palaeoclimatology, Palaeoecology* 182: 241–258.

Diester-Haass, L., 1995. Middle Eocene to early Oligocene paleoceanography of the Antarctic Ocean (Maud Rise, ODP Leg 113, Site 689): change from a low to a high productivity ocean. *Palaeogeography, Palaeoclimatology, Palaeoecology* 113: 311–334.

Diester-Haass, L., 1996. Late Eocene-Oligocene paleoceanography in the southern Indian Ocean (ODP Site 744). *Marine Geology* 130: 99–119.

Diz, P., Hall, I.R., Zahn, R., Molyneux, E.G., 2008. Paleoceanography of the southern Agulhas Plateau during the last 150 ka: Inferences from benthic foraminiferal assemblages and multispecies epifaunal carbon isotopes. *Paleoceanography* 22(4):

Domack, E.W., 1988. Biogenic facies in the Antarctic glacimarine environment: Basis for a polar glacimarine summary. *Palaeogeography, Palaeoclimatology, Palaeoecology* 63: 357–372.

Domack, E.W., Duran, D., Leventer, A., Ishman, S., Doane, S., McCallum, S., Amblas, D., Ring, J., Gilbert, R., Prentice, M., 2005. Stability of the Larsen B ice shelf on the Antarctic Peninsula during the Holocene epoch. *Nature* 436: 681–685.

Domack, E.W., Jull, A.J.T., Anderson, J.B., Linick, T.W., Williams, C.R., 1989. Application of tandem accelerator mass-spectrometer dating to late Pleistocene-Holocene sediments of the East Antarctic continental shelf. *Quaternary Research* 31: 277–287.

Domack, E.W., Taviani, M., Rodriguez, A., 1999. Recent sediment remolding on a deep shelf, Ross Sea: implications for radiocarbon dating of Antarctic marine sediments. *Quaternary Science Reviews* 18: 1445–1451.

Duplessy, J.C., Labeyrie, L., Waelbroeck, C., 2002. Constraints on the ocean oxygen isotopic enrichment between the last glacial maximum and the Holocene: Paleoceanographic implications. *Quaternary Science Reviews* 21: 315–330.

Echols, R.J., 1971. Distribution off foraminifera in the sediments of the Scotia Sea area, Antarctic waters. *Antarctic Research Series, American Geophysical Union* 15: 93–168.

Ehrmann, W., 1994. *Cenozoic glacial history of Antarctica. Reports on polar research.* Bremen University, Bremen, 137, 152 pp.

Elderfield, H., Greaves, M., Barker, S., Hall, I.R., Tripati, A., Ferretti, P., Crowhurst, S., Booth, L., Daunt, C., 2010. A record of bottom water temperature and seawater δ (18)O for the Southern Ocean over the past 440 kyr based on Mg/Ca of benthic foraminiferal Uvigerina spp. *Quaternary Science Reviews* 29(1–2): 160–169.

Elderfield, H., Rickaby, R.E.M., 2000. Oceanic Cd/P ratio and nutrient utilization in the glacial Southern Ocean. *Nature* 405: 305–310.

Fillon, R.H., 1972. Evidence from the Ross Sea for widespread submarine erosion. *Nature* (Phys.Sci.) 238: 40–42.

Finger, K.L., 1975. Benthic foraminifera from Deception Island. *Antarctartica Journal of United States of America* 10: 134.

Flower, B.P., Kennett, J.P., 1994. The middle Miocene climatic transition: East Antarctic ice sheet development, deep ocean circulation and global carbon cycling. *Palaeogeography, Palaeoclimatology, Palaeoecology* 108: 537–555.

Franck, L., Philippe, E., Jan, P., 2014. Patchiness of deep-sea benthic Foraminifera across the Southern Ocean: Insights from high-throughput DNA sequencing. *Deep-Sea Research* (Part II, Topical Studies in Oceanography), 108: 17–26.

Friederike, E., Philipp, A., Patrick, M. Isabelle, S., Sina, W., Eva-Maria, N., 2014. Particle flux characterisation and sedimentation patterns of protistan plankton during the iron fertilisation experiment LOHAFEX in the Southern Ocean. *Deep-Sea Research* (Part I, Oceanographic Research Papers), 89: 94–103.

Friedrich, O., Meier, K.J.S., 2006. Suitability of stable oxygen and carbon isotopes of calcareous dinoflagellate cysts for paleoclimatic studies: Evidence from the Campanian/Maastrichtian cooling phase. *Palaeogeography, Palaeoclimatology, Palaeoecology* 239: 456–469.

Galeotti, S., Coccioni, R., Gersonde, R., 2002. Middle Eocene-early Pliocene subantarctic planktic foraminiferal biostratigraphy of site 1090, Agulhas ridge. *Marine Micropaleontology* 45: 357–381.

Gallagher, S.J., Smith, A.J., Jonasson, K., Wallace, M.W., Holdgate, G.R., Daniels, J., 2001. The Miocene palaeoenvironmental and palaeoceanographic evolution of the Gippsland Basin, southeast Australia: A record of Southern Ocean change. *Palaeogeography, Palaeoclimatology, Palaeoecology* 172: 53–80.

Gallagher, S.J., Greenwood, D.R., Taylor, D., Smith, A.J., Wallace, M.W., Holdgate, G.R., 2003. The Pliocene climatic and environmental evolution of southeastern Australia: evidence from the marine and terrestrial realm. *Palaeogeography, Palaeoclimatology, Palaeoecology* 193: 349–382.

Gallagher, S.J., Taylor, D., Apthorpe, M., Stilwell, J.D., Boreham, C.J., Holdgate, G.R., Wallace, M.W., Quilty, P.G., 2005. Late Cretaceous dysoxia in a southern high latitude siliciclastic succession, the Otway Basin, southeastern Australia. *Palaeogeography, Palaeoclimatology, Palaeoecology* 223: 317–348.

Gallardo, V.A., Castillo, J.G., 1968. Mass mortality in the benthic infauna of Port Foster resulting from the eruptions in Deception Island, South Shetland Islands. 16: 1–13.

Garrison, D.L., Kurt, R., Buck, Marcia, M., 1991. Gowing. Plankton assemblages in the Ice edge zone of the Weddell Sea during the austral winter. *Journal of Marine Systems* 2: 123–130.

Gersonde, R., Spie, V., Flores, J.A., Hagen, R.A., Kuhn, G., 1998. The sediments of Gunnerus Ridge and Kainan Maru Seamount (Indian sector of Southern Ocean). *Deep-Sea Research Part I: Oceanographic Research Papers* 45: 1515–1540.

Gersonde, R., Wefer, G., 1987. Sedimentation of biogenic siliceous particles in Antarctic waters from the Atlantic sector. *Marine Micropaleontology* 11: 311–332.

Gilbert, I.M., Pudsey, C.J., Murray, J.W., 1998. A sediment record of cyclic bottom-current variability from the northwest Weddell Sea. *Sedimentary Geology* 115: 185–214.

Gooday, A.J., Bowser, S.S., Bernhard, J.M., 1994. The foraminifera of Explorers Cove, Antarctica: a deep-sea assemblage in shallow water. *Antarctic Journal of the United States* 29: 149–151.

Gooday, A.J., Bowser, S.S., Bernhard, J.M., 1996. Benthic foraminiferal assemblages in Explorers Cove, Antarctica: A shallow-water site with deep-sea characteristics. *Progress in Oceanography* 37: 117–166.

Gooday, A.J., Holzmann, M., Guiard, J., Pawlowski, J., 2004. A new monothalamous foraminiferan from 1000–6300 m water depth in the Weddell Sea: morphological and molecular characterisation. *Deep-Sea Research* II 51: 1603–1616.

Gooday, A.J., Pawlowski, J., 2004. *Conqueria laevis* gen. and sp. nov., a new soft-walled, monothalamous foraminiferan from the deep Weddell Sea. *Journal of the Marine Biological Association of the United Kingdom* 84: 919–924.

Gooday, A.J., Rathburn, A.E., 1999. Temporal variability in living deep-sea benthic foraminifera: a review. *Earth-Science Reviews* 46: 187–212.

Gore, D.B., Colhoun, E.A., Bell, K., 1994. Derived constituents in the glacial sediments of the Vestfold Hills, East Antarctica. *Quaternary Science Reviews* 13: 301–307.

Gostin, V.A., Belperio, A.P., Cann, J.H., 1988. The Holocene non-tropical coastal and shelf carbonate province of southern Australia. *Sedimentary Geology* 60: 51–70.

Gowing, M.M., Garrison, D.L., 1991. Austral winter distributions of large tintinnid and large sarcodinid protozooplankton in the ice-edge zone of the Weddell/Scotia seas. *Journal of Marine Systems* 2: 131–141.

Gray, S.C., Sturz, A., Bruns, M.D., Marzan, R.L., Dougherty, D., Law, H.B., 2003. Composition and distribution of sediments and benthic foraminifera in a submerged caldera after 30 years of volcanic quiescence. *Deep-Sea Research II, Topical Studies in Oceanography* 50: 1727–1751.

Guptha, M.V.S., 1983. *Benthic foraminifera from the continental margin off Queen Maud Land, Antarctica.* Department of Ocean Development, New Delhi (India).

Harloff, J., Mackensen, A., 1997. Recent benthic foraminiferal associations and ecology of the Scotia Sea and Argentine Basin. *Marine Micropaleontology* 31: 1-29.

Harwood, D.M., Scherer, R.P., Webb, P.N., 1989. Multiple Miocene marine productivity events in West Antarctica as recorded in upper Miocene sediments beneath the Ross Ice Shelf (Site J-9). *Marine Micropaleontology* 15: 91–115.

Hayward, B.W., Neil, H., Carter, R., Grenfell, H.R., Hayward, J.J., 2002. Factors influencing the distribution patterns of Recent deep-sea benthic foraminifera, east of New Zealand, Southwest Pacific Ocean. *Marine Micropaleontology* 46: 139–176.

Hayward, B.W., Grenfell, H.R., Carter, R., Hayward, J.J., 2004a. Benthic foraminiferal proxy evidence for the Neogene palaeoceanographic history of the Southwest Pacific, east of New Zealand. *Marine Geology* 205: 147–184.

Hayward, B.W., Grenfell, H.R., Sabaa, A.T., Neil, H.L., 2007. Factors influencing the distribution of sub-Antarctic deep-sea benthic foraminifera, Campbell and Bounty Plateaux, New Zealand. *Marine Micropaleontology*, 62(3): 141–166.

Hayward, B.W., Scott, G.H., Grenfell, H.R., Carter, R., Lipps, J.H. 2004b. Techniques for estimation of tidal elevation and confinement (~salinity) histories of sheltered harbours and estuaries using benthic foraminifera: examples from New Zealand. *The Holocene* 14(2): 218–232.

Healy-Williams, N. 1992. Stable isotope differences among morphotypes of Neogloboquadrina pachyderma (Ehrenberg): Implications for high-latitude palaeoceanographic studies. *Terra Nova*, 4(6): 693–700.

Healy-Williams, N., Williams, D.F., 1981. Fourier analysis of test shape of planktonic foraminifera. *Nature* 289: 485–487.

Hebbeln, D., Marchant, M., Freudenthal, T., Wefer, G., 2000. Surface sediment distribution along the Chilean continental slope related to upwelling and productivity. *Marine Geology* 164: 119–137.

Hebbeln, D., Marchant, M., Wefer, G., 2002. Paleoproductivity in the southern Peru-Chile Current through the last 33 000 yr. *Marine Geology* 186: 487–504.

Herb, R., 1971. Distribution of recent benthonic foraminifera in the Drake Passage. In: Llano, G.A., Wallen, I.E. (Eds.), Biology of the Antarctic Seas IV. *American Geophysical Union, Washington, DC*, pp. 251–300.

Heron-Allen, E., Earland, A., 1932. Foraminifera Part I. The ice-free area of the Falkland Islands and adjacent seas. *Discovery Reports* 4: 291–460.

Hillenbrand, C.D., Baesler, A., Grobe, H., 2005. The sedimentary record of the last glaciation in the western Bellingshausen Sea (West Antarctica): Implications for the interpretation of diamictons in a polar-marine setting. *Marine Geology* 216: 191–204.

Hillenbrand, C.D., Grobe, H., Diekmann, B., Kuhn, G., Futterer, D.K., 2003. Distribution of clay minerals and proxies for productivity in surface sediments of the Bellingshausen and Amundsen seas (West Antarctica)—Relation to modern environmental conditions. *Marine Geology* 193: 253–271.

Hodell, D.A., Charles, C.D., Ninnemann, U.S., 2000. Comparison of interglacial stages in the south Atlantic sector of the Southern Ocean for the past 450 kyr: Implications for marine isotope stage (MIS) 11. *Global and Planetary Change* 24: 7–26.

Hodell, D.A., Warnke, D.A., 1991. Climatic evolution of the Southern Ocean during the Pliocene epoch from 4.8 to 2.6 million years ago. *Quaternary Science Reviews* 10: 205–214.

Hofmann, E.E., 1985. The large-scale horizontal structure of the Antarctic circumpolar current from FGGE drifters. *Journal of Geophysical Research* 90: 7087–7097.

Holbourn, A., Kuhnt, Wolfgang, Simo, J.A. (Toni), Qianyu, Li., 2004, Middle Miocene isotope stratigraphy and paleoceanographic evolution of the northwest and southwest Australian margins (Wombat Plateau and Great Australian Bight). *Palaeogeography, Palaeoclimatology, Palaeoecology* 208: 1–22.

Hovan, S.A., Rea, D.K., 1992. Paleocene/Eocene boundary changes in atmospheric and oceanic circulation: A southern hemisphere record. *Geology* 20: 15–18.

Howard, W.R., Prell, W.L., 1984. A comparison of radiolarian and foraminiferal paleoecology in the Southern Indian Ocean: New evidence for the interhemispheric timing of climatic change. *Quaternary Research* 21: 244–263.

Howe, J.A., Harland, Rex, Pudsey, C.J., 2002. Dinoflagellate cyst evidence for Quaternary palaeoceanographic change in the northern Scotia Sea, South Atlantic Ocean. *Marine Geology* 191: 55–69.

Huber, B.T., 1992a. Paleobiogeography of Campanian-Maastrichtian foraminifera in the southern high latitudes. *Palaeogeography, Palaeoclimatology, Palaeoecology* 92: 3–4.

Huber, B.T., 1992b. Upper cretaceous planktic foraminiferal biozonation for the austral realm. *Marine Micropaleontology* 20: 107–128.

Huber, B.T., Leckie, R.M., 2011. Planktic foraminiferal species turnover across deep-sea Aptian/Albian boundary sections. *Journal of Foraminiferal Research* 41(1): 53–95.

Huddestun, P., 1971. Pleistocene paleoclimates are based on radiolaria from subantarctic deep-sea cores. *Deep Sea Research and Oceanographic Abstracts* 1811: 1141–1143.

Hunt, B.P.V., Hosie, G.W., 2005. Zonal structure of zooplankton communities in the Southern Ocean south of Australia: Results from a 2150 km continuous plankton recorder transect. *Deep-Sea Research I, Oceanographic Research Papers* 52: 1241–1271.

Hunt, B.P.V., Hosie, G.W., 2006a. The seasonal succession of zooplankton in the Southern Ocean south of Australia, part I: The seasonal ice zone. *Deep-Sea Research I* 53: 1182–1202.

Hunt, B.P.V., Hosie, G.W., 2006b. The seasonal succession of zooplankton in the Southern Ocean south of Australia, part II: The Sub-Antarctic to Polar Frontal Zones. *Deep-Sea Research I* 53: 1203–1223.

Igarashi, A., Numanami, H., Tsuchiya, Y., Fukuchi, M., 2001. Bathymetric distribution of fossil foraminifera within marine sediment cores from the eastern part of lutzow-holm bay, east Antarctica, and its paleoceanographic implications. *Marine Micropaleontology* 42: 125–162.

Ishman, S.E., Domack, E.W., 1994. Oceanographic controls on benthic foraminifers from the Bellingshausen margin of the Antarctic Peninsula. *Marine Micropaleontology* 24: 119–155.

Jacob, N.W.H., Alexander, M P., Delia, W.O., Kuo-Fang, H., Stefan, M., Cristiano, M.C., Jurek, B., 2016. Antarctic intermediate water circulation in the South Atlantic over the past 25,000years. *Paleoceanography* 31(10): 1302–1314.

Jan, P., Delia, F., Ana Aranda, da S., Jackie, G., 2011. Novel lineages of Southern Ocean deep-sea foraminifera revealed by environmental DNA sequencing. *Deep-Sea Research* (Part II, Topical Studies in Oceanography), 58(19–20): 1996–2003.

Jennings, A.E., Xiao, J., Licht, K.J., Andrews, J.T., 1995. Benthic foraminiferal assemblages from the western Ross Sea: Approximately 30,000 years ago to present. *Antarctic Journal of the United States* 30: 26–28.

Jones, R.W., Pudsey, C.A., 1994. Recent benthonic foraminifera from the Western Antarctic Ocean. *Journal of Micropaleontology* 13: 17–23.

Joseph, A.S., Tao, L., Peter, T.S., Andrea, B., Tianyu, C., Jenny, R., James, W.B.R., Victoria, P., Sev, K., Qian, L., Laura, F.R., 2021. Productivity and Dissolved Oxygen Controls on the Southern Ocean Deep-Sea Benthos During the Antarctic Cold Reversal. *Paleoceanography and Paleoclimatology* 36(10), [e2021PA004288]. https://doi.org/10.1029/2021PA004288

Julie, M., Salome, F., Philippe, K., Claire, L., Cedric, C., Graham, W.H., Sophie, S., Helene, H., 2016. Planktonic foraminiferal biogeography in the Indian sector of the Southern Ocean: Contribution from CPR data. *Deep-Sea Research*, (Part I), 110: 75–89.

Kanfoush, S.L., Hodell, D.A., Charles, C.D., Janecek, T.R., Rack, F.R., 2002. Comparison of ice-rafted debris and physical properties in ODP Site 1094 (South Atlantic) with the Vostok ice core over the last four climatic cycles. *Palaeogeography, Palaeoclimatology, Palaeoecology* 182: 329–349.

Keany, J., Kennett, J.P., 1972. Pliocene-early Pleistocene paleoclimatic history recorded in Antarctic-Subantarctic deep-sea cores. *Deep-Sea Research and Oceanographic Abstracts* 19: 529–538.

Keller, G., 1993. The Cretaceous-Tertiary boundary transition in the Antarctic Ocean and its global implications. *Marine Micropaleontology* 21: 1–45.

Kennett, J.P., 1978. The development of planktonic biogeography in the Southern Ocean during the Cenozoic. *Marine Micropaleontology* 3: 301–345.

Kennett, J.P., 1980. Paleoceanographic and biogeographic evolution of the Southern Ocean during the Cenozoic, and Cenozoic microfossil datums. *Palaeogeography, Palaeoclimatology, Palaeoecology* 31: 123–152.

Khare, N., Chaturvedi, S.K., 2007. Latitudinal variations in the abundance of planktonic foraminifera along Indian Sector of the Southern Ocean. *Journal Palaeontological Society of India* 52(1): 45–49.

Khare, N., Chaturvedi, S.K., 2012 Signatures of different water masses in δ 18O of G. bulloides and total planktonic foraminiferal assemblage across the Indian Ocean Sector of the Southern Ocean: Initial Results. *Oceanology* 52(3): 372–379.

Khare, N., Chaturvedi, S.K., 2012. Tracing the signature of various frontal systems in stable isotope (oxygen and carbon) of planktic foraminiferal species Globigerina bulloides in the Southern Ocean (Indian Sector). *Oceanologia* 54(2): 311–323.

Khare, N., Chaturvedi, S.K., 2006a. Size variations of planktonic foraminiferal population in Indian Ocean sector of Southern Ocean. *Indian Journal of Marine Sciences* 35: 221–226.

Khare, N., Chaturvedi, S.K., 2006b. Mechanical stress of planktonic foraminiferal assembleages in Indian Ocean sector of Southern Ocean. *Indian Journal of Earth Sciences* 33(1–4): 15–24.

Khare, N., Chaturvedi, S.K., Saraswat, R., 2009a. Oxygen Isotope Records of Globigerina Bulloides Across a North-south Transect in the Southwestern Indian Ocean. *Ocean Science Journal* 44(2): 117–123.

Khare, N., Mazumder, A., Govil, P., 2010. Abundance and coiling direction in planktic species Neogloboquadrina pachyderma (Ehrenberg) as indicators of hydrological conditions: evidence from N—S transect of Indian Ocean, *Current Science* 98(8): 1108–1112

Khare, N., Mazumder, A., Govil, P., 2012. Do changes in planktonic foraminifera correspond to dimorphic reproduction? *Oceanologi*a 52(3): 1–8.

Khare, N., Mazumder, A., Pawan, G., Singh, V.P., 2009c. Environmental implication on chamber accretion of Neogloboquadrina Pachyderma (Ehrenberg) in South Indian Ocean. *Geological Society of India* 73: 379–385.

Khare, N., Pawan, G., Mazumder, A., 2009b. Latitudinal trends in morphological characteristics of Neogloboquadrina pachyderma (Ehrenberg) along a north-south transect in the south-western Indian Ocean. *Geo-Marine Letters* 29: 61–69.

Khare, N., Saraswat, R., Chaturvedi, S.K., 2007. On the intriguing relationship between seawater salinity and δ18O of Globigerina bulloides at higher latitudes. *Journal Indian Geophysical Union* 11(2): 73–78.

Khare, N., Saraswat, R., Chaturvedi, S.K., 2008. Assessing the applicability of δ18O Globigerina bulloides to estimate paleotemperature from the southwestern Indian Ocean. *Current Science* 94: 255–258.

Kim, J. H. Crosta, X., Michel, E., Schouten, S., Duprat, J., Damste, J. S. S., 2009. Impact of lateral transport on organic proxies in the Southern Ocean. *Quaternary Research* 71(2): 246–250.

Klaas, C., 2001. Spring distribution of larger (>64 μm) protozoans in the Atlantic sector of the Southern Ocean. *Deep-Sea Research I* 48: 1627–1649.

Krasheninnikov, V.A., Basov, I.A., 1986. Late Mesozoic and Cenozoic stratigraphy and geological history of the South Atlantic high latitudes. *Palaeogeography, Palaeoclimatology, Palaeoecology* 55: 145–188.

Kucera, M., Malmgren, B.A., Sturesson, U., 1997. Foraminiferal dissolution at shallow depths of the Walvis Ridge and Rio Grande Rise during the latest Cretaceous: Inferences for deep-water circulation in the South Atlantic. *Palaeogeography, Palaeoclimatology, Palaeoecology* 129(3–4): 195–212.

Kurihara, K., Kennett, J.P., 1992. Paleoceanographic significance of Neogene benthic foraminiferal changes in a southwest Pacific bathyal depth transect. *Marine Micropaleontology* 19: 181–189.

Labeyrie, L.D., Pichon, J.J., Labracherie, M., Ippolito, P., Duprat, J., Duplessy, J.C., 1986. Melting history of Antarctica during the past 60,000 years. *Nature* 322: 701–706.

Labeyrie, L., Duplessy, J., Grojean, M., 1984a. *The Antarctic ocean at the last glacial maximum: Analysis of $^{18}O/^{16}O$ in fossil diatoms and foraminifera. Reunion des Sciences de la Terre*. Institut Geologie Bassin Aquitaine, Talence (France), and S.G.F., Paris.

Labeyrie, L., Juillet, A., Duplessy, J., 1984b. Oxygen isotopic stratigraphy: Fossil diatoms vs foraminifera. In: Mann D. G. (Ed.), *Proceedings of the seventh international diatom symposium*. Koenigstein: Otto Koeltz, pp. 477–491.

Leckie, R.M., Webb, P.N., 1983. Late oligocene-early Miocene glacial record of the Ross Sea, Antarctica: Evidence from DSDP site 270. *Geology* 11: 578–582.

Lena, H., 1980. Benthonic foraminifera from the north-west sector of the Antarctic peninsula. *Physis. Secciones A, B, C* 39: 9–20.

Lena, H., 1981. Benthonic foraminifera collected by the R/V hero near Adelaide, Anvers, Brabant Islands, 1972–73. *Antarctic Journal of the United States* 15: 135–136.

Lena, H., 1981. Preliminary studies of planktonic foraminifera in surface sediments from the South Atlantic Ocean. *Antarctic Journal of the United States* 15: 101–102.

Li, B., Yoon, H.I., Park, B.K., 2000. Foraminiferal assemblages and CaCO3 dissolution since the last deglaciation in the Maxwell Bay, King George Island, Antarctica. *Marine Geology* 169: 239–257.

Li, W., Chen, Y., 1987. Preliminary study on the surface sediment foraminifera from the western sea area of Antarctic Peninsula. Marine geology and Quaternary geology/Haiyang Dizhi Yu Disiji Dizhi. *Qingdao* 7: 67–80.

Licht, K., 2000. *Investigations into the late quaternary history of the Ross Sea, Antarctica*. Ph.D. Thesis, University of Colorado, Boulder, 216 pp. Referenced in Dissertation Abstracts International Part B: Science and Engineering [Diss. Abst. Int. Pt. B—Sci. and Eng.]. 60: 3162.

Lindenberg, H.G., Auras, A., 1984. Distribution of arenaceous foraminifera in-depth profiles of the Southern Ocean (Kerguelen Plateau area). *Palaeogeography, Palaeoclimatology, Palaeoecology* 48: 61–106.

Loutit, T.S., 1981. Late Miocene palaeoclimatology: Subantarctic water mass, Southwest Pacific. *Marine Micropaleontology* 6: 1–27.

Loutit, T.S., Kennett, J.P., Savin, S.M., 1983. Miocene equatorial and southwest Pacific paleoceanography from stable isotope evidence. *Marine Micropaleontology* 8: 215–233.

Lovell, L.L., Trego, K.D., 2003. The epibenthic megafaunal and benthic infaunal invertebrates of port foster, Deception Island (South Shetland Islands, Antarctica). *Deep-Sea Research* (Part II, Topical Studies in Oceanography) 50: 1799–1819.

Lower, C., Cann, J., Haynes, D., 2013. Microfossil evidence for salinity events in the Holocene Coorong Lagoon, South Australia. *Australian Journal of Earth Sciences* 60(5): 573–587.

Lu, G., Keller, G., 1993. The palaeocene-eocene transition in the Antarctic Indian Ocean: Inference from planktic foraminifera. *Marine Micropaleontology* 21: 101–142.

Lucchi, R., Rebesco, G., Camerlenghi, M., Busetti, A., Tomadin, M.L., Villa, G., 2002. Mid-late Pleistocene glacimarine sedimentary processes of a high-latitude, deep-sea sediment drift (Antarctic Peninsula Pacific margin). *Marine Geology* 189: 343–370.

Luz, B., 1977. Late Pleistocene paleoclimates of the South Pacific based on statistical analysis of planktonic foraminifers. *Palaeogeography, Palaeoclimatology, Palaeoecology* 22: 61–78.

Mackensen, A., Douglas, R.G., 1989. Down-core distribution of live and dead deep-water benthic foraminifera in box cores from the Weddell Sea and the California continental borderland. *Deep-Sea Research Part A. Oceanographic Research Papers* 36: 879–900.

Mackensen, A., Ehrmann, W.U., 1992. Middle Eocene through early Oligocene climate history and paleoceanography in the Southern Ocean: Stable oxygen and carbon isotopes from ODP sites on Maud Rise and Kerguelen Plateau. *Marine Geology* 108: 1–27.

Mackensen, A., Fu¨tterer, D.K., Grobe, H., Schmiedl, G., 1993. Benthic foraminiferal assemblages from the eastern South Atlantic Polar Front region between 35° and 57°S: Distribution, ecology and fossilization potential. *Marine Micropaleontology* 22: 33–69.

Mackensen, A., Grobe, H., Hubberten, H.W., Spiess, V., Fütterer, D.K., 1989. Stable isotope stratigraphy from the Antarctic continental margin during the last one million years. *Marine Geology* 87: 315–321.

Mackensen, A., Grobe, H., Kuhn, G., Futterer, D.K., 1990. Benthic foraminiferal assemblages from the eastern Weddell Sea between 68 and 73°S: Distribution, ecology and fossilization potential. *Marine Micropaleontology* 16: 241–283.

Mackensen, A., Rudolph, M., Kuhn, G., 2001. Late Pleistocene deep-water circulation in the subantarctic eastern Atlantic. *Global and Planetary Change* 30: 197–229.

Mackensen, A., Schmiedl, G., Harloff, J., Giese, M., 1995. Deep-sea foraminifera in the South Atlantic Ocean: Ecology and assemblage generation. *Micropaleontology* 41: 342–358.

MacLeod, N., 1993. The Maastrichtian-Danian radiation of triserial and biserial planktic foraminifera: Testing phylogenetic and adaptational hypotheses in the (micro)fossil record. *Marine Micropaleontology* 21: 47–100.

Majewski, W., 2002. Mid-Miocene invasion of ecological niches by planktonic foraminifera of the Kerguelen Plateau, Antarctica. *Marine Micropaleontology* 46: 59–81.

Majewski, W., 2003. Water–depth distribution of Miocene planktonic foraminifera from ODP Site 744, Southern Indian Ocean. *Journal of Foraminiferal Research* 33: 144–154.

Malmgren, B.A., 1983. Ranking of dissolution susceptibility of planktonic foraminifera at high latitudes of the South Atlantic Ocean. *Marine Micropaleontology* 8: 183–191.

Malmgren, B.A., Kennett, J.P., 1973. Recent planktonic foraminiferal distribution in high latitudes of the South Pacific: A multivariate statistical study. *Palaeogeography, Palaeoclimatology, Palaeoecology* 14: 127–136.

Malmgren, B.A., Kennett, J.P., 1978. Test size variation in globigerina bulloides in response to quaternary palaeoceanographic changes. *Nature* 275: 123–124.

Marchant, M., Guzman-Alvis, A., 2002. Bibliographic index on aquatic biodiversity of chile: Recent planktonic foraminifera. *Ciencia y Tecnologia del Mar* (Valparaiso) 25: 163–165.

Margolis, S.V., Kroopnick, P.M., Goodney, D.E., 1977. Cenozoic and late Mesozoic paleoceanographic and paleoglacial history recorded in circum-Antarctic deep-sea sediments. *Marine Geology* 25: 131–146.

Maria, G., Fanny, M.M., Jamie, W., Andy, R., Daniela, N.S., 2022. Exploring the impact of climate change on the global distribution of non-spinose planktonic foraminifera using a trait-based ecosystem model. *Global Change Biology* 28(3): 1063–1076.

Marsh, K.V., Buddemeier, R.W., 1984. *Marine plankton as an indicator of low-level radionuclide contamination in the Southern Ocean.* Lawrence Livermore National Lab., CA (USA), NTIS Order No.: DE84016306/GAR; Contract W-7405-ENG-48.

Martınez, J.I., Deckker, De, P.T., Barrows, T., 1999. Palaeoceanography of the last glacial maximum in the eastern Indian Ocean: planktonic foraminiferal evidence. *Palaeogeography, Palaeoclimatology, Palaeoecology* 147: 73–99.

Matsumoto, K., Lynch-Stieglitz, J., Anderson, R.F., 2001. Similar glacial and Holocene Southern Ocean hydrography. *Paleoceanography* 16: 1–10.

Mazaud, A., Sicre, M.A., Ezat, U., Pichon, J.J., Duprat, J., Laj, C., Kissel, C., Beaufort, L., Michel, E., Turon, J.L., 2002. Geomagnetic-assisted stratigraphy and sea surface temperature changes in core MD94–103 (Southern Indian Ocean): possible implications for North-South climatic relationships around H4. *Earth and Planetary Science Letters* 201: 159–170.

Mazumder, A., Khare, N., Govil, P., 2009. Cosmopolitanism of the planktic foraminiferal species Globigerinita glutinata—A testimony by Q-mode cluster analysis. *International Journal of Geology* 3(1): 1–7.

Mazumder, A., Khare, N., Govil, P. 2014. *The Interdependency of the Morphological Variations of the Planktonic Foraminiferal Species Globigerina bulloides in Surface Sediments on the Environmental Parameters of the Southwestern Indian Ocean.* International Scholarly Research Notices, Article ID 621479, 9 pages. http://dx.doi.org/10.1155/2014/621479

Mazumder, A., Khare, N., Govil, P., 2016. Statistical Approach to Correlate the Morphological Variations in Foraminifera Neogloboquadrina Pachyderma with the Hydrological Parameters in Surface Sediments along NS Transect in SW Indian Ocean. International *Journal of Earth Sciences and Engineering* 09(03): 966–972.

McGowran, B., Li, Q., Cann, J., Padley, D., McKirdy, D.M., Shafik, S., 1997. Biogeographic impact of the Leeuwin current in southern Australia since the late middle Eocene. *Palaeogeography, Palaeoclimatology, Palaeoecology* 136: 19–40.

McKnight, W.M.J., 1962. The distribution of foraminifera off parts of the Antarctic coast. *Bulletin of American Paleontology* 44: 65–158.

Mead, G.A., 1985. Recent benthic foraminifera in the polar front region of the southwest Atlantic. *Micropaleontology* 31: 221–248.

Mead, G.A., Hodell, D.A., 1995. Controls on the $^{87}Sr/^{86}Sr$ composition of seawater from the middle Eocene to Oligocene: Hole 689B, Maud Rise, Antarctica. *Paleoceanography* 10: 327–346.

Mead, G.A., Kennett, J.P., 1987. The distribution of recent benthic foraminifera in the polar front region, southwest Atlantic. *Marine Micropaleontology* 11: 343–361.

Meir, A., Jonathan, E., 2017. The onset of modern-like Atlantic meridional overturning circulation at the Eocene-Oligocene transition: Evidence, causes, and possible implications for global cooling. *Geochemistry, Geophysics, Geosystems* 18(6): 2177–2199.

Melguen, M., Debrabant, P., Chamley, H., Maillot, H., Hoffert, M., Courtois, C., 1978. Influence of deep currents on sedimentary features of the Vema channel (south Atlantic Ocean) at the end of the Cenozoic era. *Bulletin de la Société Géologique de France* 20: 121–136.

Melguen, M., Thiede, J., 1974. Facies distribution and dissolution depths of surface sediment components from the Vema channel and the Rio Grande rise (Southwest Atlantic Ocean). *Marine Geology* 17: 341–353.

Melis, R., Colizza, E., Pizzolato, F., Rosso, A., 2002. Preliminary study of the calcareous taphocoenoses in Late Quaternary glacial marine sequences of the Ross Sea (Antarctica). *Geobios, Mémoire Special* 24.

Melles, M., Kuhn, G., Fuetterer, D.K., Meischner, D., 1994. Processes of modern sedimentation in the southern Weddell sea, Antarctica: evidence from surface sediments. *Polarforschung, Bremerhaven* 64: 45–74.

Milam, R.W., Anderson, J.B., 1981. Distribution and ecology of recent benthonic foraminifera of the Adelie-George V continental shelf and slope, Antarctica. *Marine Micropaleontology* 6: 297–325.

Mohan, R., Shetye, S.S., Tiwari, M., Anilkumar, N., 2015. Secondary calcification of planktic foraminifera from the Indian sector of Southern Ocean. *Acta Geologica Sinica* 89 (1): 27–37.

Morigi, C., Capotondi, L., Giglio, F., Langone, L., Brilli, M., Turi, B., Ravaioli, M., 2003. A possible record of the Younger Dryas event in deep-sea sediments of the Southern Ocean (Pacific sector). *Palaeogeography, Palaeoclimatology, Palaeoecology* 198: 265–278.

Moy, A.D., Howard, W.R., Bray, S.G., Trull, T.W., 2009. Reduced calcification in modern Southern Ocean planktonic foraminifera. *Nature Geoscience* 2(4): 276–280.

Murphy, L., Warnke, D.A., Andersson, C., Channell, J., Stoner, J., 2002. History of ice rafting at south Atlantic ODP site 177–1092 during the Gauss and late Gilbert chrons. *Palaeogeography, Palaeoclimatology, Palaeoecology* 182: 183–196.

Murray, J.W., Pudsey, C.J., 2004. Living (stained) and dead foraminifera from the newly ice-free Larsen Ice Shelf, Weddell Sea, Antarctica: Ecology and taphonomy. *Marine Micropaleontology* 53: 67–81.

Naik, D.K., Saraswat, R., Khare, N., Pandey, A.C., Nigam, R., 2014. Hydrographic changes in the Agulhas Recirculation Region during the late Quaternary. *Climate of the Past* 10 (2): 745.

Nancy, H.W., 1983. Fourier shape analysis of globorotalia truncatulinoides from late quaternary sediments in the southern Indian Ocean. *Marine Micropaleontology* 8: 1–15.

Nelson, C.S., Hendy, I.L., Neil, H.L., Hendy, C.H., Weaver, P.P.E., 2000. Last glacial jetting of cold waters through the Subtropical Convergence Zone in the Southwest Pacific off eastern New Zealand, and some geological implications. *Palaeogeography, Palaeoclimatology, Palaeoecology* 156: 103–121.

Niebler, H., 1995. *Reconstruction of paleo-environmental parameters using stable isotopes and faunal assemblages of planktonic foraminifera in the south Atlantic ocean.* Reports on polar research, 1995.

Niebler, H.S., Gersonde, R., 1998. A planktic foraminiferal transfer function for the southern South Atlantic Ocean. *Marine Micropaleontology* 34: 213–234.

Nomura, R., 1995. Paleogene to Neogene deep-sea paleoceanography in the eastern Indian ocean: Benthic foraminifera from ODP sites 747, 757 and 758. *Micropaleontology* 41: 251–290.

Northcote, L.C., Neil, H.L., 2005. Seasonal variations in foraminiferal flux in the Southern ocean, Campbell Plateau, New Zealand. *Marine Micropaleontology* 56: 122–137.

Nowlin Jr, W.D., Clifford, M., 1982. The kinematic and thermohaline zonation of the Antarctic Circumpolar Current at Drake Passage. *Journal of Marine Research* 40: 481–507.

Okada, H., Wells, P., 1997. Late Quaternary nannofossil indicators of climate change in two deep-sea cores associated with the Leeuwin Current off Western Australia. *Palaeogeography, Palaeoclimatology, Palaeoecology* 131: 413 432

Oppo, D.W., Horowitz, M., 2000. Glacial deep water geometry: South Atlantic benthic foraminiferal Cd/Ca and δ13C evidence. *Paleoceanography* 15: 147–160.

Orsi, A.H., Whiteworth III, T., Nowlin Jr, W.D., 1995. On the meridional extent and fronts of the Antarctic Circumpolar Current. *Deep-sea Research I* (40): 641–673.

Oslick, J., 1998. *Reconstructing hydrographic gradients and the carbon isotopic composition of surface waters using multiple species of planktonic foraminifera.* Ph.D. Thesis, Brown University, 210 pp.

Pandey, P. C., Khare, N., Sudhakar, M. 2006. Oceanographic research: Indian efforts and preliminary results from the Southern Ocean. *Current Science* 90(7): 978–984.

Parker, F.L., Berger, W.H., 1971. Faunal and solution patterns of planktonic foraminifera in surface sediments of the South Pacific. *Deep-Sea Research and Oceanographic Abstracts* 18: 73–107.

Passchier, S., Krissek, L.A., 2008. Oligocene-Miocene Antarctic continental weathering record and paleoclimatic implications, Cape Roberts Drilling Project, Ross Sea, Antarctica. *Palaeogeography, Palaeoclimatology, Palaeoecology* 260(1–2): 30–40.

Pawlowski, J., Cornelius, N., Lecroq, B., Longet, D., Fahrni, J., Cedhagen, T., Gooday, A., 2006. Bipolar gene flow in deep-sea foraminifera. In Thatje, S; Tyler, P; Talbot, P; Horton, T; Maclaren, L; Murty, S; Tothe, N; Billett, D (Eds.), *International Deep-Sea Biology Symp., Southampton*, (UK), 9–14, July 2006. National Oceanography Center, Southampton.

Pawlowski, J., Fahrni, J.F., Guiard, J., Conlan, K., Hardecker, J., Habura, A., 2005. Allogromiid foraminifera and gromiids from under the Ross Ice Shelf: Morphological and molecular diversity. *Polar Biology* 28: 514–522.

Pawlowski, J., Fahrni, J.I., Brykczynska, U., Habura, A., Bowser, S.I., 2002. Molecular data reveal high taxonomic diversity of allogromiid foraminifera in Explorers cove (McMurdo Sound, Antarctica). *Polar Biology* 25: 96–105.

Pena, L.D., Goldstein, S.L., Hemming, Jones, K.M., Calvo, E., Pelejero, C., Cacho, I., 2013. Rapid changes in meridional advection of Southern Ocean intermediate waters to the tropical Pacific during the last 30kyr. *Earth and Planetary Science Letters* 368: 20–32.

Perry, P.J., 2000. *Quaternary benthic foraminiferal distribution in the Ross Sea, Antarctica, and its relationship to oceanography.* Ph.D. Thesis, Louisiana State University and Agricultural and Mechanical Col., LA, USA. *Referenced in Dissertation Abstracts International Part B: Science and Engineering* [Diss. Abst. Int. Pt. B—Sci. and Eng.]. 60: 4486.

Petersen, S.V., Schrag, D.P., 2015. Antarctic ice growth before and after the Eocene-Oligocene transition: New estimates from clumped isotope paleothermometry. *Paleoceanography* 30 (10): 1305–1317.

Peterson, L.C., Lohmann, G.P., 1982. Major change in Atlantic Deep and bottom waters 700,000 yr ago: Benthonic foraminiferal evidence from the South Atlantic. *Quaternary Research* 17: 26–38.

Petrizzo, M.R., Kenneth, G.M., David, K.W., 2022. Late Cretaceous Paleoceanographic Evolution and the Onset of Cooling in the Santonian at Southern High Latitudes (IODP Site U1513, SE Indian Ocean). *Paleoceanography and Paleoclimatology* 37 (1). http://dx.doi.org/10.1029/2021PA004353

Petrizzo, M.R., 2000. Upper Turonian—lower Campanian planktonic foraminifera from southern mid-high latitudes (Exmouth Plateau, NW Australia): biostratigraphy and taxonomic notes. *Cretaceous Research* 21: 479–505.

Petrizzo, M.R., 2001, Late Cretaceous planktonic foraminifera from Kerguelen Plateau (ODP Leg 183): new data to improve the Southern Ocean biozonation. *Cretaceous Research* 22: 829–855.

Pharr, R.B., Douglas, Williams, F., 1987. Shape changes in Globorotalia truncatulinoides as a function of ontogeny and paleobiogeography in the Southern Ocean. *Marine Micropaleontology* 12: 343–355.

Pichon, J., Sikes, E.L., Hiramatsu, C., Robertson, L., 1998. Comparison of U^{k}_{37} and diatom assemblage sea surface temperature estimates with atlas derived data in Holocene sediments from the southern west Indian ocean. *Journal of Marine Systems* 17: 541–554.

Pilskaln, C.H., Manganini, S.J., Trull, T.W.L., Armand, W., Howard, V.L., Asper, R., 2004. Massom Geochemical particle fluxes in the southern dan ocean seasonal ice zone: Prydz Bay region, East Antarctica. *Deep-Sea Research* I 51: 307–332.

Piscova, A., Roman, M., Bulínová, M., Pokorný, M., Sanderson, D., Cresswell, A., Lirio, J.M., Coria, S.H., Nedbalová, L., Lami, A., Musazzi, S., Vijver, B.V., Nývlt, D., Kopalová, K., 2019. Late-Holocene palaeoenvironmental changes at Lake Esmeralda (Vega Island, Antarctic Peninsula) based on a multi-proxy analysis of laminated lake sediment. *The Holocene* 29(7): 1155–1175.

Pollard, R. T., Lucas, M. I., Read, J. F., 2002. Physical controls on biogeochemical zonation in the Southern Ocean. *Deep Sea Research, Part II* 49, 3289–3305. https://doi.org/10.1016/S0967-0645(02)00084-X.

Pudsey, C.J., Murray, J.W., Appleby, P., Evans, J., 2006. Ice shelf history from petrographic and foraminiferal evidence, Northeast Antarctic Peninsula. *Quaternary Science Reviews* 25: 2357–2379.

Pudsey, C.J., 2000. Sedimentation on the continental rise west of the Antarctic Peninsula over the last three glacial cycles. *Marine Geology* 167: 313–338.

Pushkin, A.F., Smirnov, I.S., Neelov, A.B., 1991. A contribution to the study of bottom fauna from the fish tail bay of the bunger oasis (East Antarctica). Informatsionnyj byulleten' Sovetskoj Antarkticheskoj ehkspeditsii. *St. Petersburg* 116: 24–27.

Quilty, P.G., 2001. Reworked Paleocene and Eocene foraminifera, Mac. Robertson Shelf, East Antarctica: Paleoenvironmental implications. *Journal of Foraminiferal Research* 31: 369–384.

Rathburn, A.E., De Deckker, P., 1997. Magnesium and strontium compositions of recent benthic foraminifera from the Coral Sea, Australia and Prydz Bay, Antarctica. *Marine Micropaleontology* 32: 231–248.

Rathburn, A.E., Pichon, J.J., Ayress, M.A., Deckker, De P., 1997. Microfossil and stable-isotope evidence for changes in Late Holocene palaeoproductivity and palaeoceanographic conditions in the Prydz Bay region of Antarctica. *Palaeogeography, Palaeoclimatology, Palaeoecology* 131: 485–510.

Rebecca, T.M., Wojciech, M., John, B.A., Yusuke, Y., Rodrigo, F., Martin, J., 2017. Oceanographic influences on the stability of the Cosgrove Ice Shelf, Antarctica. *Holocene* 27 (11): 1645–1658.

Rickaby, R.E.M., Elderfield, H., 1999. Planktonic foraminiferal Cd/Ca: Paleonutrients or paleotemperature? *Paleoceanography* 14: 293–303.

Rintoul, S.R., Wolff, J.O., Griffiths, F.B., Bindoff, N.L., Church, J.A., Tilbrook, B.D., Parslow, J.S., Rosenberg, M., 2002. Southern Ocean processes and climate: Recent progress by ANARE. In: Marchant, H. J. Lugg, D. J. and Quilty, P. G. (Eds), *Australian Antarctic science: The first 50 years of ANARE.* Australian Antarctic Division, Kingston, Tasmania, pp. 519–540.

Robert, C., Diester-Haass, L., Paturel, J., 2005. Clay mineral assemblages, siliciclastic input and paleoproductivity at ODP Site 1085 off Southwest Africa: A late Miocene-early Pliocene history of Orange River discharges and Benguela current activity, and their relation to global sea-level change. *Marine Geology* 216: 221–238.

Robert, C., Haass, D.L., Chamley, H., 2002.Late Eocene-Oligocene oceanographic development at southern high latitudes, from terrigenous and biogenic particles: a comparison of Kerguelen Plateau and Maud Rise, ODP Sites 744 and 689. *Marine Geology* 191: 37–54.

Roberts, D., Howard, W.R., Moy, A.D., Roberts, J.L., Trull, T.W., Bray, S.G., Hopcroft, R.R., 2008. Interannual variability of pteropod shell weights in the high-CO2 Southern Ocean. *Biogeosciences Discussions* 5(6): 4453–4480.

Roberts, D., van Ommen, T.D., McMinn, A., Morgan, V., Roberts, J.L., 2001. Late-Holocene East Antarctic climate trends from ice-core and lake-sediment proxies. *The Holocene* 11(1): 117–120.

Robinson, R.S., Mix, A., Martinez, P., 2007. Southern Ocean control on the extent of denitrification in the southeast Pacific over the last 70 ka. *Quaternary Science Reviews* 26: 201–212.

Sabbatini, A., Morigi, C., Ravaioli, M., Negri, A., 2004. Abyssal benthic foraminfera in the polar front region (pacific sector): Faunal composition, standing stock and size structure. *Chemistry and Ecology* 20: S117-S129.

Saidova, K., 1998. Benthic foraminifera communities of the Southern Ocean. *Okeanologiya* 38: 561–567.

Salvi, C., Busetti, M., Marinoni, L., Brambati, A., 2006. Late Quaternary glacial marine to marine sedimentation in the Pennell Trough (Ross Sea, Antarctica). *Palaeogeography, Palaeoclimatology, Palaeoecology* 231: 199–214.

Sandi, M.S., Sarah, E.F., Haojia, R., Ralf, S., Emily, M.T., Martínez-García, A., Luca, S., Alakendra, R., Gerald, H.H., Daniel, M.S., 2020. The Nitrogen Isotopic Composition of Tissue and Shell-Bound Organic Matter of Planktic Foraminifera in Southern Ocean Surface Waters. *Geochemistry, Geophysics, Geosystems* 21 (2). http://dx.doi.org/10.1029/2019GC008440

Saraswat, R., Khare, N., Chaturvedi, S.K., Rajkumar, A., 2007. Seawater pH and planktic foraminiferal abundance: Preliminary observations from the western Indian Ocean. *Current Science* 93(5): 703–706.

Scheibnerova, V., 1977. Synthesis of the cretaceous benthonic foraminifera recovered by the deep sea drilling project in the Indian Ocean. In Heirtzler J.R., Bolli H.M., Davies T.A., Saunders J.B. and Sclater J.G. (Eds.), *Indian Ocean geology and biostratigraphy: Studies following Deep-Sea Drilling Legs* 22–29. Washington, DC (USA).: Publ. by American Geophysical Union.

Scher, H. D., Delaney, M. L., 2010. *Evaluating the Source of Nd to the Southern Ocean During the Eocene Oligocene Transition.* American Geophysical Union, URL:www.agu.org

Scherer, R.P., 1991. Quaternary and Tertiary microfossils from beneath Ice Stream B: Evidence for a dynamic West Antarctic Ice Sheet History. *Global and Planetary Change* 4: 395–412.

Schrag, D.P., Adkins, J.F., McIntyre, K., Alexander, J.L., Hodell, D.A., Charles, C.D., 2002. The oxygen isotopic composition of seawater during the last glacial maximum. *Quaternary Science Reviews* 21: 331–342.

Schroder, C.J., 1990. High latitude agglutinated foraminifera: Prydz Bay (Antarctica) vs. Lancaster Sound (Canadian Arctic). In: Hemleben, C. (Ed.), *Paleoecology, Biostratigraphy, Paleoceanography and Taxonomy of Agglutinated Foraminifera.* Kluwer Academic Publishers, Netherlands, pp. 315–343.

Schumacher, S., 2001. Microhabitat preferences of benthic foraminifera in South Atlantic Ocean sediments. Bremerhaven (FRG): Alfred-Wegener-Institut fuer Polar- und Meeresforschung. *Ber. Polarforsch./Rep. Polar Res.* 403: 151, 2001.

Schumacher, S., Lazarus, D., 2004. Regional differences in pelagic productivity in the late Eocene to early Oligocene-a comparison of southern high latitudes and lower latitudes. *Palaeogeography, Palaeoclimatology, Palaeoecology* 214: 243–263.

Scott, G.H., Kennett, J.P., Wilson, K.J., Hayward, B.W., 2007. *Globorotalia puncticulata*: population divergence, dispersal and extinction related to Pliocene-Quaternary water masses. *Marine Micropaleontology* 62: 235–253

Shchedrina, Z.G., 1982. Vertical distribution of agglutinated foraminifera in the southern hemisphere from material collected by the soviet antarctic expedition, 1955–1966. Antarktika. *Doklady Komissii* 21: 133–156.

Shemesh, A., Charles, C.D., Fairbanks, R.G., 1992. Oxygen isotopes in biogenic silica: Global changes in ocean temperature and isotopic composition. *Science (Washington)* 256: 1434–1436.

Shemesh, A., Macko, S.A., Charles, C.D., Rau, G.H., 1993. Isotopic evidence for reduced productivity in the glacial Southern Ocean. *Science* 262: 407–410.

Sicre, M.A., Labeyrie, L., Ezat, U., Duprat, J., Turon, J.L., Schmidt, S., Michel, E., Mazaud, A., 2005. Mid-latitude Southern Indian Ocean response to Northern Hemisphere Heinrich events. *Earth and Planetary Science Letters* 240: 724–731.

Sikes, E., King, A., Howard, W., 2001. *Sediment trap evidence for seasonality-based differences in alkenone and foraminiferal temperature estimates in the Southern Ocean.* EOS Trans. Am. Geophys. Union, v. 82, suppl., [np]. suppl.

Sissingh, W., 1973. Foraminifera of the Breid Bay area (Antarctica). *Netherlands Journal of Sea Research* 6: 355–364.

Smith, J., Bentley, A., Michael, J., Hodgson, D.A., Roberts, S.J., Leng, M.J., Lloyd, J.M., Barrett, M.S., Bryant, C.S., David, E., 2007. Oceanic and atmospheric forcing of early Holocene ice shelf retreat, George VI Ice Shelf, Antarctica Peninsula. *Quaternary Science Reviews* 26(3–4): 500–516.

Stott, L.D., Kennett, J.P., 1989. New constraints on early Tertiary palaeo-productivity from carbon isotopes in foraminifera. *Nature* 342: 526–529.

Suhr, S.B., Pond, D.W., 2006. Antarctic benthic foraminifera facilitates rapid cycling of phytoplankton-derived organic carbon. *Deep-Sea Research II* 53: 895–902.

Suhr, S.B., Pond, D.W., Gooday, A.J., Smith, C.R., 2003. Selective feeding by benthic foraminifera on phytodetritus on the western Antarctic Peninsula shelf: Evidence from fatty acid biomarker analysis. *Marine Ecology Progress Series* 262: 153–162.

Takahashi, K.T., Hosie, G.W., Kitchener, J.A., McLeod, D.J., Odate, T., Fukuchi, M., 2010 August. Comparison of zooplankton distribution patterns between four seasons in the Indian Ocean sector of the Southern Ocean. *Polar Science* 4(2): 317–331.

Tapia, R., Nuernberg, D., Ronge, T., Tiedemann, R., 2015. Disparities in glacial advection of Southern Ocean Intermediate Water to the South Pacific Gyre. *Earth and Planetary Science Letters* 40: 152–164.

Temnikow, N.K., Lipps, J.H., 1975. Foraminiferal ecology: R/V Hero cruise 75–1a. *Antarct. J.U.S* 10: 132–133.

Theissen, K.M., Robert, B., Dunbar, A.K., Cooper, Mucciarone, D.A., Hoffmann, D., 2003. The Pleistocene evolution of the East Antarctic Ice Sheet in the Prydz Bay region: stable isotopic evidence from ODP Site 1167. *Global and Planetary Change* 39: 227–256.

Theyer, F., 1971. Benthic foraminiferal trends, Pacific-Antarctic basin. Deep-Sea Research and Oceanographic Abstracts 18: 723–738.

Thomas, A.R., Silke, S., Ralf, T., Matthias, P., Ute, M., Dirk, N., Gerhard, K., 2015. Pushing the boundaries: Glacial/interglacial variability of intermediate and deep waters in the southwest Pacific over the last 350,000 years. *Paleoceanography* 30(2): 23–38.

Thomas, J.W., Hillenbrand, C., Alexander M P., Claire S A., Thomas, F., James A S., Werner, E., David A H., 2019. Paleocirculation and Ventilation History of Southern Ocean Sourced Deep Water Masses During the Last 800,000 Years. *Paleoceanography and Paleoclimatology* 34 (5): 833–852.

Thomas, W.T., Abraham, P., Diana, M.D., Tim, S., Kate, B., Bronte, T., 2018. Distribution of planktonic biogenic carbonate organisms in the Southern Ocean south of Australia: a baseline for ocean acidification impact assessment. *Biogeosciences* 15 (1): 31–49.

Timothy, J, B., Katrin, J.M., Kaitlin, A., Deborah, J.T., 2014. The dynamics of global change at the Paleocene-Eocene thermal maximum: A data-model comparison. *Geophysics, Geosystems* 15(10): 3830–3848.

Tiwari, M., Mohan, R., Meloth, T., Naik, S .S., Sudhakar, M., 2011. Effect of varying frontal systems on stable oxygen and carbon isotopic compositions of modern planktic foraminifera of Southern Ocean. *Current Science* 100(6): 881–887.

Trenberth, K.E., Large, W.G., Olson, J.G., 1990. The mean annual cycle in global ocean wind stress. *Journal of Physical Oceanography* 30: 1742–1760.

Vargas, De., Pawlowski, C., Renaud, J.S., Hilbrecht, H., 2001. Pleistocene adaptive radiation in globorotalia truncatulinoides: Genetic, morphologic, and environmental evidence. *Paleobiology* 27: 104–125.

Villa, G., Persico, D., Bonci, M.C., Lucchi, R.G., Morigi, C., Rebesco, M., 2003. Biostratigraphic characterization and quaternary microfossil palaeoecology in sediment drifts west of the Antarctic peninsula—implications for cyclic glacial-interglacial deposition. *Palaeogeography, Palaeoclimatology, Palaeoecology* 198: 237–263.

Volbers, A.N.A., Henrich, R., 2002. Present water mass calcium carbonate corrosiveness in the eastern South Atlantic inferred from the ultrastructural breakdown of Globigerina bulloides in surface sediments. *Marine Geology* 186: 471–486.

Volbers, A.N.A., Henrich, R., 2004. Calcium carbonate corrosiveness in the South Atlantic during the Last Glacial Maximum as inferred from changes in the preservation of Globigerina bulloides: A proxy to determine deep-water circulation patterns? M*arine Geology* 204: 43–57.

Vorobeva, L., 1998. Meiobenthos of the Ukrainian Antarctic station Akademik Vernadsky region. In Gozhik P. F. (Ed.), *Second Ukrainian Antarctic Expedition Report 1997/1998*, no. 2 Ukrainian Antarctic Centre [Byul. Ukr. Antarkt. Tsent./Bull. Ukr. Antarct. Cent.], Kiev (Ukraine), pp. 191–197.

Waelbroeck, C., Labeyrie, L., Michel, E., Duplessy, J.C., McManus, J.F., Lambeck, K., et al., 2002. Sea-level and deep water temperature changes derived from benthic foraminifera isotopic records. *Quaternary Science Reviews* 21: 295–305.

Ward, B.L., Barrett, P.J., Vella, P., 1987. Distribution and ecology of benthic foraminifera in McMurdo Sound, Antarctica. *Palaeogeography, Palaeoclimatology, Palaeoecology* 58: 139–153.

Webb, P.N., Harwood, D.M., McKelvey, B.C., Mercer, J.H., Stott, L.D., 1984. Cenozoic marine sedimentation and ice-volume variation on the East Antarctic craton. *Geology* 12: 287–291.

Webb, P.N., Strong, C., 2006. Foraminiferal biostratigraphy and palaeoecology in upper oligocene-lower miocene glacial marine sequences 9, 10, and 11, CRP-2/2A drill hole, victoria land basin, Antarctica. *Palaeogeography, Palaeoclimatology, Palaeoecology* 231: 71–100.

Webb, P.N., Harwoodb, D.M., Mabin, M.G.C., McKelveyd, B.C., 1996. A marine and terrestrial Sirius Group succession, middle Beardmore Glacier-Queen Alexandra Range, Transantarctic Mountains, Antarctica. *Marine Micropaleontology* 272: 273–297.

Wei, R., Abouchami, W., Zahn, R., Masque, P., 2016. Deep circulation changes in the South Atlantic since the Last Glacial Maximum from Nd isotope and multi-proxy records. *Earth and Planetary Science Letters* 434: 18–29.

Wells, P., Okada, H., 1997. Response of nannoplankton to major changes in sea-surface temperature and movements of hydrological fronts over site DSDP 594 (south chatham rise, southeastern New Zealand), during the last 130 kyr. *Marine Micropaleontology* 32: 341–363.

Whiteworth, T., 1980. Zonation and geostrophic flow of the Antarctic Circumpolar Current at Drake Passage. *Deep-Sea Research A* (27): 497–507.

Widmark, J.G.V., 1995. Multiple deep-water sources and trophic regimes in the latest cretaceous deep-sea: Evidence from benthic foraminifera. *Marine Micropaleontology* 26: 361–384.

Williams, D.F., Keany, J., 1978. Comparison of radiolarian/planktonic foraminiferal paleoceanography of the subantarctic Indian Ocean. *Quaternary Research* 9: 71–86.

Williams, M., Haywood, A.M., Taylor, S.P., Paul, J., Valdes, B., Sellwood, W., 2005. Claus-Dieter Hillenbrand. Evaluating the efficacy of planktonic foraminifer calcite δ18O data for sea surface temperature reconstruction for the Late Miocene. *Geobios* 38: 843–863.

Woelders, L., Vellekoop, J., Kroon, D., Smit, J., Casadio, S. Pramparo, M. B., Dinares-Turell, J., Peterse, F., Sluijs, A., Lenaerts, J.T.M., Speijer, R.P., 2017. Latest Cretaceous climatic and environmental change in the South Atlantic region. *Paleoceanography*, 32(5): 466–483.

Yoon, H.I., Park, B.K., Kim, Y., Kang, C.Y., 2002. Glaciomarine sedimentation and its paleoclimatic implications on the Antarctic Peninsula shelf over the last 15 000 years. *Palaeogeography, Palaeoclimatology, Palaeoecology* 185: 235–254.

Yoon, H.I., Park, B.K., Kim, Y., Kim, D., 2000. Glaciomarine sedimentation and its paleoceanographic implications along the fjord margins in the South Shetland Islands, Antarctica during the last 6000 years. *Palaeogeography, Palaeoclimatology, Palaeoecology* 157: 189–211.

Zachos, J.C., Breza, J.R., Wise, S.W., 1992. Early oligocene ice-sheet expansion on Antarctica: Stable isotope and sedimentological evidence from Kerguelen plateau, southern Indian ocean. *Geology* 20: 569–573.

Zheng, L., Wu, Z., 1989. *Relationship between the distribution of foraminifera and environmental factors in the northwestern sea area on the Antarctic Peninsula.* China First Symp. on Southern Ocean Expedition, Hangzhou (China).

Zhu, G., 1989. Diet analysis of Antarctic krill Euphausia superba dana. *Acta Oceanologica Sinica/Haiyang Xuebao* 8: 457–462.

Ziegler, Martin, Z., Paula, D., Ian R.H., Rainer, Z., 2013. Millennial-scale changes in atmospheric CO2 levels linked to the Southern Ocean carbon isotope gradient and dust flux. *Nature Geoscience* 6: 457–461.

Zwaan, G.J.V., Duijnstee, I.A.P., Dulk, den M., Ernst, S.R., Jannink, N.T., Kouwenhoven, T.J., 1999. Benthic foraminifers: proxies or problems? A review of paleocological concepts. *Earth-Science Reviews* 46: 213–236.

6 Tracking the Ionospheric Responses over Antarctica during the December 4, 2021, Total Solar Eclipse

A. S. Sunil,[1] K. K. Ajith,[2] P. S. Sunil,[1] and Dhanya Thomas[3]

[1] CUSAT-NCPOR Centre for Polar Sciences, Department of Marine Geology and Geophysics, School of Marine Sciences, Cochin University of Science and Technology, Kochi, India

[2] National Atmospheric Research Laboratory, Gadanki, India

[3] CSIR Fourth Paradigm Institute, Bangalore, India

CONTENTS

DOI: 10.1201/9781003284413-6

6.1 INTRODUCTION

Solar eclipses occur when the Moon aligns between Sun and Earth, blocks the solar radiation, and casts a shadow on Earth's surface. During solar eclipses, the Sun's photosphere (bright surface of the Sun) gets covered by Moon and its corona (the outermost part of the Sun's atmosphere) becomes visible. The alignment of the Sun, Moon and Earth required for the occurrence of a solar eclipse is depicted in Figure 6.1a. As shown in this schematic diagram, the alignment of these three celestial bodies should be in a perfect or near-perfect straight line for an eclipse to take place. This alignment generally happens during the new moon phase of every lunar cycle. However, solar eclipses do not occur at every new moon due to the peculiar orientation of the Moon's orbital plane. As shown in the figure, the Moon's orbital plane is tilted ~5° relative to

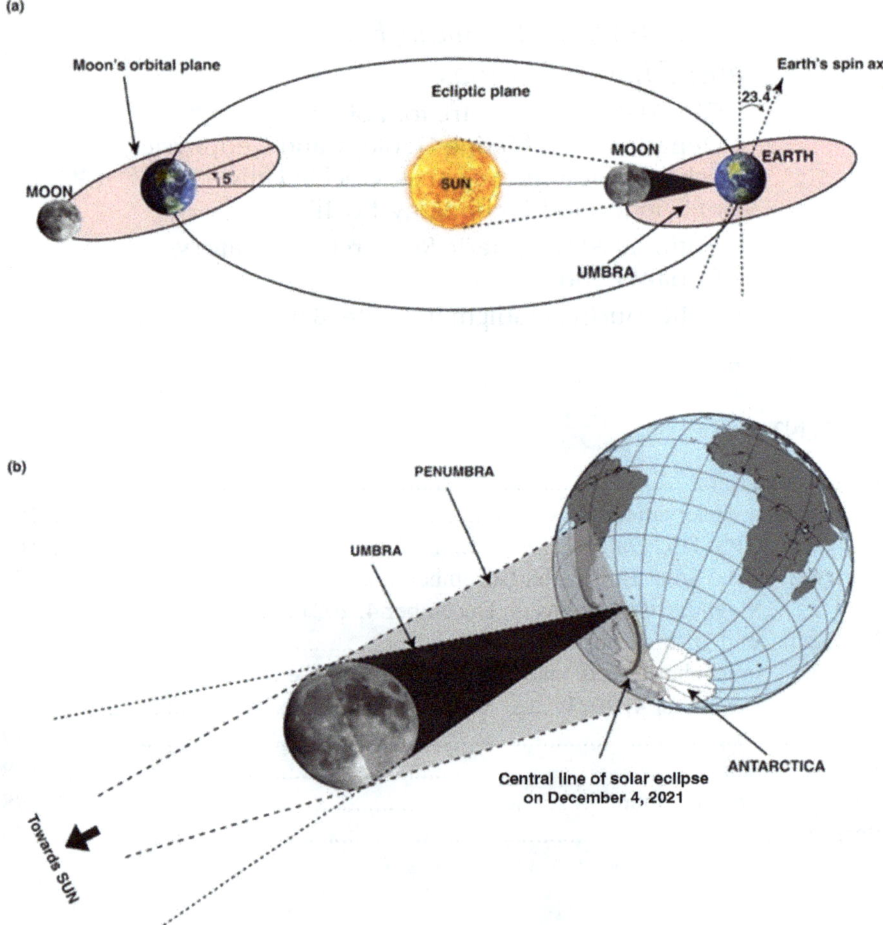

FIGURE 6.1 (a) Schematic diagram showing the mechanism of the solar eclipse. The orbital plane of the Moon is tilted ~5° to the ecliptic plane. (b) The central line of the total solar eclipse occurred on December 4, 2021, and passed through the Antarctic region.

the ecliptic plane (plane of Earth's orbit around the Sun). Lunar nodes are the points where these two planes intersect. The specific alignment needed for the occurrence of solar eclipse happens, only when the new Moon is near the lunar node. The Moon's shadow has a central region known as the umbra and an outer region called penumbra. Umbra is the darkest region of the shadow where the solar radiation gets completely blocked and penumbra is the region where the sunlight is partially obscured.

Solar eclipses are mainly classified into three categories which are total solar eclipse, annular solar eclipse, and partial solar eclipse. A total solar eclipse takes place when the Sun, Moon and Earth align in a direct line. During a total solar eclipse, the Moon orbits closest to the Earth and casts its umbra on the Earth's surface. Likewise, annular eclipses also possess the same geometry as total solar eclipses except for Moon's orbital position. During an annular eclipse, Moon orbits farthest from the Earth and the umbra does not reach the Earth's surface. As a result, the outer edge of the Sun remains visible like an annular ring. Unlike the discussed cases, a partial solar eclipse occurs when these three celestial bodies do not align in a perfectly straight line. The Moon only casts its penumbra on Earth's surface and partially covers the Sun's disk during a partial solar eclipse.

During solar eclipses, the absence of solar radiation produces dynamic changes in the Earth's ionosphere. The Ionosphere is the partially ionized region of the atmosphere where free electrons and ions are present and it ranges from ~50 km to ~1000 km altitude. The photo ionization by solar UV and EUV radiation results in the production of electrons and ions in the ionosphere (Rishbeth and Garriott, 1969; Hargreaves, 1992). During the total solar eclipse period, the blockage of solar radiation causes a sudden decrease in electron density in the shadow region of Earth's ionosphere. After the eclipse, the ionospheric electron density returns to its normal daytime conditions. Thus, the solar eclipse is a wonderful opportunity to study the ionospheric electron density variation in the absence of solar radiation for a short duration.

The dynamical changes in the ionosphere during the solar eclipse have been studied fairly well in the past (Rishbeth, 1968; Cohen, 1984; Pradipta et al., 2018; Goncharenko et al., 2018; Coster et al., 2017). Many researchers studied the ionospheric signatures of solar eclipses using various observation techniques such as ionosondes, incoherent radar systems, and satellite measurements (Zhang et al., 2020; Farges et al., 2001; Goncharenako et al., 2018). However, depending upon its occurrence time, number of events, geographical location, and availability of GPS data, very limited studies have been carried out so far in the Antarctic region (Kameda et al., 2009). In the present study, an investigation has been carried out to detect the ionospheric electron density variation over the Antarctic region during the total solar eclipse on December 4, 2021.

6.2 DATA AND OBSERVATIONS

6.2.1 GPS-TEC Data

This study mainly uses Global Positioning System (GPS) derived Total Electron Content (TEC) data to investigate the ionospheric response over Antarctica during the December 4, 2021, solar eclipse. The TEC is the total number of electrons integrated from satellite to receiver along a cylinder having a cross-sectional area of 1 m^2

(Klobuchar, 1987). That is, TEC is an integral measure of electron density from the GPS satellite to the receiver. It is measured in the unit of electrons per square meter (1 TEC unit or TECU = 10^{16} electrons/m^2). The TEC estimated along the satellite Line of Sight (LOS) is termed slant TEC (STEC). The STEC can be converted into vertical TEC (VTEC) using a standard mapping function (Mannucci et al., 1993). A total of 35 GPS stations over the Antarctic region were used in the present study (Table 6.1). The

TABLE 6.1
List of GPS Stations Used in This Study

Station code	Latitude	Longitude
BACK	74.43° S	102.50° W
BEAN	75.96° S	069.40° W
BENN	84.79° S	116.50° W
BSA1	67.82° S	067.30° W
CAPF	66.01° S	060.60° W
CLRK	77.34° S	141.90° W
CRDI	82.86° S	053.20° W
DUPT	64.80° S	062.90° W
GLDK	72.23° S	100.60° W
HAAG	77.03° S	078.30° W
HOWN	77.53° S	086.80° W
HTON	74.08° S	061.80° W
LPLY	73.11° S	090.30° W
MAJK	81.66° S	021.90° W
MCAR	76.32° S	144.40° W
MCRG	73.67° S	094.70° W
MRTP	74.18° S	115.20° W
PAL2	64.78° S	064.10° W
PIRT	81.10° S	085.20° W
PRPT	66.00° S	065.40° W
RMBO	83.87° S	066.40° W
ROBN	65.23° S	059.50° W
SGP1	65.55° S	061.80° W
SGP4	66.69° S	062.50° W
SGP5	67.28° S	064.90° W
SLTR	75.09° S	113.90° W
SPGT	64.29° S	061.10° W
STEW	84.19° S	086.30° W
THUR	72.53° S	097.60° W
TOMO	75.80° S	114.70° W
TRVE	69.99° S	067.60° W
UTHW	77.57° S	109.10° W
WHTM	82.68° S	104.40° W
WILN	80.04° S	080.60° W
WLRD	80.54° S	096.70° W

geographic locations of these stations are listed in Table 6.1 and represented as green triangles in Figure 6.2. The VTEC is estimated from the GPS data using the GPS-TEC analysis application methodology developed at Boston College (Seemala, 2011).

6.2.2 Total Solar Eclipse on December 4, 2021

The December 4, 2021, eclipse falls under the category of the total solar eclipse. Figure 6.1b shows the path of totality and its central line which moves across West Antarctica from east to west. For most the eclipses, the central line moves from

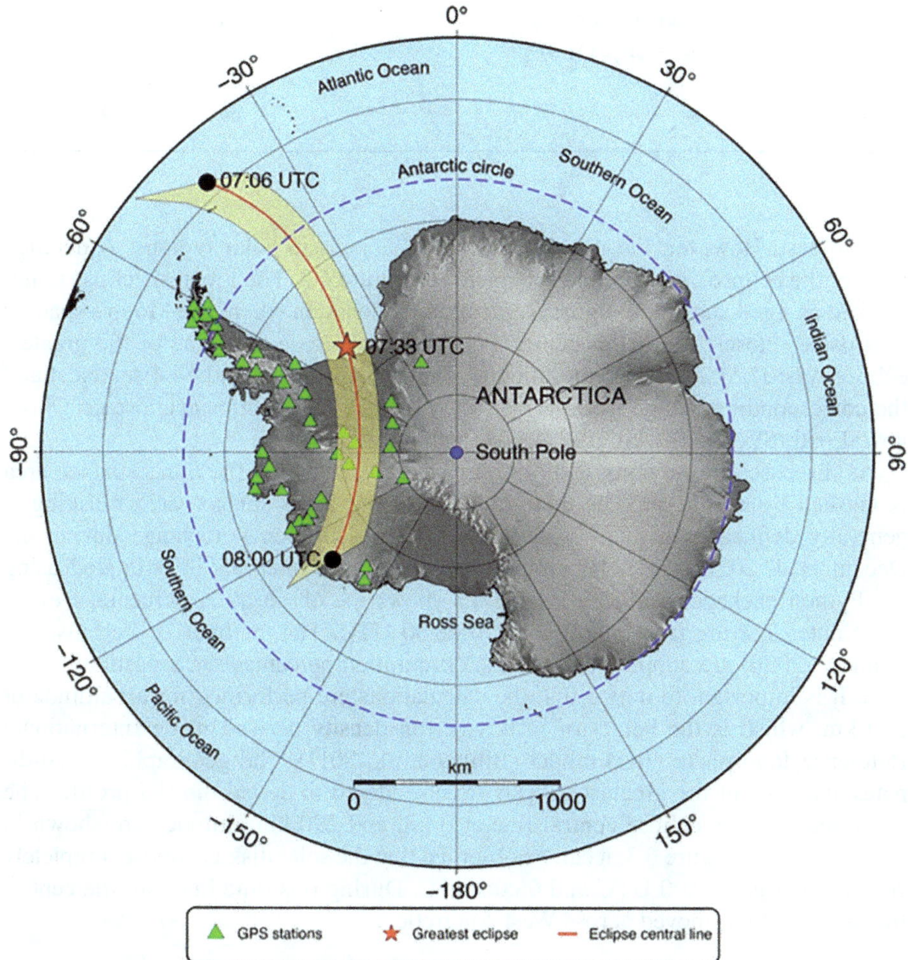

FIGURE 6.2 Trajectory of eclipse central line is shown as the red curve. The red star represents the greatest eclipse. The yellow shaded region shows the northern and southern limits of the umbra. Green triangles are GPS stations.

Eclipse data source: https://eclipse.gsfc.nasa.gov/SEpath/SEpath2001/SE2021Dec04Tpath.html.

TABLE 6.2
Coordinates of Eclipse Central Line at Altitudes 0 Km and 290 Km

Time (UTC)	Eclipse central line coordinates at 0 km altitude (latitude, longitude)	Eclipse central line coordinates at 290 km altitude (latitude, longitude)
06:50	–	(–50.20°, –35.56°)
07:00	–	(–59.20°, –24.11°)
07:10	(–62.57°, –38.36°)	(–65.29°, –19.14°)
07:20	(–69.33°, –36.59°)	(–70.56°, –15.89°)
07:30	(–75.11°, –40.53°)	(–76.03°, –13.86°)
07:40	(–80.01°, –55.63°)	(–81.35°, –13.83°)
07:50	(–81.99°, –95.27°)	(–86.80°, –25.78°)
08:00	(–76.84°, –131.69°)	(–86.39°, –156.20°)
08:10	–	(–78.72°, –164.90°)
08:17	–	(–69.22°, –158.89°)

west to east. However, the direction reversal happens in polar regions. A detailed view of the eclipse's central line is depicted in Figure 6.2. The greatest eclipse point, marked as a red star in the figure, represents the location where the Moon's shadow axis passes closest to Earth's center. For the present case, the time of the greatest eclipse was 07:33:28.2 UTC (duration of totality 01 minutes and 54.4 seconds) and the corresponding location was at 46.16° W and 76.78° S (https://eclipse.gsfc.nasa.gov/SEpath/SEpath2001/SE2021Dec04Tpath.html).

As discussed in previous sections, during a solar eclipse, the Sun's surface area is masked by the moon. The quantification of the Sun's surface area masking is generally defined as eclipse obscuration. The obscuration percentage during the December 4, 2021, total solar eclipse at various time intervals is estimated using the Python package *pyephem*. Figure 6.3 shows the obscuration percentage every 30 minutes starting from 06:00 UTC to 08:30 UTC. The methods of Verhulst and Stankov (2020) are adopted to estimate obscuration percentage at a particular altitude. It is important to notice that the calculations are performed for an altitude of 290 km, which is the height of peak electron density derived using International Reference Ionosphere (IRI) model (Bilitza et al., 2017). The geographical coordinates and time of the greatest eclipse are considered to derive the IRI profile. The coordinates of the eclipse central line at 0 km and 290 km altitudes are shown in Table 6.2. From Figure 6.3, it can be observed that the solar disk is almost completely covered between 07:00 UTC and 08:00 UTC. During this time interval, the central line of the eclipse moved across West Antarctica.

6.2.3 Magnetic Conditions on December 4, 2021

Apart from solar radiation, geomagnetic activity plays an important role in the distribution of ionospheric plasma (Coster et al., 2007; Immel and Mannucci, 2013). Hence, to exclude the influence of magnetic activity and to understand the effect of the solar

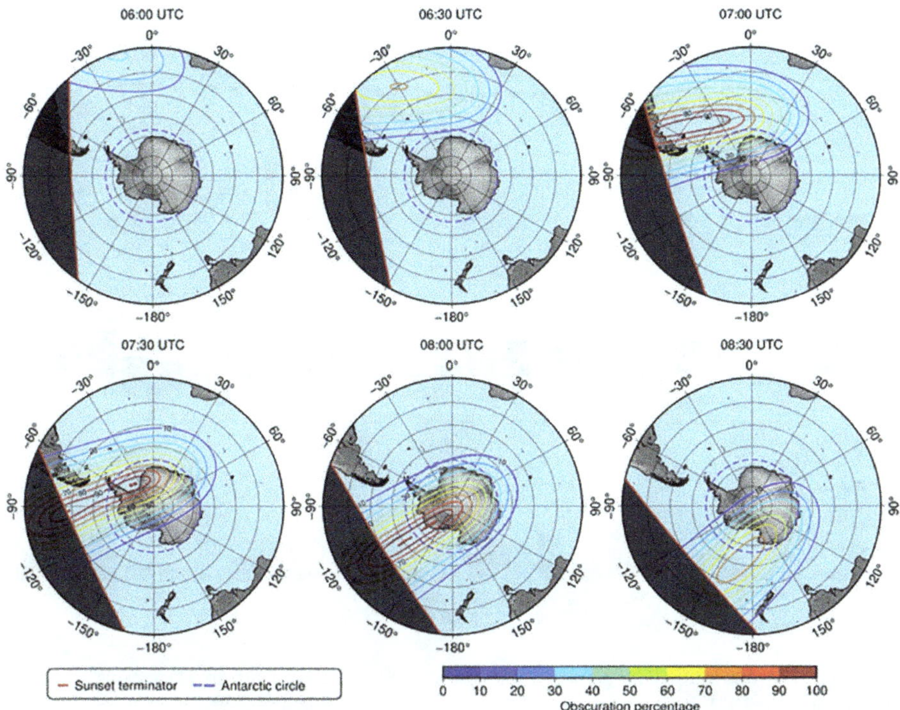

FIGURE 6.3 Eclipse obscuration calculated using the Python package pyephem and the methods of Verhulst and Stankov (2020). The colored contours depict the obscuration percentage. The red line shows the sunset terminator, and the shaded region represents the nightside. The sunset terminator and obscuration percentage are estimated for an altitude of 290 km (height of peak electron density derived from the IRI model).

eclipse on the electron density variation it's important to make sure that the observation period is geomagnetically quiet. Figure 6.4 shows the variations in geomagnetic indices Sym-H, Kp and ap respectively during 3–5 December 2021. The Sym-H values during this period is varying between –25 to –5 nT indicating the magnetically quiet conditions. Similarly, the values of the Kp index (Kp < 3.5) and ap index (ap < 18) further confirm the quiet geomagnetic conditions during this period.

6.3 RESULTS AND DISCUSSIONS

6.3.1 Ionospheric Response of Total Solar Eclipse on December 4, 2021

To understand the effect of the total solar eclipse on ionospheric electron density over Antarctica, GPS-TEC data were used. Since TEC is the integrated ionospheric electron density, a 2D-TEC map can give information about the ionization loss due to the absence of solar radiation associated with the solar eclipse. Figure 6.5 shows

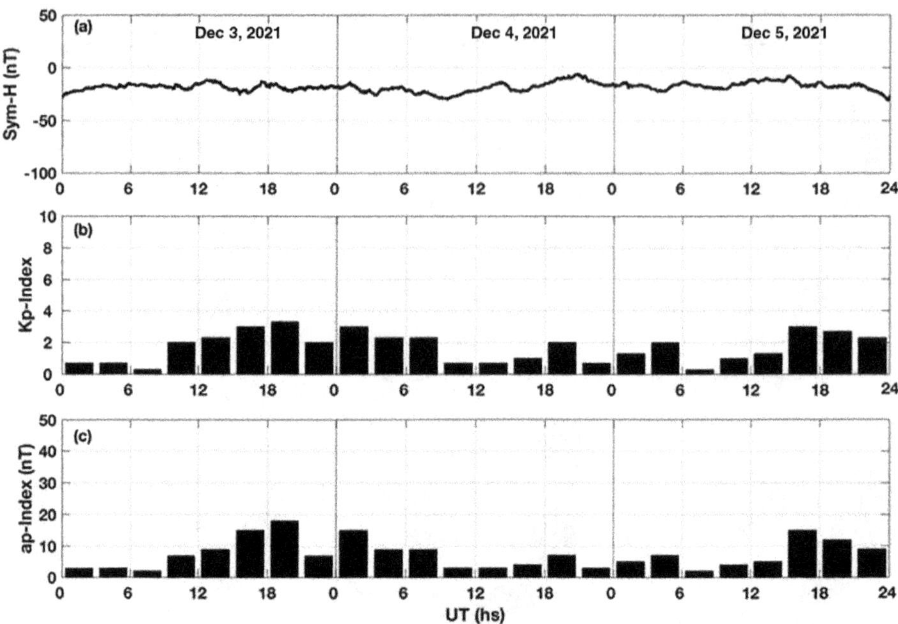

FIGURE 6.4 Variations in geomagnetic indices Sym-H, Kp and ap respectively during December 3–5, 2021.

the 2D-TEC maps generated using the VTEC data from 35 GPS stations over the Antarctic region at 06:30, 07:00, 07:30, 08:00 and 08:30 UTC. The VTEC values are averaged in 1° latitude x 1° longitude grids. To understand the variation in the electron density, the mean 2D-TEC map was constructed by averaging the VTEC values on December 3, 2021 (One day before the eclipse) and December 5, 2021 (One day after the eclipse) and shown in Figure 6.5a. Figure 6.5b presents the VTEC distribution over the Antarctic region on the solar eclipse day (December 4, 2021). The red solid and dashed curves indicate the movement of the eclipse totality path at 0 and 290 km with respect to time. Finally, Figure 6.5c shows the difference in VTEC on the eclipse day from the average values of one day before and after the eclipse day. From Figure 6.5c it is clear that VTEC values are significantly reduced over the regions close to the obscuration path during the eclipse period. On the eclipse day VTEC values show a deviation of 5 to 10 TECU from the average VTEC of one day before and after the eclipse.

To investigate the continuous temporal evolution of ionospheric electron density variations during the eclipse, the vertical TEC variations from various GPS stations located along the path of eclipse totality are analyzed. Figure 6.6 summarizes the results of this analysis. In Figure 6.6a, the red dashed curve represents the path of the eclipse central line at an altitude of 290 km. The latitude and longitude correspond to the starting point of the central line at 06:50 UTC are –50.20° and –35.56° respectively (Table 6.2). As seen from the figure, the totality path crosses Antarctica

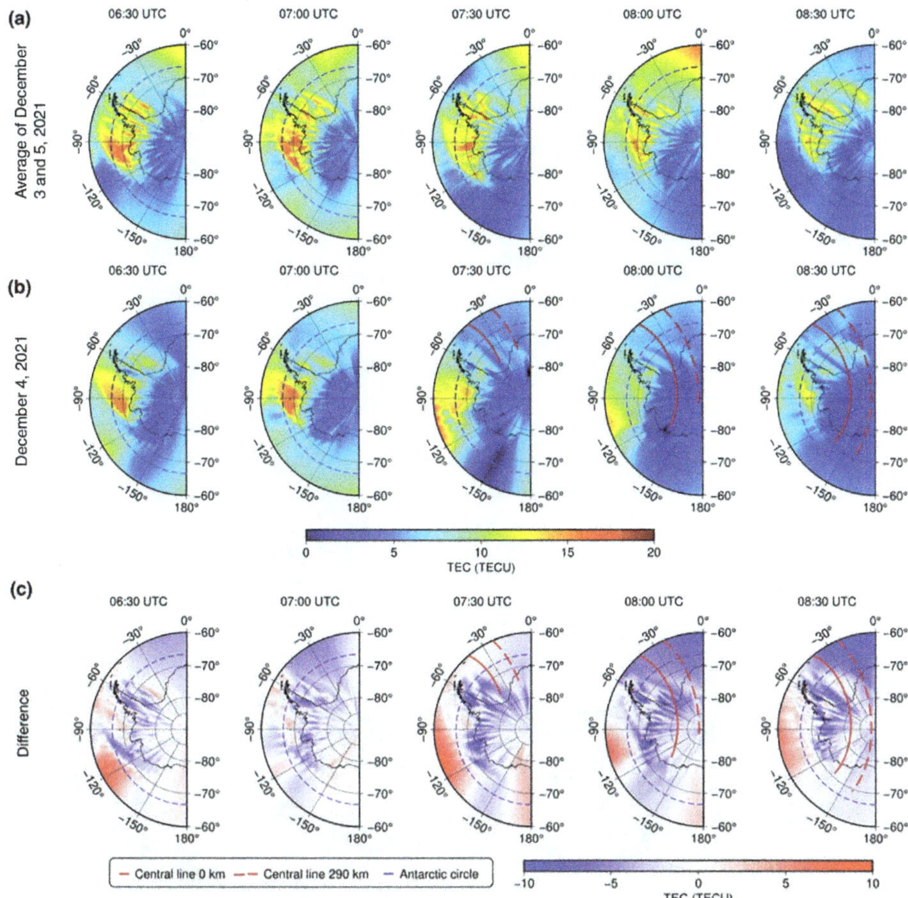

FIGURE 6.5 2D-TEC maps over the Antarctic region at 06:30, 07:00, 07:30, 08:00 and 08:30 UTC. (a) average TEC map obtained using the TEC values of one day before and after eclipse day (i.e., December 3 and 5, 2021) (b) on the solar eclipse day (December 4, 2021). (c) The absolute TEC deviation from two-day average values during the solar eclipse. The red solid and dashed curves show the eclipse center line at 0 km and 290 km, respectively.

and ends at 08:17 UTC. The corresponding geographical coordinates are –69.22° latitude and –158.89° longitude. 8 GPS sites are selected (majk, crdi, rmbo, stew, whtm, wiln, pirt and wlrd) located near the eclipse central line to analyze the TEC variations. The IPP (Ionospheric Pierce Point) trajectories of PRN 14 observed from these stations are plotted as black curves in the figure. It can be noted from the figure that, the satellite IPPs move towards the central line of eclipse totality.

The time series of vertical TEC variations observed in PRN 14 from the selected 8 GPS stations are illustrated in Figure 6.6b. The red curves are the TEC observations on the eclipse day (December 4, 2021). To compare it with non-eclipse days, averaged values of TEC variations on one day prior (December 3, 2021) and one

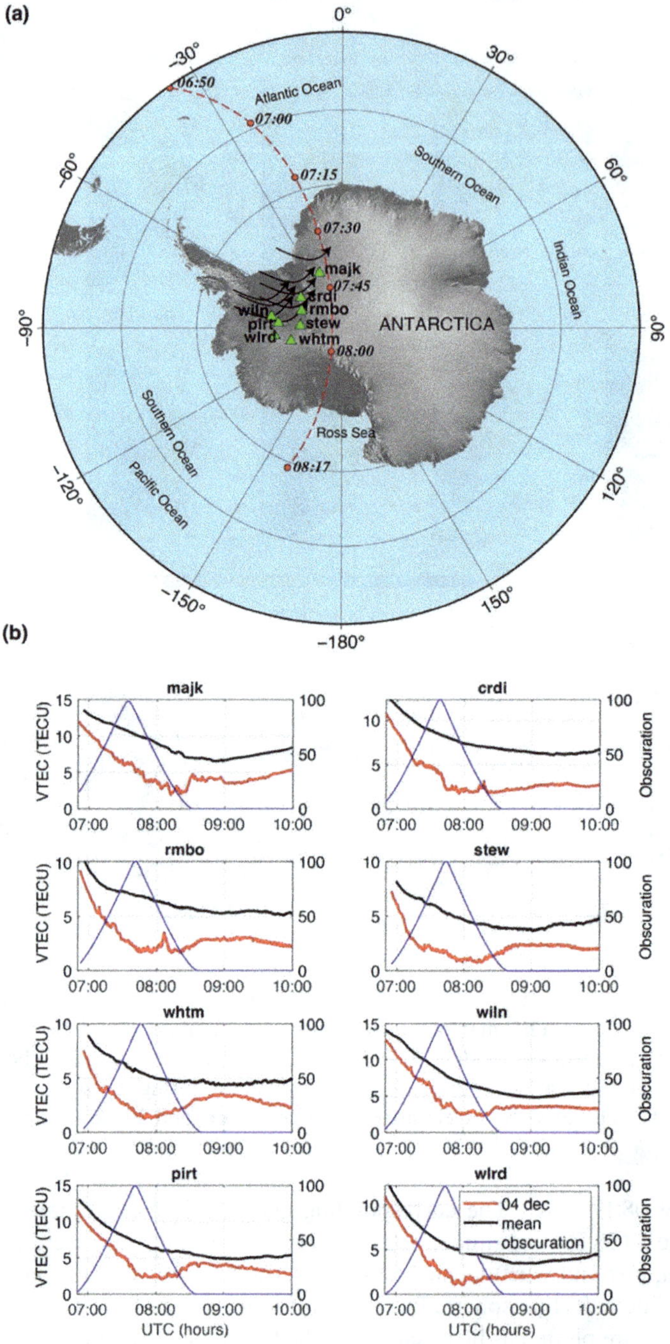

FIGURE 6.6 (a) Path of eclipse central line and IPP trajectories. The red dashed curve represents the track of the eclipse central line at 290 km. Total obscuration starts at 06:50 UTC and ends at 08:17 UTC at this altitude. The black curves show the IPP trajectories of PRN 14. (b) Red curves show the TEC variations from PRN 14 corresponding to December 4, 2021. Black curves are the average TEC of December 3–5. Obscuration percentages along the IPP trajectories are represented as blue curves.

day after (December 5, 2021) the eclipse is used. The black curves in each subplots represent the averaged time series. To quantify the effect of the eclipse, the obscuration percentage along each IPP track is derived and shown as blue curves in the figure. As seen from the figure, TEC depletions are observed in all the stations on the eclipse day. The TEC values gradually reduce and reach a minimum as the eclipse progresses. Later, the TEC values are recovered and return to the normal level. From Figure 6.6b, complete obscurations along the selected IPPs are observed between 07:30 UTC and 08:00 UTC. TEC minimum values in all the time series are recorded after the solar radiations are completely blocked (100% obscuration). The solar disk becomes fully visible again after ~08:30 UTC. In most of the stations, the TEC values returned to their normal condition after ~08:30 UTC.

Many studies have addressed the ionospheric responses of solar eclipses (Rishbeth, 1968; Evans, 1965; Klobuchar and Whitney, 1965; Nayak and Yiğit, 2018; Verhulst and Stankov, 2020; Maurya et al., 2020; Shrivastava et al., 2021). A depletion in ionospheric electron density during the eclipse period and the recovery after the completion of the eclipse are the general trends observed. In the present case, as shown in Figure 6.6b, a gradual decrease in vertical TEC is observed in all the stations followed by a recovery phase. At the same time, there exists a time lag between complete obscuration and TEC minimum. The drop of solar EUV flux during the eclipse is responsible for the reduction of electron density at the E layer, F1 layer and F2 layer, which in turn reduces the TEC (an integrated measure of electron density) along the satellite-receiver line of sight (Cherniak and Zakharenkova, 2018). The major reason behind the time lag is usually attributed to the lifetime of charged particles at various ionospheric layers (Nayak and Yiğit, 2018; Adeniyi et al., 2007). Ions/electrons in the E layer and F1 layer have a lesser life than that of the F2 layer. As a result, charged particle density in the lower ionospheric layers (E and F1) reduces immediately during the eclipse. On the other hand, the electron density of the F2 layer drops after a time delay.

CONCLUSION

The totality path of the total solar eclipse occurred on December 4, 2021, travelled across West Antarctica from east to west. To study the ionospheric response associated with this event, the TEC observations from 35 GPS stations distributed across West Antarctica are analyzed. To better correlate the effects of the eclipse on ionospheric electron density distribution, the eclipse obscuration percentage along each satellite IPP trajectories is derived. Compared to non-eclipse days, clear depletions in the TEC values are observed in all the stations after the eclipse onset. Apart from that, a gradual recovery to normal TEC values is recorded after the completion of the eclipse. The TEC depletion and recovery are well associated with the eclipse obscuration. Interestingly, the minimum TEC values and eclipse obscuration are not perfectly correlating. The time lags are attributed to the lifetime of charged particles at various ionospheric layers.

ACKNOWLEDGMENT

A. S. Sunil sincerely acknowledges the University Post-Doctoral Fellowships 2021 of Cochin University of Science and Technology, Kochi, India for the research fellowship. The authors also thank the Editor, Dr. Neloy Khare, for the constant support and encouragement.

DATA AVAILABILITY STATEMENT

The pyephem Python package is provided by B. Rhodes through the website http://rhodesmill.org/pyephem/. The SYM-H, Kp and ap indices were obtained from SPDF, NASA, USA (https://omniweb.gsfc.nasa.gov). GPS RINEX data used in this study are available at SOPAC (http://sopac.ucsd.edu).

REFERENCES

Adeniyi, J. O., Radicella, S. M., Adimula, I. A., Willoughby, A. A., Oladipo, O. A., & Olawepo, O. (2007). Signature of the 29 March 2006 eclipse on the ionosphere over an equatorial station. *Journal of Geophysical Research*, 112, A06314, https://doi.org/10.1029/2006JA012197.

Bilitza, D., Altadill, D., Truhlik, V., Shubin, V., Galkin, I., Reinisch, B., & Huang, X. (2017). International Reference Ionosphere 2016: From ionospheric climate to real-time weather predictions. *Space Weather*, 15, 418–429. https://doi.org/10.1002/2016SW001593

Cherniak, I., & Zakharenkova, I. (2018). Ionospheric total electron content response to the great American solar eclipse of 21 August 2017. *Geophysical Research Letters*, 45, 1199–1208. https://doi.org/10.1002/2017GL075989

Cohen, E. A. (1984). The study of the effect of solar eclipses on the ionosphere based on satellite beacon observations. *Radio Science*, 19, 769–777. https://doi.org/10.1029/RS019i003p00769

Coster, A. J., Colerico, M. J., Foster, J. C., Rideout, W., & Rich F. (2007). Longitude sector comparisons of storm enhanced density. *Geophysical Research Letters*, 34, L18105. https://doi.org/10.1029/2007GL030682

Coster, A. J., Goncharenko, L., Zhang, S.-R., Erickson, P. J., Rideout, W., & Vierinen, J. (2017). GNSS observations of ionospheric variations during the 21 August 2017 solar eclipse. *Geophysical Research Letters*, 44, 12041–12048. https://doi.org/10.1002/2017GL075774

Evans, J. V. (1965). An F region eclipse. *Journal of Geophysical Research*, 70, 131–142. https://doi.org/10.1029/JZ070i001p00131

Farges, T., Jodogne, J. C., Bamford, R., Le Roux, Y., Gauthier, F., Vila, P. M., . . . Miro, G. (2001). Disturbances of the western European ionosphere during the total solar eclipse of 11 August 1999 measured by a wide ionosonde and radar network. *Journal of Atmospheric and Solar-Terrestrial Physics*, 63, 915–924. https://doi.org/10.1016/S1364-6826(00)00195-4

Goncharenko, L. P., Erickson, P. J., Zhang, S.-R., Galkin, I., Coster, A. J., & Jonah, O. F. (2018). Ionospheric response to the solar eclipse of 21 August 2017 in Millstone Hill (42N) observations. *Geophysical Research Letters*, 45, 4601–4609. https://doi.org/10.1029/2018GL077334

Hargreaves, J. K. (1992). *The Solar-Terrestrial Environment.* Cambridge Press.

Immel, T. J., & Mannucci, A. J. (2013), Ionospheric redistribution during geomagnetic storms. *Journal of Geophysical Research-Space Physics*, 118, 7928–7939. https://doi.org/10.1002/2013JA018919

Kameda, T., Fujita, K., Sugita, O., Hirasawa, N., & Takahashi, S. (2009). Total solar eclipse over Antarctica on 23 November 2003 and its effects on the atmosphere and snow near the ice sheet surface at Dome Fuji. *Journal of Geophysical Research*, 114, D18115. https://doi.org/10.1029/2009JD011886

Klobuchar, J. A., & Whitney, H. E. (1965). Ionospheric electron content measurements during a solar eclipse. *Journal of Geophysical Research*, 70, 1254–1257.

Klobuchar, J. A. (1987). Ionospheric time-delay algorithm for single-frequency GPS users. *IEEE Transactions on Aerospace and Electronic Systems*, 3, 325–331.

Mannucci, A. J., Wison, B. D., & Edwards, C. D. (1993). A new method for monitoring the Earth's ionospheric total electron content using GPS global networks. *Proceedings of ION GPS-93*, 93, 1323–1332.

Maurya, A. K., Shrivastava, M. N. & Kumar, K. N. (2020). Ionospheric monitoring with the Chilean GPS eyeball during the South American total solar eclipse on 2nd July 2019. *Scientific Reports*, 10, 19380. https://doi.org/10.1038/s41598-020-75986-7

Nayak, C., & Yiğit, E. (2018). GPS-TEC observation of gravity waves generated in the ionosphere during 21 August 2017 total solar eclipse. *Journal of Geophysical Research: Space Physics*, 123, 725–738. https://doi.org/10.1002/2017JA024845

Pradipta, R., Yizengaw, E., & Doherty, P. (2018). Ionospheric density irregularities, turbulence, and wave disturbances during the total solar eclipse over North America on 21 August 2017. *Geophysical Research Letters*, 45, 7909–7917. https://doi.org/10.1029/2018GL079383

Rishbeth, H. (1968). Solar eclipses and ionospheric theory. *Space Science Reviews*, 8, 543–554. https://doi.org/10.1007/BF00175006

Rishbeth, H., & Garriott, O. K. (1969). *Introduction to Ionospheric Physics*. Academic Press.

Seemala, G. K. (2011). GPS-TEC analysis application. *Tech Rep, Institute for Science Research, U.S.A.* https://seemala.blogspot.com

Shrivastava, M. N., Maurya, A. K., & Kumar, K. N. (2021). Ionospheric perturbation during the South American total solar eclipse on 14th December 2020 revealed with the Chilean GPS eyeball. *Scientific Reports*, 11, 20324. https://doi.org/10.1038/s41598-021-98727-w

Verhulst, T. G. W., & Stankov, S. M. (2020). Height dependency of solar eclipse effects: The ionospheric perspective. *Journal of Geophysical Research: Space Physics*, 125, e2020JA028088. https://doi.org/10.1029/2020JA028088

Zhang, R., Le, H., Li, W., Ma, H., Yang, Y., Huang, H., et al. (2020). Multiple technique observations of the ionospheric responses to the 21 June 2020 solar eclipse. *Journal of Geophysical Research: Space Physics*, 125, e2020JA028450. https://doi.org/10.1029/2020JA028450

7 Isotope Hydrochemistry of Lakes and Transient Ponds of East Antarctica with Varying Degree of Environmental Condition

T. R. Resmi,[1] *Girish Gopinath,*[2] *P. S. Sunil,*[3] *M. Praveenbabu,*[1] *and Rahul Rawat*[4]

[1] Ecology and Environment Research Group, Centre for Water Resources Development and Management, Kozhikode, India

[2] Department of Climate Variability and Aquatic Ecosystems, Kerala University of Fisheries and Ocean Studies, India

[3] CUSAT-NCPOR Centre for Polar Sciences, Department of Marine Geology and Geophysics, School of Marine Sciences, Cochin University of Science and Technology, Kochi, India

[4] Indian Institute of Geomagnetism, Mumbai, India

CONTENTS

DOI: 10.1201/9781003284413-7

7.1 INTRODUCTION

The postglacial retreat of the ice cap and isostatic rebound of the Earth's crust following the most recent deglaciation have resulted in making 1–2% of the Antarctic continent ice-free regions. Among the ice-free regions, 1–2% contain coastal oasis and hill regions such as Schirmacher Oasis and Larsemann Hills of East Antarctica. During the austral summer, melt water accumulates in the land depressions, and lakes and ponds are formed in this oasis. The lakes are varying in-depth, quality, water spread area, and water availability (seasonal or perennial). Radiation processes of the lakes are decided by many factors; mainly by the differences in the albedo of new and old ice, the amount of suspended matter and secondarily by the differences in areas covered by snow (Kaup, 1982). According to Hodgson (2012), the increase of radiation in water is mainly determined by the irregularities at the lower surface of the ice. Accordingly, the radiation regime determines seasonal variations in water temperature and the cooling and warming effects are determined by the glaciers and rocks respectively. Cooling of the lake water by the inflowing melt water has the same effect as the turbulent heat exchange with the atmosphere in summer. Thus the variation in snow and ice cover changes all ecological variables in Antarctica lakes.

In this paper, we discuss the chemical and isotopic evolution of some lakes and transient ponds in East Antarctica situated in two different environmental settings. The Schirmacher Oasis hosts more than 100 lakes and the Larsemann Hills about 150 lakes. Earlier studies have shown that the chemical composition of lakes in these two regions is distinctly different (Haberlandt et al., 1987; Warrier et al., 2016; Srivastava et al., 2012; Phartiyal, 2014; Smith et al., 2006; Mojib et al., 2009; Sinha and Chatterjee, 2000; Gopinath et al., 2020; Resmi et al., 2022). As these lakes are fed by glacial melt water, the chemical composition depends mainly on the precipitation and regional bedrock geology. However, continuous cycles of freezing, thawing and evaporation alter the primary composition giving distinct chemical signatures. Similar to the chemical composition, stable isotope ratios are also varied in these lakes. The water bodies of Antarctica have been investigated for stable isotopes by many investigators (Haberlandt et al., 1987; Strauch et al., 2011; Haendel et al., 2011; Wand et al., 2011; Richter and Strauch, 1983; Gopinath et al., 2020; Resmi et al., 2022). The vital role played by the Antarctica lakes in the water balance of the Schirmacher Oasis, especially in the estimation of the contribution of the melt water from the glaciers and the snow/ice fields of the oasis through stable isotopic composition.

7.2 MATERIALS AND METHODS

7.2.1 Schirmacher Oasis

The Schirmacher Oasis (70^{0}44'30" S to 70^{0}46'30" and 11^{0}22'40" E to 11^{0}54'00" E) is a narrow strip of the ice-free region covering an area of about 35 km^2 laying in the east-west direction (Figure 7.1). The central part of the oasis has a maximum width of about 2.7 km and has undulating topography with small hills (~200 m elevation) and depressions formed by glacial valleys and lakes. The region is characterized by deglaciated landscape interspersed with more than 100 freshwater lakes (Mathur, 2006). The lakes can be classified as proglacial, landlocked, or inland and epiglacial lakes that occur in well-defined lineament and glacial basins formed during the late Pleistocene-Holocene period.

The Schirmacher Oasis region is geomorphologically distinguishable into three units: 1) polar ice sheet, 2) Schirmacher mainland including lakes, and 3) coastal-shelf area. All the three units extend in east-west directions paralleling the coastline (Figure 7.1). The polar ice sheet, which contains abundant sand and silt-size sediments, covers a large area in the south. The exposed mainland unit mostly represented by high-grade metamorphic rocks (Sengupta, 1986), shows an undulating topography with low altitude hills of 50 to 200 m elevation and inland lakes formed due to glacier erosion.

7.2.2 Larsemann Hills

The Larsemann Hills is the ice-free coastal oasis in the Ingrid Christensen Coast of Princess Elizabeth Land in East Antarctica (69°30'S, 76°19'58''E). It is spread over approximately 50 km^2 and includes about 150 lakes with different ionic characters (Figure 7.2). Stornes and Broknes are the two significant peninsulas in the Larsemann Hills. The Indian research station 'Bharti' is located in the central part of the Larsemann Hills, which is about ~12 km^2 in area. The introductory geology of the Larsemann Hills consists of supracrustal volcanogenic and sedimentary rocks metamorphosed under granulite facies conditions. The supracrustal rocks are intruded by several generations of pegmatites and granites and are underlain by and possibly derived from the Proterozoic orthopyroxene-bearing orthogneiss basement

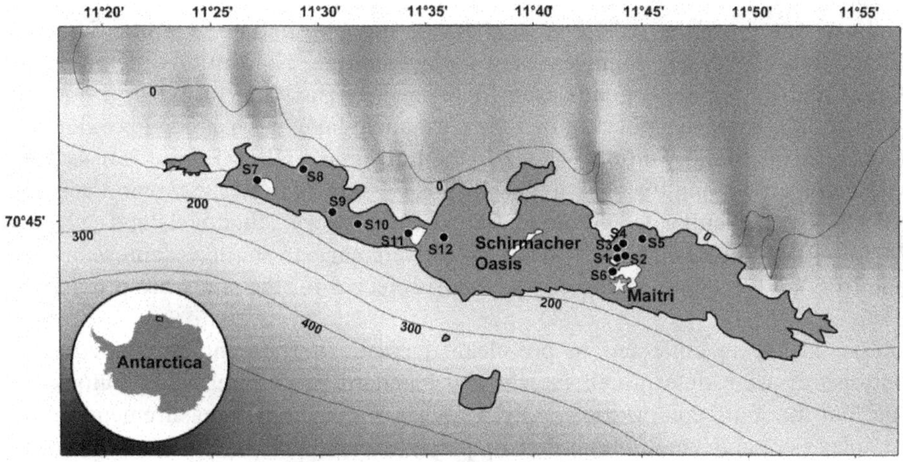

FIGURE 7.1 Sampling locations map of the Schirmacher Oasis, East Antarctica in black circles. The inset map shows the location of Schirmacher Oasis in a rectangle.

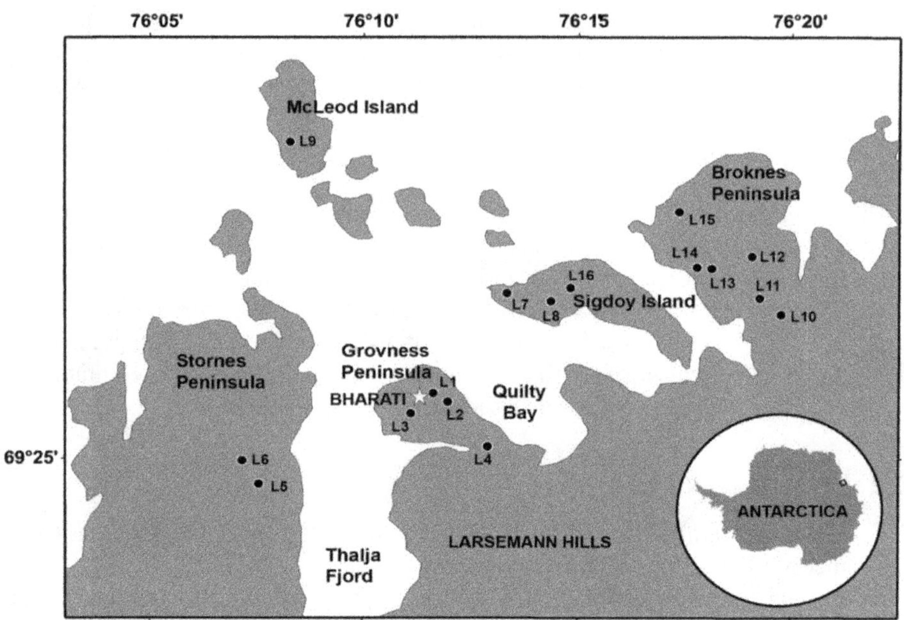

FIGURE 7.2 Sampling locations map of the Larsemann Hills, East Antarctica in black circles. The black circles indicate the sample location. The inset map shows the location of Larsemann Hills in the rectangle.

(ATCM XXXVII Final Report, 2014). Similar to the lakes in Schirmacher Oasis, the lakes in Larsemann Hills are formed in the land depressions caused by glacial erosion.

Lake water samples were collected during the 29th Indian Antarctic Scientific Expedition (2009–10). Samples were collected from 12 locations including the Lakes and the shallow ponds from Schirmacher Oasis during austral summer (Figure 7.1). Samples collected include permanent lakes such as the Priyadarshini Lake near the Indian Station, Maitri. In the Larsemann Hills (Figure 7.2), samples were collected from 16 lakes in the Broknes Peninsula (BP), Storens Peninsula (SP), Grovnes Peninsula (GP, near Bharti Promontory), Sigdoy Island (SI), and McLeod Island (MI). Most of the samples were surface shallow depression storage from both regions. Details of the lakes selected for the study are provided in Table 7.1.

Samples were collected in pre-cleaned plastic bottles and physicochemical parameters were determined *in situ* and standard procedures were followed to find out the ionic composition (APHA, 1995). For isotopic measurements, water samples were collected in high-density polyethylene (HDPE) bottles and the ratios were determined in an off-axis integrated cavity output spectroscopy system (IWA-45EP, Los Gatos research), with an external precision of ±0.1‰ for $\delta^{18}O$ and ±0.5‰ for δD.

TABLE 7.1
Location Details of the Sampled Lakes and Ponds in Schirmacher Oasis and Larsemann Hills, East Antarctica

Sl. No.	Code	Location Name	Type
1	S 1	SO1	Surface Shallow Depression Storage
2	S 2	SO2	Precipitation Dominant
3	S 3	S3	Surface Shallow Depression Storage
4	S 4	S4	Surface Shallow Depression Storage
5	S 5	S5	Precipitation Dominant
6	S 6	S6	Precipitation Dominant
7	S 7	S7	Surface Shallow Depression Storage
8	S 8	S8	Surface Shallow Depression Storage
9	S 9	S9	Surface Shallow Depression Storage
10	S 10	S10	Surface Shallow Depression Storage
11	S 11	S11	Surface Shallow Depression Storage
12	S 12	S12	Precipitation Dominant
13	L1	Grovness Peninsula	Surface Shallow Depression Storage
14	L2	Grovness Peninsula	Surface Shallow Depression Storage
15	L3	Grovness Peninsula	Surface Shallow Depression Storage
16	L4	Grovness Peninsula	Surface Shallow Depression Storage
17	L5	Stornes Peninsula	Precipitation Dominant
18	L6	Stornes Peninsula	Precipitation Dominant
19	L7	Sigdoy Island	Surface Shallow Depression Storage
20	L8	Sigdoy Island	Surface Shallow Depression Storage
21	L9	McLeod Island	Precipitation Dominant
22	L10	Broknes Peninsula	Surface Shallow Depression Storage
23	L11	Broknes Peninsula	Precipitation Dominant
24	L12	Broknes Peninsula	Precipitation Dominant
25	L13	Broknes Peninsula	Precipitation Dominant
26	L14	Broknes Peninsula	Precipitation Dominant
27	L15	Broknes Peninsula	Precipitation Dominant
28	L16	Sigdoy Island	Surface Shallow Depression Storage

7.3 RESULTS AND DISCUSSION

7.3.1 Physicochemical Characteristics

The physicochemical characteristics of the lake water of the two study regions are given in Table 7.2. pH of the lakes and ponds in the Schirmacher Oasis and Larsemann Hills were slightly acidic (Ave: 6.6 and 6.8 respectively). In Larsemann Hills, pH > 7 was noted in the Sigdoy and McLeod Island lakes compared to the peninsular lakes (Groveness, Broknes, and Stornes). However electrical conductivity (EC) was lower in samples from Schirmacher Oasis (Ave: 83.6 μS/cm) than from Larsemann Hills (Ave: 494.5 μS/cm). In both these places, a large variation in EC among the samples was noted. In general, lakes from Schirmacher Oasis can be classified into two as (1) EC < 100 μS/cm and (2) EC > 200 μS/cm. Only three lakes (S3, S5 & S12) recorded EC > 200 μS/cm, which were precipitation dominant except M6, and in the rest of the samples, EC was below 100 μS/cm.

Three classes can be considered in the case of lakes of Larsemann Hills, such as EC < 500 μS/cm (L1–L7, L12–L13) and EC = 500–1000 μS/cm (L11) and EC > 1000 μS/cm (L8–L10), based on the EC. Lakes in the GP and SP belonged solely to the first category, whereas the lakes in the MI and one each from SI and BP belonged to the higher EC category. Only one lake in the BP was in the second category. The lakes in the BP showed large variations in ionic content falling in the three categories.

TABLE 7.2
Chemical Composition of Lake Waters of Schirmacher Oasis and Larsemann Hills

code	pH	EC (μS/cm)	Na (mg/l)	K (mg/l)	Ca (mg/l)	Mg (mg/l)	Cl (mg/l)	SO_4 (mg/l)	HCO_3 (mg/l)
S 1	6.4	15.7	0.8	0.5	1.3	1.3	3.8	4.8	2.5
S 2	7.1	9.0	1.7	0.0	1.4	1.3	3.8	1.9	10.1
S 3	7.2	105	6.7	3.1	8.4	2.4	11.5	29.0	24.5
S 4	6.3	18.9	1.7	1.0	1.5	1.3	3.8	4.8	1.8
S 5	7.1	511	20.2	15.3	38.4	11.5	53.4	67.6	54.8
S 6	6.6	14.2	1.7	0.0	1.2	1.3	5.7	2.1	2.2
S 7	6.4	14.4	1.7	1.0	1.1	1.3	1.9	9.7	2.1
S 8	5.9	76.6	1.7	1.0	8.7	1.4	15.3	4.8	2.3
S 9	6.2	8.2	1.7	0.0	1.0	1.3	B.D.L.	2.4	4.3
S 10	6.1	8.2	1.7	0.0	1.0	1.3	B.D.L.	4.8	2.1
S 11	6.8	4.1	1.7	0.0	0.9	1.2	B.D.L.	4.8	2.3
S 12	7.3	218	13.5	7.2	16	5.3	22.9	77.3	5.7
L1	6.38	360	3.4	5.1	48.5	2.1	80.2	12.9	9.7
L2	6.99	383	5.0	3.1	53.5	2.9	84.0	13.1	19.3
L3	6.86	205	1.7	4.1	29.4	1.9	47.7	10.6	14.5
L4	7.27	234	6.7	5.1	29.9	1.8	51.5	9.8	29.0
L5	6.29	274	5.0	3.1	32.4	1.9	57.3	6.7	9.7
L6	6.22	13.8	1.7	0.0	1.5	1.2	1.9	4.0	4.8
L7	6.4	141	3.4	1.0	17	1.5	22.9	3.3	14.5
L8	7.66	1210	25.2	23.5	130	7.3	274.8	3.9	120.8
L9	7.27	1330	10.1	21.5	170	8.7	324.5	35.9	33.8
L10	7.06	1210	15.1	21.5	140	6.6	284.4	39.9	38.6
L11	6.41	532	3.4	5.1	75	3.8	110.7	19.0	24.2
L12	6.75	373	6.7	5.1	43.5	2.3	61.1	33.0	24.2
L13	6.65	163	3.4	2.0	21.2	1.6	26.7	7.3	4.8

- B.D.L. denotes below detection level

Similarly, the SI lakes belonged to one of the lowest and highest ionic content categories at the same time. On the whole, the lakes in the islands showed higher conductivity, perhaps due to the proximity to the sea or may be due to increased evaporation as being closed basins.

According to the electrical conductivity, ionic concentration in the lake water of Larsemann Hills was significantly higher than in the Schirmacher Oasis, perhaps due to the proximity to the sea or may be due to increased evaporation as being closed basins. However, the order of abundance of ions in the lakes of the two places was different. Though Ca was the dominant cation in both the oasis, SO_4 was the predominant anion in Schirmacher Oasis and chloride ion was dominant in Larsemann Hills.

The ionic composition of Antarctic lakes is determined by lake-ice thawing which can either result in more diluted lake water due to the addition of melt water of ionic content or, more concentrated lake water if marine aerosols are accumulated in the snow and ice. Earlier studies (Phartiyal et al., 2011) have shown that the lakes in Schirmacher Oasis are with minimal marine influence due to the distance from the sea and those which were close to the ice sheet were characterized by a higher Ca to Na ratio. In addition, in Schirmacher Oasis, all lakes are in elevations above the Holocene marine limit that had no hydrological contact with the ocean in the past (Hodgson, 2012) and have a glacial origin, in contrast to the saline lakes and ponds in East Antarctica. The lakes in East Antarctica are isolation basins of low elevations. Many of the lakes in Schirmacher Oasis are of Holocene age and are thus relatively young (Phartiyal et al., 2011). The lakes are flushed by a continuous supply of melt water in the summer by the streams from the East Antarctic Ice Sheet or permanent snow banks upstream.

7.3.1.1 Interionic Relationships

The inter-ionic relations of the lake waters of the study area were determined with Pearson's product momentum correlation (Table 7.3). Most of the ions in the lakes of both the oasis were correlated with each other either strongly or moderately. In general, the strong correlation among the ions of the lakes can be attributed to their common origin. However, in the lakes of the Larsemann Hills, bicarbonate ions did not show any relationship with any other ions except with Ca.

Gibb's classification was used to find out the origin of ions in the lake water (Figure 7.3), which, can differentiate between the three major sources such as (1) evaporation/seawater, (2) rock-water interaction/weathering, and (3) precipitation. Though the ionic concentrations were relatively low in both the oasis, the processes determining the presence of ions in the lakes were distinctly different. In Schirmacher Oasis, most of the lakes have precipitation origin, and in the Larsemann Hills, interaction with base rocks was evident.

To discern the ions generated by weathering reactions, major hydrochemical facies present in these lakes were found using Piper's classification (Table 7.4). The water types (hydrochemical facies) obtained were mixed types with no dominant ions for the lakes in Schirmacher Oasis and Ca-Cl type showing a high degree of rock—water interaction with a concentration of ions by evaporation.

TABLE 7.3
Pearson's Product Momentum Correlation among the Major Ions in the Lake Waters of Schirmacher and Larsemann Hills Oasis

	pH	EC	Ca	Mg	Na	K	Cl	HCO_3	SO_4
Schirmacher Oasis									
pH	1								
EC	0.50	1							
Ca	0.47	***1.00***	1						
Mg	0.51	***0.99***	***0.98***	1					
Na	0.63	***0.97***	***0.96***	***0.97***	1				
K	0.53	***0.99***	***0.98***	***0.99***	***0.98***	1			
Cl	0.46	***0.99***	***0.99***	***0.97***	***0.94***	***0.97***	1		
HCO_3	0.53	***0.89***	***0.88***	***0.88***	***0.83***	***0.87***	***0.86***	1	
SO_4	***0.67***	***0.86***	***0.84***	***0.85***	***0.95***	***0.89***	***0.82***	0.64	1
Larsemann Hills									
	pH	**EC**	**Ca**	**Mg**	**Na**	**K**	**Cl**	**HCO_3**	**SO_4**
pH	1								
EC	***0.69***	1							
Ca	***0.66***	***0.99***	1						
Mg	***0.69***	***0.99***	***0.99***	1					
Na	***0.77***	***0.81***	***0.74***	***0.78***	1				
K	***0.75***	***0.98***	***0.95***	***0.96***	***0.88***	1			
Cl	***0.69***	***1.00***	***0.99***	***0.99***	***0.79***	***0.98***	1		
HCO_3	0.27	***0.61***	***0.65***	0.57	0.22	0.53	0.60	1	
SO_4	***0.77***	***0.70***	0.63	***0.68***	***0.94***	***0.77***	***0.67***	0.03	1

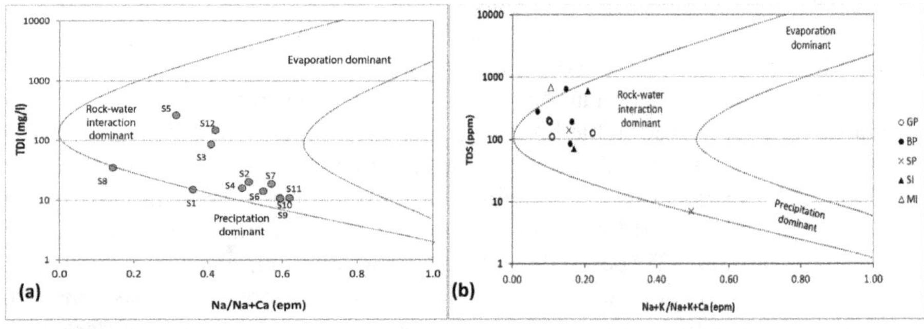

FIGURE 7.3 Gibb's diagram of lake water of (a) Schirmacher Oasis and (b) Larsemann Hills.

For understanding the major weathering reactions that are responsible for the ionic composition of the lake waters of the study area, different ionic relations were plotted (Figure 7.4 and 7.5).

The relation of total alkalinity which, is mainly due to the bicarbonate alkalinity, with total cations (TZ^+) (Figures. 7.4 a & 7.5 a) showed an excess of cations, which

TABLE 7.4
Major Hydrochemical Facies of Lake Waters of the Study Area

Code	Water Type	L1	Ca-Cl
S1	Mg-Ca-Cl-SO_4	L2	Ca-Cl
S2	Mg-Na-Ca-HCO_3-Cl	L3	Ca-Cl
S3	Ca-Na-SO_4-HCO_3-Cl	L4	Ca-Cl-SO_4
S4	Mg-Ca-Na-Cl-SO_4	L5	Ca-Cl
S5	Ca-Mg-Na-Cl-SO_4-HCO_3	L6	Mg-Ca-Na-SO_4-HCO_3-Cl
S6	Mg-Na-Ca-Cl	L7	Ca-Cl-SO_4
S7	Mg-Na-SO_4	L8	Ca-Cl-SO_4
S8	Ca-Cl	L9	Ca-Cl
S9	Mg-Na-Ca-HCO_3-SO_4	L10	Ca-Cl
S10	Mg-Na-Ca-SO_4	L11	Ca-Cl
S11	Mg-Na-Ca-SO_4-HCO_3	L12	Ca-Cl
S12	Ca-Na-Mg-SO_4-Cl	L13	Ca-Cl

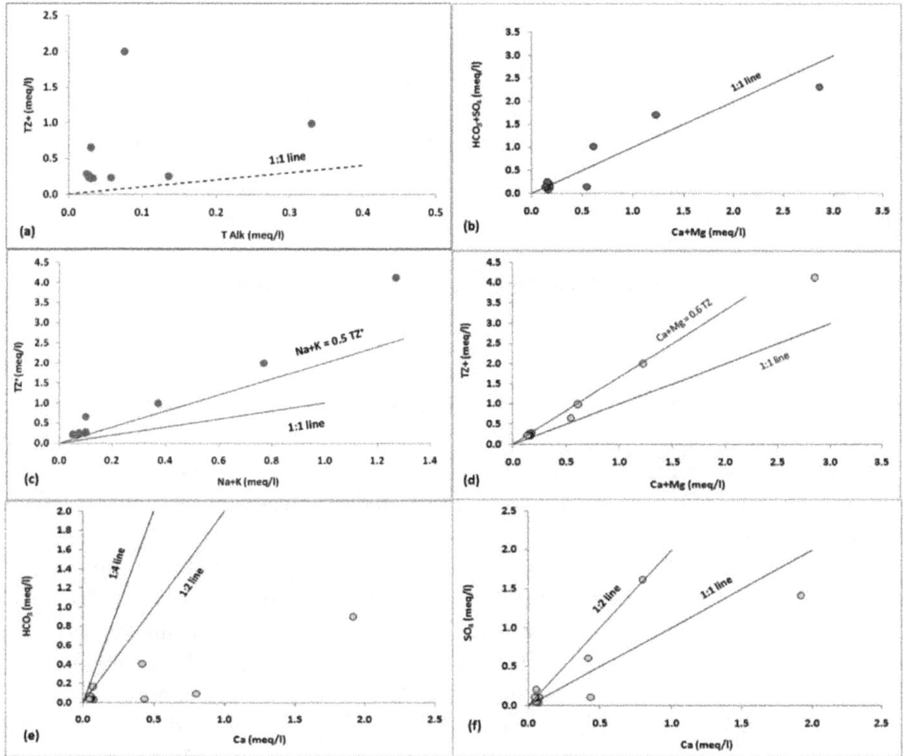

FIGURE 7.4 (a–f) Variation of interionic relations in Schirmacher Oasis

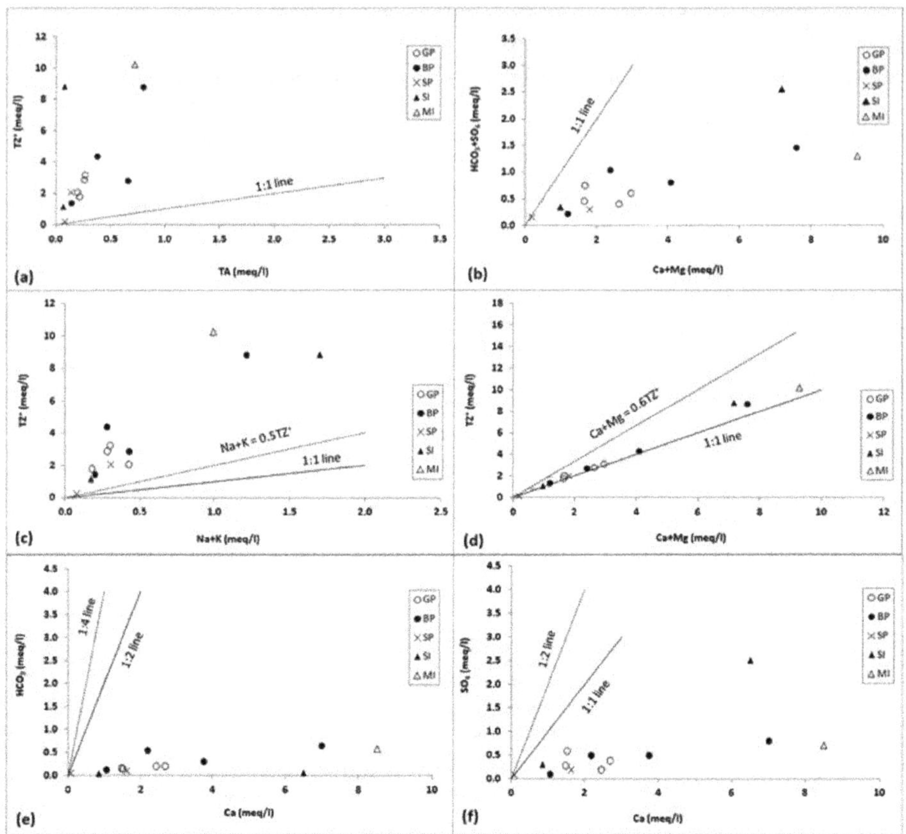

FIGURE 7.5 (a–f) Variation of interionic relations in Larsemann Hills

cannot be balanced by the bicarbonate ion alone in the lakes of both the oasis. It has been reported earlier that the rate of weathering reactions is very high in the peninsulas of Larsemann Hills (Kiernen et al., 2009), and, as a consequence, the dissolved carbon dioxide was used up and accordingly bicarbonate ion concentrations are lesser in these lakes.

In the Ca+Mg *vs* HCO_3+SO_4 plot (Figure 7.5 a) of the Schirmacher Oasis lakes, samples falling on the 1:1 equiline suggested that these ions are derived from the carbonate and sulphate minerals. In the Larsemann Hills, there is an excess of Ca and Mg, which points to additional sources of these ions. Since the geology of the area is mostly determined by silicate rocks, silicate weathering can be the source of ions in the lake. To find out the silicate weathering, Na + K and Ca + Mg were plotted against total cations separately (Figures 7.4 c, d & 7.5 c, d). In the lakes of the Schirmacher Oasis, samples were found mostly near the Na+K = 0.5 TZ^+ line, indicating weathering of silicates by the inflowing melt water, contributing Na and K ions mainly, to the lake water. In addition, the Ca+Mg *vs* total cations (TZ^+) plot

(Figure 7.4 d) showed most of the samples along the 0.6:1 line (Ca+Mg = 0.6 TZ^+) indicating these ions also may be originated from silicate weathering. The Ca to Mg molar ratio of the Schirmacher Oasis lake water samples was mostly below 1, and in a few, it was greater than 2.0. If the molar ratio of these two ions is equal to one, dissolution of dolomite will occur, whereas a higher ratio indicates calcite dissolution. The molar ratio of the lake water samples in the Schirmacher Oasis is mostly less than one, indicating the dissolution of dolomite. Further, a higher Ca/Mg molar ratio (>2), in a few samples is indicative of the dissolution of silicate minerals.

However, in the case of Larsemann Hills, the lake water Na and K were not varying linearly with total cations or did not fall on the 0.5 TZ^+ line pointing to the possibility of Ca/Na exchange reactions. However, as the Ca and Mg ions were varying linearly with total cations in most of the lakes, these ions may have originated from silicate weathering reactions. The molar ratios of Ca to Mg were > 2, in most of the lakes indicating the predominance of silicate weathering reactions (Katz et al., 1998).

However, to find out sources of ions other than silicate weathering, Ca was plotted against HCO_3 and SO_4 in both the lakes (Figures. 7.4 e, f & 7.5 e, f). In Schirmacher Oasis, some lakes were following the 1:2 line indicating calcite weathering and some had an excess of Ca compared to bicarbonate. Similarly, Ca *vs* SO_4^{2-} plot, some were plotting along the 1:1 and 1:2 lines and there are excess Ca ions. Samples falling on the 1:1 line denoted gypsum and anhydrite dissolution, whereas excess Ca points for additional source. In Larsemann Hills, the lake waters, sources of Ca not be accounted due only to silicate minerals as these are not balanced either by the bicarbonate or sulphate ion. The excess calcium in the lake water may be due to additional geochemical processes such as exchange reactions contributing to these ions.

Chloroalkali indices (CAI 1 & 2; see Figure 7.6) devised by Schoeller (1965, 1967) were used to determine the ion exchange reactions in the lakes. If both the indices are negative, there is an exchange of Ca and Mg of the water with Na and K of the Earth material. If reverse ion-exchange reactions are occurring, the indices will be positive and accordingly, the Ca and Mg of the rock material will get exchanged with Na and K ions in water.

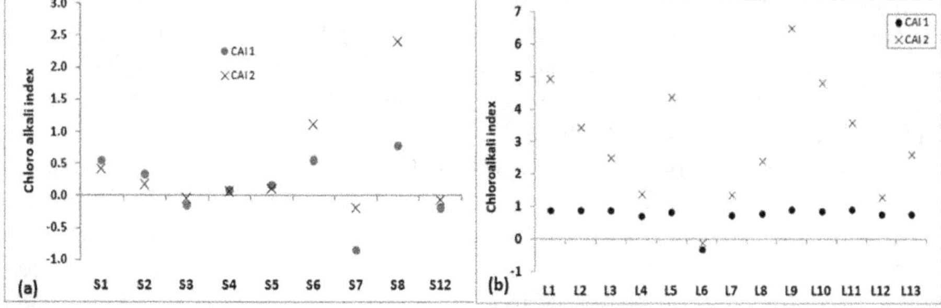

FIGURE 7.6 Chloro alkali indices in the lake water of Schirmacher Oasis and Larsemann Hills

The CAI were positive for most of the lakes in Schumacher as well as Larsemann Hills oasis indicating reverse ion exchange reactions are also contributing Ca and Mg ions to these lakes. In Schirmacher Oasis, since the concentration of chloride ions was very low in some of the lakes, the indices were not computed. The excess Ca present in some of the lakes is explained by the reverse ion exchange processes. In Larsemann Hills particularly, reverse ion-exchange reactions were very prominent. Except for L6 of SP, all the lakes showed reverse ion-exchange reactions, contributing Ca and Mg to the lake water (Figure 7.5). Hence apart from silicate weathering, Ca ions are added to the water by the reverse ion exchange process accounting for the excess Ca present in the water. A negative ion exchange reaction was observed in L6, in Stornes Peninsula, and the main source of Na and Ca in this lake water is the weathering of silicate minerals.

7.3.2 Isotopic Characterization of Lakes of Schirmacher Oasis and Larsemann Hills

The stable isotope composition of the lake water of the 12 lakes in Schirmacher Oasis and 16 lakes in the Larsemann Hills is provided in Table 7.5. Similar to the chemical composition, large variation is observed in the isotopic ratios of these lake waters.

Schirmacher Oasis: $\delta^{18}O$= −8‰ to −31.2‰; δD = −167.2‰ to −250.2‰
Larsemann Hills: $\delta^{18}O$ = −7.6 ‰ and −17.9 ‰; δD = −108.7 ‰ and −146.5‰.

The lakes in Schirmacher Oasis were depleted in heavier isotopes than in Larsemann Hills and the samples collected from precipitation dominant lakes in the Schirmacher were relatively enriched than the depression storage. In Larsemann Hills, the opposite distribution, with heavier isotopes in depression storage was found.

The observed δ values are typical for the Polar lakes, which receive substantial glacial melt water input. During the formation of ice, the heavier water molecules are preferentially added to the solid phase. Experimental studies have shown that for $^{18}O/^{16}O$, the isotopic enrichment factor of ice with respect to water ($\alpha_{ice\text{-}water}$) ranges from 1.00291 to 1.0048 and for ($^{2}H/^{1}H$), it ranges from 1.0171 to 1.0212 (Horita et al., 2008). When ice is formed fast, the equilibrium fractionation values are still lower, particularly for $^{2}H/^{1}H$ (1.017) (Wang et al., 2000). Slightly lower values of $^{2}\alpha_{ice\text{-}water}$ (1.013–1.017) were observed in ice covers of lakes and rivers (Gibson and Prowse, 1999; Michel, 1986; Miller and Aiken, 1996). According to Souchez et al. (1987), the apparent fractionation factor for $^{2}\alpha_{ice\text{-}water}$ is roughly half the equilibrium value for water and ice. Horita et al. (2008) have shown that the stagnant diffusive layer in the liquid phase favoring the transport of lighter water molecules controls the transport of water molecules onto the bottom of the ice. Accordingly, the melt water will be depleted in heavier isotopes. Consequently, the equilibrium isotope fractionation at the ice-water interface is partly shifted favoring the heavy water molecules. This shifting depends on the rate of freezing and the extent of mixing of the water column. Hence, molecular diffusion controls the transport of the water molecules in the boundary layer, the thickness of which determines the apparent ice-water isotope fractionation factor (Ferrik et al., 2002).

TABLE 7.5
Stable Isotope Data of Lake Waters of Schirmacher Oasis and Larsemann Hills, East Antarctica

Code	Location Name	$\delta^{18}O$ (‰)	$\delta^{2}H$ (‰)	d-excess
Schirmacher Oasis				
S1	SO1	−25.3	−207.9	−5.15
S2	SO2	−24.3	−201.2	−7.07
S3	S3	−24.0	−202.0	−10.2
S4	S4	−24.7	−206.9	−9.18
S5	S5	−18.0	−167.2	−23.3
S6	S6	−25.1	−213.8	−12.8
S7	S7	−21.2	−192.5	−23.2
S8	S8	−31.2	−250.2	−0.97
S9	S9	−30.9	−249.1	−1.73
S10	S10	−26.6	−220.1	−7.64
S11	S11	−28.6	−237.1	−8.37
S12	S12	−19.3	−184.5	−30.5
Larsemann Hills				
L1	GP	−17.4	−138.7	0.08
L2	GP	−12.7	−110.6	−9.08
L3	GP	−13.4	−113.3	−6.02
L4	GP	−14.8	−122.4	−4.09
L5	SP	−17.9	−146.5	−3.12
L6	SP	−16.8	−141.7	−7.06
L7	SI	−15	−130.2	−10.4
L8	SI	−11.8	−111	−16.3
L9	MI	−7.6	−108.7	−48.3
L10	BP	−12.7	−112	−10.5
L11	BP	−12.8	−117	−14.6
L12	BP	−11	−104.7	−16.6
L13	BP	−17.3	−138.7	−0.66
L14	BP	−17.9	−142.5	0.98
L15	BP	−14	−116.5	−4.19
L16	SI	−13.3	−110.3	−4.31

Compared to other lakes such as in the Schirmacher Oasis in East Antarctica, the lakes in Larsemann Hills have an enriched isotopic composition (Figure 7.7). L9 located on the McLeod Island was the most enriched ($\delta^{18}O$ = −7.6‰). The lakes in the Groveness Peninsula (GP) have an average isotopic composition of $\delta^{18}O$ = −14.6‰, δD= −121.3‰; lakes in Stornes Peninsula (SP) have $\delta^{18}O$ = −16.8‰, δD = −141.7‰; Sigdoy Island (SI) has $\delta^{18}O$ = −12.5‰, δD = −110.6‰, and the Broknes Peninsula (BP) has $\delta^{18}O$ = −14.3‰, δD = −121.9‰. Lakes located in the islands were enriched and on average, a difference of 5‰ was noted as that of the peninsular lakes. As expected, most of the lakes in the precipitation dominant category were depleted in heavier isotopes than the shallow depression storage lakes. However, lake L1 (GP), though being

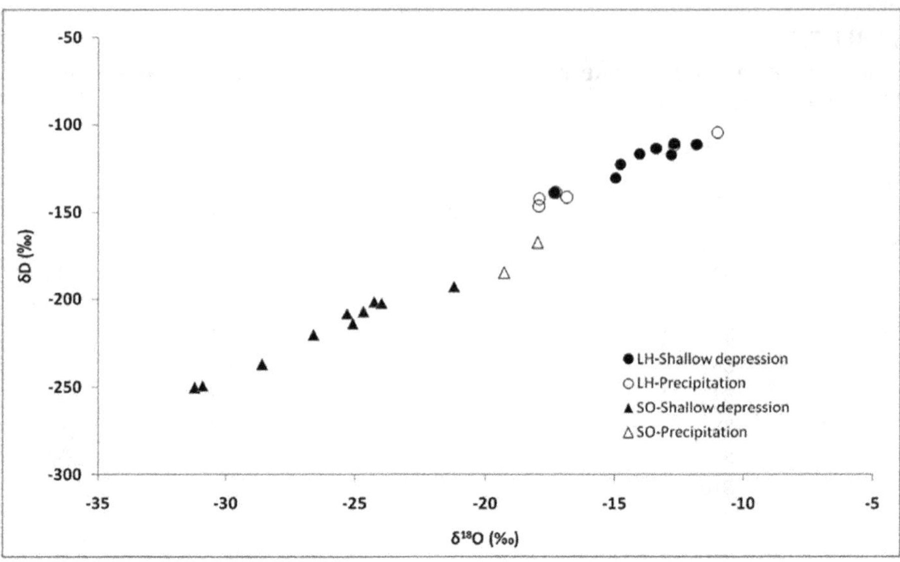

FIGURE 7.7 Isotopic composition of the precipitation dominant lakes and surface shallow depression lakes in Schirmacher Oasis and Larsemann Hills, East Antarctica

depression storage, was depleted in heavier isotopes may be because of the inflow of depleted melt eater. Similarly, L9 (MI) and L12 (BP) had isotopic enrichment though they were precipitation dominant. The contrasting behavior can be attributed to the evaporation processes occurring in the lakes.

7.3.3 δD-δ^{18}O Relationship

The δD and δ^{18}O values of lake water of Schirmacher Oasis were found to plot below the GMWL with a slope of 6 in the regression plot (Figure 7.8). As seen earlier, due to the diffusion-controlled kinetic effect at the liquid-ice interface, the formation of ice in lakes have a smaller isotope fractionation for $^{2}H/^{1}H$ (1.013–1.017), which results in a slope of 5–6. In addition, in some lakes, evaporative enrichment of water is evident from the higher ionic content. This is also confirmed by their relatively enriched isotopic composition. The evaporation of lake water can also lower the slope of the δD-δ^{18}O regression line. The lake water samples can be grouped into two based on the isotopic composition as $\delta^{18}O < -20‰$ and δ^{18}O between –15‰ and –20‰. Most of the samples were found in the first group. Relatively enriched δ values were obtained for S5 and S12, in which ionic concentrations were also higher. Since the sampled lakes were shallow, evaporation in the austral summer may have concentrated the solutes and enriched the water in these lakes.

The lake waters of Larsemann Hills have a slope of 4.54 close to that observed for systems dominant in evaporation. Horita et al. (2008) have shown that in a naturally freezing system, the remaining water will be depleted in heavier isotopes and will have a slope of 5–6 in the regression plot of δ^{18}O-δD that are similar to the slopes due to evaporation processes. The intersection point of the lake water line (LWL) with

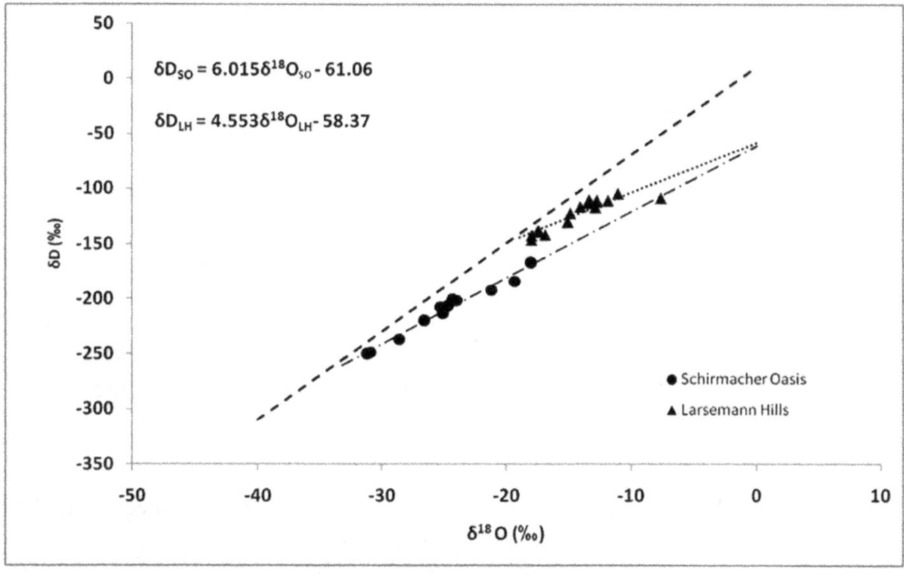

FIGURE 7.8 Regression plot $\delta^{18}O$ *vs* δD of the lake waters. The evaporation line or lake water line is also given

the global meteoric water line represents the isotopic composition of the precipitation in the region. Thus $\delta^{18}O = -20‰$ and $\delta D = -144‰$ can be considered as the initial isotopic composition of precipitation. A few lakes in the BP and SP have δ values close to this value but are more positive probably due to the kinetically controlled ice-water fractionation processes. However, the slope of LWL is mainly defined by other lakes in which evaporation can be the dominant process. The d-excess values obtained also are in accordance. The evaporation process enriches the water body lowering the d-excess of the residual water. The island lakes have the lowest d-excess values being shallow basins with entrapped seawater.

7.4 CONCLUSIONS

The lakes of the Schirmacher Oasis and Larsemann Hills region of the East Antarctic region were explored to find the chemical and stable isotope compositions. In Schirmacher Oasis, the lakes were fresh water in character, which receives input from glacial melt water. The major geochemical processes can be the weathering of silicate and carbonate minerals along with reverse ion-exchange reactions and in some lakes, evaporation was found to control the ionic makeup. The lakes were isotopically depleted and are characteristic of the polar lakes receiving glacial melt water. The kinetic fractionation effect at the ice-liquid boundary/evaporation of lake water determines the slope of the regression line of the lake water. Isotope data also affirmed the effect of evaporation from the shallow lakes during austral summer. In the Larsemann Hills in East Antarctica, weathering and reverse ion exchange processes were the source of ions. The lakes have an enriched isotopic composition than the lakes in the Schirmacher Oasis. The isotopic ratios could identify the lakes as precipitation dominant and evaporation dominant. The kinetic controlled ice-water fractionation and evaporation are the processes controlling the isotopic composition of the lake water in the Larsemann Hills.

ACKNOWLEDGMENTS

The authors gratefully acknowledge the support rendered by the Director, Indian Institute of Geomagnetism (IIG) and the Director, National Centre for Polar and Ocean Research (NCPOR) for the permission and support to carry out this study. The authors express their sincere thanks to the Executive Director, CWRDM (Centre for Water Resources Development and Management), Kozhikode, for its facilities. The authors also thank the Editor, Dr. Neloy Khare, for the constant support and encouragement.

REFERENCES

APHA. *Standard Methods for the analysis of water and waste water*, 19th edn. American Public Health Association, Washington, D.C., 1995.

ATCM XXXVII Final Report. *Larsemann Hills, East Antarctica Antarctic specially managed area management plan*, 2014. Secretariat of the Antarctic Treaty Buenos Aires 2014, https://documents.ats.aq/ATCM37/fr/ATCM37_fr002_e.pdf

Ferrik MG, Calkins DJ, Perron NM, Cragin JH, Kendall C. Diffusion model validation and interpretation of stable isotopes in river and lake ice. *Hydrol. Process.* 2002; 16(4): 851–872.

Gibson JJ, Prowse TD. Isotopic characteristic of ice cover in a large northern river basin. *Hydrol. Process.* 1999; 13(16): 2537–2548.

Gopinath G, Resmi TR, Praveenbabu M, Pragath M, Sunil PS, Rawat R. Isotope hydrochemistry of lakes in Schirmacher Oasis, East Antarctica. *Indian J. Geo-Mar. Sci.* 2020; 49 (6): 947–953.

Haberlandt R, Haendel D, Hermichen W-D, Strauch G, Wand U. *The lakes of central queen maud land (East Antarctica): Their recent isotope hydrologic characteristics, origin and evolution.* International Atomic Energy Agency, Vienna, 1987.

Haendel D, Hermichen W-D, Höfling R, Kowski P. Hydrology of the lakes in Central Wohlthat Massif, East Antarctica: new results. *Isot. Environ. Healt. S.* 2011; 47 (4): 402–406.

Hodgson DA. Antarctic lakes. In: Bengtsson L, Herschy RW, Fairbridge R, editors. *Encyclopedia of lakes and reservoirs.* Springer, Berlin, 2012, pp. 26–31.

Horita J, Rozanski, K, Cohen S. Isotope effects in the evaporation of water: a status report of the Craig-Gordon model. *Isot. Environ. Healt. S.* 2008; 44(1): 23–49.

Katz BG, Coplen TB, Bullen TD, Davis JH. Use of chemical and isotopic tracers to characterize the interaction between groundwater and surface water in mantled karst. *Ground Water*, 1998; 35: 1014–1028.

Kaup EB. The schirmacher ponds: General description, temperature and radiation regime. *Polar Geog. Geol.* 1982; 6 (3): 219–229.

Kiernan K, Gore D, Fink D, White D, McConnell A, Sigurdsson I. Deglaciation and weathering of Larsemann Hills, East Antarctica. *Antarct. Sci.* 2009; 21 (4): 373–382. doi:10.1017/S0954102009002028

Mathur AK, Ashana R, Ravindra R. Arcellaceans (thecamoebians) from core sediments of Priyadarshini Lake, Schirmacher Oasis, Eastern Antarctica. *Curr. Sci.* 2006; 90 (12): 1603–1605.

Michel FA. Isotope geochemistry of Frost-Blister Ice, North Fork Pass, Yukon, Canada. *Can. J. Earth Sci.* 1986; 23(4): 543–549.

Miller LG, Aiken GR. Effects of glacial melt water inflows and moat freezing on mixing in an ice-covered Antarctic Lake as interpreted from stable isotope and tritium distributions. *Limnol. Oceanogr.* 1996; 41(5): 966–976.

Mojib N, Huang J, Hoover RB, Pikuta EV, Pikuta EV, Storrie-Lombardi M et al. Diversity of bacterial communities in the lakes of Schirmacher Oasis, Antarctica. In: *Proceedings of the SPIE 7441, Instruments and Methods for Astrobiology and Planetary Missions XII*, 74410J, 2009. doi:10.1117/12.831289

Phartiyal B. Holocene paleoclimatic variation in the Schirmacher Oasis, East Antarctica: A mineral magnetic approach. *Polar Sci.* 2014; 8 (4): 357–369.

Phartiyal B, Sharma A, Bera SK. Glacial lakes and geomorphological evolution of Schirmacher Oasis, East Antarctica, during late quaternary. *Quat. Internat.* 2011; 235 (1–2): 128–136.

Resmi TR, Gopinath G, Praveenbabu M, Pragath M, Sunil PS, Arjun P, Rawat R. Chemical and isotopic characterization of lakes in the Larsemann Hills, East Antarctica. In: Khare N, editors. *Assessing the Antarctic environment from a climate change perspective, earth and environmental sciences library.* Springer Nature, Switzerland AG, 2022, pp. 153–166.

Richter W, Strauch G. Deuterium and 18O variations in lakes of the Schirmacher Oasis (East Antarctica). *Isotopenpraxis*, 1983; 19 (5): 145–153.

Schoeller H. Qualitative evaluation of groundwater resources. In: *Methods and techniques of groundwater investigations and development.* UNESCO, Paris, 1965, pp. 54–83.

Schoeller H. *Geochemistry of groundwater—An international guide for research and practice.* UNESCO, Paris, 1967, pp. 1–18.

Sengupta SM. Geology of Schirmacher Range (Dakshin Gangotri), East Antarctica. Scientific report of the 3rd Indian science expedition to Antarctica. *DOD, Govt. India Pub. Techn. Pub.* 1986; 3: 187–217.

Sinha R, Chatterjee A. 2000. Thermal structure, sedimentology, and hydro-geochemistry of lake Priyadarshini, Schirrmacher oasis, Antarctica. *Sixteenth Indian Expedition to Antarctica, Sci. Rep. Dep. Ocean Develop. Techn. Pub.* 2000; 14: 1–36.

Smith JA, Hodgson DA, Bentley MJ, et al. Limnology of two Antarctic epishelf lakes and their potential to record periods of ice shelf loss. *J. Paleolimnol.* 2006; 35 (2): 373–394.

Souchez R, Tison JL, Jouzel J. Freezing rate determination by the isotopic composition of the ice. *Geophy. Res. Lett.* 1987; 14(6): 599–602.

Srivastava AK, Randive KR, Khare N. Mineralogical and geochemical studies of glacial sediments from Schirmacher Oasis, East Antarctica. *Quatern. Int.* 2012; 292: 205–216.

Strauch G, Haendel D, Maaß I, Mühle K, Runge A. Isotope variations of hydrogen, carbon and nitrogen in florae from the Schirmacher Oasis, East Antarctica. *Isot. Environ. Healt. S.* 2011; 47(3): 280–285.

Wand U, Hermichena W-D, Brüggemannb E, Zierath R, Klokov VD. Stable isotope and hydrogeochemical studies of Beaver Lake and Radok Lake, MacRobertson Land, East Antarctica. *Isot. Environ. Healt. S.* 2011; 47(4): 407–414.

Wang YQ, Chen XB, Meng GL, Wang S, Wang Z. On changing trends of δD during seawater freezing and evaporation. *Cold Reg. Sci. Technol.* 2000; 31(1): 27–31.

Warrier AK, Pednekar H, Mahesh BS, Mohan R, Gazi S. Sediment grain size and surface textural observations of quartz grains in late quaternary lacustrine sediments from Schirmacher Oasis, East Antarctica: Paleoenvironmental significance. *Polar Sci.* 2016; 10 (1): 89–100.

8 Limnological Assessment of Water Bodies of Extreme Antarctic Climatic Conditions

Rajni Khare,[1] Ashwani Wanganeo,[1] and Rajni Wanganeo[2]

[1] Department of Environmental Sciences & Limnology, Bakatullah University, Bhopal (Formerly)

[2] Department of Zoology, Government Benzeer College of Commerce & Science, Bhopal (Formerly)

CONTENTS

DOI: 10.1201/9781003284413-8

8.1 INTRODUCTION

The Antarctic, a region though remotely situated and hostile in terms of conditions is an integral and highly significant component of the Earth's climate system. These regions are the first to suffer from global warming and constitute an archive of vital climatic variations. Antarctica is the world's last unspoiled wilderness. It drives the world's climate and has a central role in regulating the Earth's environmental processes, including the world's atmospheric and oceanic systems, global tides, and sea levels. Antarctica's diverse ecosystem is dependent on a healthy environment. Remote, inaccessible, and inhospitable, Antarctica was the last continent to be discovered.

One of the Earth's last great frontiers, Antarctica presents a foreboding environment to life. Freezing temperatures are a challenge to biological metabolic functions. Freezing which keeps in its solid, ice state, makes the continent the driest "desert" on the planet, which is another challenge to living in this unforgiving environment (Figure 8.1).

Nonetheless, Antarctica is not a sterile, lifeless environment. Micro-organisms have been found to exist in the upper layers of the ice sheet and microbes and microfossils have been found in deep ice cores. Some microbes in ice cores may even date back to more than 200,000 years. Colonies of bacteria can survive the long winters of extreme cold by a form of hibernation. In the Antarctic "summer," the warmth of sunlight is sufficient to melt small quantities of liquid water around dust particles acting as tiny solar collectors. It is at these sites that the bacteria complete their life cycles. Algae have also been discovered coexisting in these colonies, whose seeds over-winter and then "sprout" in the south polar summer.

Interestingly, the ice-free areas of Antarctica display a sharp contrast to most other ecosystems in the world, which exist under far more moderate environmental conditions. Such ecosystems are dominated by microorganisms, mosses, lichens, and relatively few groups of invertebrates; higher forms of life are virtually nonexistent.

The icy continent "Antarctica" holds answers to many scientific riddles and thus offers tremendous scope for carrying out research in various scientific fields. Since the discovery of oil and other mineral resources in the polar regions, significant interest has been generated in exploring these regions in the interest of seeking knowledge and potential resources. And awareness of the environment. The untrammeled nature of the Antarctic also offers a unique opportunity for nature's singular laboratory where scientists can measure the effects of changes in the environment. The unique location, climatic conditions and pristine nature makes this region ideal to study biological and environmental phenomena. Antarctica, being a storehouse of 90% of the world's ice and 75% of freshwater locked in the form of ice, provides a unique platform to study continental glaciers and sea ice in a pristine environment.

Antarctica is particularly vulnerable to some types of environmental change, notably those that would require a biological activity for reversal or amelioration. Pollutants that would be readily biodegradable elsewhere can have very long lifetimes in the Antarctic environment, increasing the possibility of long-term alteration through human activities. Its preservation and well-being are therefore vital to the

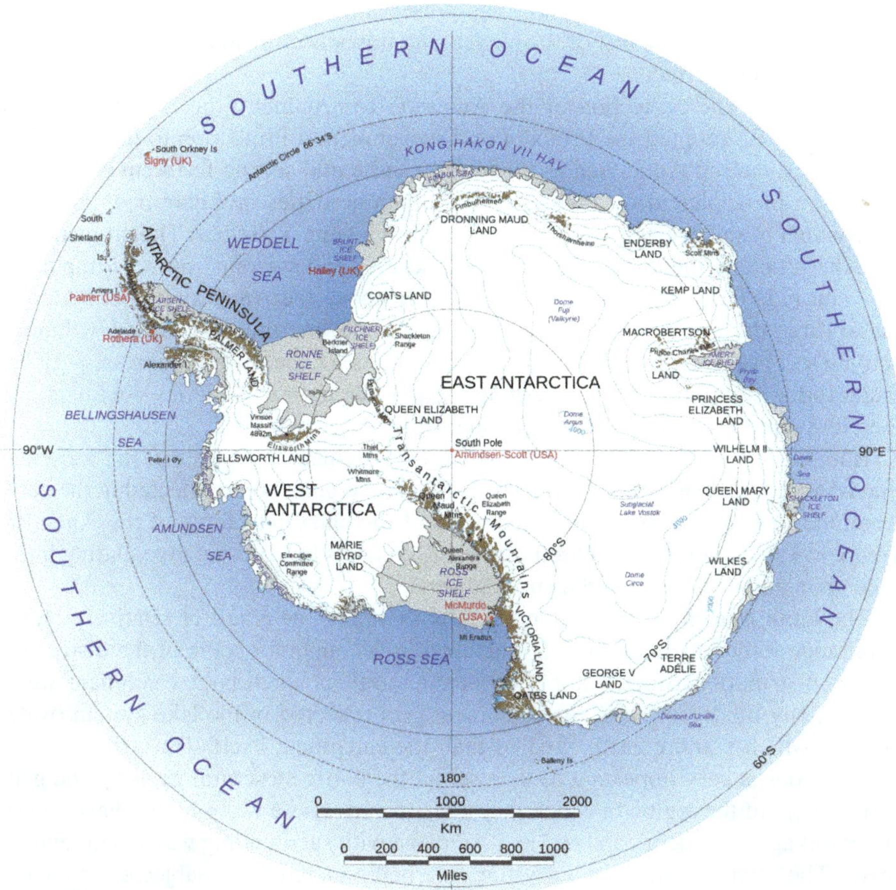

FIGURE 8.1 The Antarctic continent.

Source: https://commons.wikimedia.org/wiki/File:Antarctica.svg.

health of the rest of the planet, and impacts on Antarctica's environment could have global effects.

Antarctica, as a global "Hot Spot" of biodiversity, is dominated by extremophilic bacteria, fungi and algae which are well adapted to the extreme habitats of Antarctica, such as low temperature, and low humidity, high salt content and high pressures.

The preservation of Antarctica's environment and fragile ecosystem, the open and free exchange of scientific knowledge, and international cooperation rest on a very fragile political base and are governed by the Antarctic Treaty System. Antarctica is one of the few remaining nearly pristine sites in the world and is certainly by far the largest such site. Antarctica is particularly vulnerable to some types of environmental change, notably those that would require a biological activity for reversal or amelioration. Pollutants that would be readily biodegradable elsewhere, can have very long lifetimes in the Antarctic environment, increasing the possibility of long-term alteration through human activities. Its preservation and well-being are therefore vital to the health of the rest of the planet, and impacts on Antarctica's environment could have global effects.

To enhance the protection of the Antarctic environment, the Antarctic Treaty parties in 1991 adopted the Protocol on Environmental Protection to the Antarctic Treaty, designating Antarctica as a natural reserve and setting forth environmental protection principles to be applied to all human activities in Antarctica, including the conduct of science, tourism, and other anthropogenic activities. The protection of the Antarctic continent, and the great Southern Ocean surrounding it, is important for mankind. The discovery of the ozone hole above the Antarctic in 1985 alerted the world to the potentially dangerous changes in the environment caused by human activities. This discovery led to the first measures to control pollution on a global scale. India is committed to preserving the pristine nature of the icy continent and therefore undertakes regular environmental Impact assessments.

Human activity in the Antarctic began as a consequence of a deep desire for scientific knowledge along with economical gains. These goals were reflected in the early Antarctic exploration in the 19th century and the entry into force of the Antarctic Treaty in 1961. Interest in Antarctica now also reflects humankind's expanding influence upon and awareness of the environment.

The biota inhabiting the region today is therefore the result of relatively recent long-range re-colonization, which undoubtedly accounts for some of the distinctive species distribution. Since water bodies are natural sumps of large catchment areas, as such, any modification of the catchment gets reflected in the lake system (water bodies), which is more easily studied than the catchment itself. Thus, the study of water bodies is very important as they reveal the entire structural geology and geomorphology of the region in which they are situated. These water bodies have a variety of linkages for energy and nutrient exchange with surrounding watersheds and air sheds. The inland freshwater ecosystems are being increasingly subjected to greater stress from various human activities (Wood and Gibson, 1974; Hemasundaram, 2003). The eutrophication of water means enrichment with nutrients and the resulting degradation of its quality is accompanied by a luxuriant growth of micro and macrophytes, which is recognized as a major worldwide problem including in polar regions. The enrichment of nutrients occurs due to the disposal of pollutants from the runoff from surrounding areas.

To understand the nature of the physicochemical and biological events taking place in the water body, it is important to relate these events to their shape and size (morphometry). The size and shape of a waterbody are, however, largely dependent on the forces that have produced these basins. Thus, it is important to understand the geomorphic aspects of these aquatic waterbodies, as their specific nature depends upon their formation type, latitudinal position, and the type of water they hold/receive.

With the opening of Antarctica, continent to human influence in the form of scientific programs and recent tourist influx, a lot of pressure is being felt on the aquatic eco-systems occurring around the fringe of the continent (Antarctica). The biological productivity of waters together with all the causal influence on the quantitative features along with its actual potential aspects need to be worked out under extreme climatic conditions. All the organisms present in a particular area have unique functions and methods to play and occupy the specific habitat. The physical-chemical conditions of the area have got a direct influence on the organisms and at the same time the habit and habitats are also influenced by them.

The various limnological parameters vary considerably in the Antarctic where the temperature remains almost below zero throughout the year as compared to tropical regions. The temperature plays a vital role in the productivity of a particular waterbody. The oxygen level is low due to low photosynthetic activity in the waterbodies of Antarctica.

In freshwater bodies, there are enormous differences in production between even geographically adjacent lakes. The increased solar energy towards summer brings about a dramatic increase in the metabolic activity in the waterbodies, which is a significant feature in Antarctic waterbodies. Though the biological population is very low in freshwater, vertical variation is supposed to be very prominent in Antarctic waterbodies.

Interestingly, peripheral Antarctica has a number of water bodies of various types, viz. (1) perennially exposed, (2) seasonally covered with an ice sheet, and (3) perennially covered with the ice sheet. A few of these lakes lie in proximity to the shoreline and thus are affected during the transgression and regression of the sea.

Over 100 small and large-sized freshwater lakes exist in the Schirmacher Oasis, yet the limnological studies in the Antarctic region are limited. Although, initial attempts were made by Ingole and Parulekar (1987, 1990, 1993) on ecological assessment of the freshwater habitats (Schirmacher Oasis) and the nutrient concentrations and primary productivity in some of the lakes by Matondkar and Gomes (1983), and SenGupta and Qasim (1983). However, the integrated study on the chemical, microbiological and ecological characterization of the freshwater lakes was started in 1983–4, during the fourth Indian Scientific Expedition to Antarctica (Ingole and Parulekar, 1987). Recently Wanganeo *et al.* (2018) studied the variations in physicochemical characteristics of water bodies placed at different geographical locations in Antarctica.

To assess the extraordinary evolutionary adaptations to the extreme environment, the limnological investigations thus gain importance. Although considerable investigations have been made by the researchers in the other fields of polar science, little information is known about the systematic limnological studies in the high latitude polar region. However, it is important to note that inadequate attention has been paid to the qualitative, quantitative, or ecological study on limnology so far from the Antarctic region. Both freshwater and meromictic lakes in the Antarctic region have received little attention from the ecological studies point of view as well as the quantitative estimation of the biological population under severe climatic conditions.

Since the Limnological study of Antarctic waterbodies is still in infancy it as such, needs to be worked out thoroughly Such studies will throw light on fundamental ecological principles in a relatively unpolluted natural laboratory with its unique ecosystems. In this chapter, we address the limnological investigation in relation to the different physicochemical characteristics of water bodies of the Schirmacher Oasis region for a period of two austral summers (2006–7 and 2007–8). This study will form a baseline for assessing the Environmental impact on pristine Antarctic terrestrial and freshwater (or saline) habitats due to anthropogenic activities.

8.2 PRIYADARSHINI WATER BODY

Priyadarshini water body, trending in ENE and WSW direction (Figure 8.2) is situated in the Schirmacher Oasis located about 100 km south of Lazarev Sea up to 228 m. The water body is perennially fed by the melting of the polar ice that stands all along the southern margin of the oasis as a fortress and drains out into smaller lakes in the north and finally onto a vast ice shelf extending tens of kilometers up to polynya.

Priyadarshini water body has been reported to have formed after the last deglaciation (Pleistocene-Holocene) and not much information is available about the geologic history of the water body since not many cores have been collected and studied to unravel the geologic history of the water body.

Priyadarshini water body (L-49) is by far the second-largest landlocked and one of the deepest lakes, out of more than a hundred lakes in the Schirmacher Oasis. It is situated in the vicinity of the second permanent Indian Station in Antarctica, i.e., Maitri, at 70°45'39.4" S latitude and 11°44'8.6" E longitude, at an elevation of 117 m AMSL. In general, the Priyadarshini water body extends in ENE and WSW directions. The water body is perennially fed by the melting of the polar ice that stands all along the southern margin of the oasis and drains out into smaller water bodies in the north and finally onto vast ice shelves extending tens of kilometers up to polynya.

A continuous cycle of freezing and thawing in winter and summer brings a large amount of changes in the physicochemical and biological properties of the water and bottom sediments. The influences of environmental and physical changes provide unique conditions for the development of biotic communities in and around the lakes. The extreme air temperature varies from a minimum of –40 ºC in winter to a maximum of +6 ºC in summer. Precipitation in the form of snowfall is quite heavy but most of it is blown by strong winds from easterly directions. Summer melting of glaciers and snow banks replenishes the water losses in these lakes due to evaporation. Due to the very harsh environmental conditions, the Antarctic flora is sparse and almost Cryptogamic, i.e., comprising Bacteria, Algae, Fungi, Lichens, and Bryophytes.

8.3 MORPHOMETRIC STUDIES

Morphometric parameters were calculated after Goldman and Horne (1983).

8.3.1 BATHYMETRY (DEPTH)

Lake Priyadarshini, the largest and one of the deeper lakes of Schirmacher Oasis, was selected to carry out a bathymetric survey of the lake. Though Ingole and Perulaker (1990) had carried out a bathymetric survey of the Priyadarshini lake by spot sounding using conventional string rope and, Swain and Chaturvedi (2017) attempted with GPR, the present bathymetric survey is a systematic approach to carry out the continuous bathymetric profiling of the lake using portable geophysical equipment called a "HydroBox" echosounder (Chaturvedi and Khare, 2006). A total of 1800 line meter area was surveyed and continuous profiles of depth data in digital and analogue forms were recorded. The data collected during the survey were analyzed and the bathymetric map of the surveyed area of the lake is prepared. The bathymetric data collected in the present study once integrated with the sub-bottom profiler data will provide detailed information about the lake and would suggest the suitable locations to retrieve the long sediment cores for paleoclimatic study. The map and data generated can be effectively utilized for various other scientific purposes for collecting samples for biological/environmental studies.

To understand the nature of the physicochemical and biological events taking place in the water body, it is important to relate these events to their shape and size (morphometry). The chemical and biological conditions of lakes are influenced by their morphometric characteristics, especially retention time, volume, surface area, and mean depth (volume/surface area). These factors affect processes related to input, output, internal circulation of substances, re-suspension, and sedimentation. Mean depth is generally considered the single best index of morphometric conditions and shows an inverse relationship to productivity at all trophic levels in large lakes.

The bathymetric survey of the Priyadarshini water body (Figure 8.2) in Schirmacher Oasis has been undertaken using a modern shallow water echo-sounder. A total of 1800 line meter areas were surveyed, and continuous bathymetric profiling data were recorded. Some important morphometric features of the Priyadarshini water body have been given in Table 8.1.

FIGURE 8.2 Areal view of Priyadarshini Lake near Maitri.

TABLE 8.1
Important Morphometric Features of Priyadarshini Water Body

Shore length	m	1024.00
Shore line development		2.78
Maximum length	m	418.00
Maximum depth	m	5.90
Mean depth	m	2.95
Area	m^2	33880.00
Relative depth	m	2.84
Volume	m^3	99946.00

8.4 LIMNOLOGICAL STUDIES

8.4.1 Physicochemical Parameters

Samples from different water bodies placed at different longitudes and latitudes were collected during the austral summer period only. Standard methods as given in American Public Health Association- APHA (1985) and Golterman and Clymo (1979), were followed for analysis purposes besides using a portable HACH field laboratory kit. Samples were brought to the laboratory to measure temperature (air and water), transparency, depth, conductivity, pH, dissolved oxygen, free CO_2, alkalinity, total dissolved solids (TDS), hardness, total hardness, calcium and magnesium hardness, phosphate, nitrogen (ammonia), nitrogen nitrite, nitrogen nitrate, sulphate, chloride, iron, silica, and manganese.

Physicochemical parameters viz. temperature, pH, conductivity, dissolved oxygen, total dissolved solids etc., were analyzed in the field itself as they are liable to changes. A total of 19 limnological parameters of water *viz.*, temperature, transparency, pH, dissolved oxygen, free carbon dioxide, total alkalinity, conductivity, $CaCO_3$ hardness, total dissolved solid, chloride, ammonia, nitrite, nitrate, and phosphate were determined. All the parameters were analyzed following the standard methods (Golterman, 1969; Michael, 1984; Trivedi and Goel, 1984; APHA, 1985, 1989; Wetzel and Likens, 1991).

8.5 ANALYTICAL TECHNIQUES

Analytical techniques as described in APHA *et al.* (2005) and Trivedi and Goel (1986) were used for the physicochemical analysis. The physicochemical parameters viz., ambient and water temperature were determined by mercury filled centigrade thermometer, depth by measuring tap graduated in meters and centimeters, transparency by Secchi disc, color by visual estimation, electrical conductivity by conductivity meter, total dissolved solids by gravimetric method, pH by digital pH meter, dissolved oxygen by Winkler's iodometric method (Lind, 1979; APHA, 1985), free carbon dioxide by titration method using phenolphthalein solution as an indicator, total alkalinity by titration method using strong acid and methyl orange and phenolphthalein solutions as indicators using standard methods described by Boyd (1979) and APHA (1992), total hardness and calcium hardness by ethylene diamine tetra acetic acid (EDTA) titration method using eriochrome black-T and murexide indicators (Lind, 1979; APHA, 1985), chlorides by argentometric method, sulphates by turbidimetric method, nitrate-nitrogen by colorimetric method using brucine sulfanilic acid, nitrite by diazotization method inorganic phosphorous by using stannous chloride method, silicate by molybdosilicate method, ammonia by volumetric method, sulfide by precipitation method, biochemical oxygen demand by Winkler's Iodometric method, chemical oxygen demand by reflux condensation method using ferroin as an indicator, magnesium by indirect method and sodium and potassium were estimated with the help of a flame photometer. Physicochemical characteristics of water, like ambient and water temperature, depth, transparency, color, pH, dissolved oxygen, free carbon dioxide, total alkalinity, total hardness, chlorides, calcium, and magnesium were determined at the sampling sites immediately after the collection of water samples while rest of the parameters were analyzed in the laboratory within six hours after the collection of water samples.

8.6 BIOTIC PARAMETERS

8.6.1 Phytoplankton

Phytoplanktons are chlorophyll bearing suspended microscopic organisms consisting of algae. The Phytoplankton samples were collected with the help of a plankton net made of bolting silk of 20μ mesh size. The collected samples were preserved with a formaldehyde solution and then the samples were identified and the results for phytoplankton have been expressed as units/liter (Wanganeo and Wanganeo, 1991, 2006). The biomass of phytoplankton was analyzed as per procedures provided by Edmondson (1959).

8.6.2 Zooplankton

The zooplanktons are the microscopic free-swimming animal components of the aquatic system which are represented by a wide array of taxonomic groups of which the members belong to Protozoa, Rotifera, Cladocera and Copepoda are the most common and dominant among the entire consumer communities. The Zooplankton samples were collected with the help of a plankton net made of bolting silk of 20μ mesh size. The collected samples were preserved with a formaldehyde solution and then the samples were identified and the results for zooplankton have been expressed as organisms/l (Wanganeo and Wanganeo, 1991, 2006). The biomass of zooplankton was analyzed in accordance with Downing and Rigler (1984).

8.7 ENVIRONMENTAL PARAMETERS

The monitoring of the Priyadarshini **water body** was carried out by Ingole *et al.* (1987) and Ingole and Dhargalkar (1998). The atmospheric temperature varies from –34.5° C to +8.2° C. The temperature of the ambient water below the surface frozen layer ranged from −1.5° C to +4.7° C. The amount of dissolved oxygen varies from 8.4 to 12.1 mg l (Ingole *et al.* 1987). The biological productivity as well as biomass show biannual peak (during the austral summer and spring season). The productivity variations have been closely related to the ice extent and atmospheric conditions above and adjacent to the lake. The pH of the water has been reported to vary between 8.1 and 8.8 (Ingole and Dhargalkar, 1998).

Dissolution of the alkaline rocks in the vicinity of the lake by the glacial water has been cited as the reason for the relatively high pH. The salinity and amount of dissolved nutrients are very low. The bottom sediments are fine black and covered by blue-green algae. In general, coarse sand with a very low amount of silt clay characterizes most of the lake floor. The sublittoral margin of the Priyadarshini water body is occupied by blue-green algae and moss (Ingole *et al.*, 1987; Khare *et al.*, 2008).

8.8 NUTRIENT ANALYSIS OF AQUATIC ECOSYSTEMS

8.8.1 In Streams

Differences in nutrient concentrations among streams have been reported by Pandey *et al.* (1995). Table 8.2 shows that NH_4^+ and NO_3—N and phosphate were the lowest in-stream WN, whereas a maximum of NO_{3-} and NH_4^+ was observed in stream EGF, close to stream WN. Stream ST was richer in phosphate than the other streams (Table 8.2). An ANOVA indicated that six streams differed significantly in physicochemical characteristics ($F_{stream} = 48.21$; $p < 0.001$).

8.8.2 In Glaciers, Lakes, and Ponds

Data regarding the chemical analyses of water samples collected from three different fresh water bodies in and around the study area have been collated from published reports (Pandey *et al.*, 1995) and are presented in Table 8.3. Stream (SEM) contains a high amount of Cl^-, SO_{4-}^{2}, NO_{3-} and a moderate amount of NH_4^+ and

TABLE 8.2
Chemical Analysis of Water Samples Collected from Middle Regions of Different Streams of Schirmacher Oasis, Antarctica

Physicochemical characteristics	Streams*					
	WN	SEM	ST	EGF	EM	NS
pH	7.9 ± 0.3	7.4 ± 0.2	7.4 ± 0.2	7.8 ± 0.5	7.5 ± 0.3	7.4 ± 0.2
Chloride (mgl^{-1})	1.4 ± 0.2	1.2 ± 0.2	1.1 ± 0.2	1.1 ± 0.2	1.2 ± 0.1	1.1 ± 0.1
Sulphate (µgl^{-1})	3.2 ± 0.6	2.9 ± 0.5	3.4 ± 0.5	3.0 ± 0.3	3.0 ± 0.3	3.0 ± 0.4
Nitrate-N (µgl^{-1})	9.5 ± 0.8	13.0 ± 2.1	14.5 ± 1.6	15.5 ± 2.4	12.1 ± 0.9	12.0 ± 1.5
Ammonium-N (µgl^{-1})	2.4 ± 0.4	6.9 ± 1.0	5.0 ± 0.3	8.5 ± 0.9	3.8 ± 0.5	4.9 ± 0.6
Orthophosphate (µgl^{-1})	0.6 ± 0.1	1.5 ± 0.1	1.6 ± 0.3	0.9 ± 0.1	1.0 ± 0.1	1.1 ± 0.2

Source: After Pandey *et al.*, 1995

TABLE 8.3
Chemical Analysis of Water Samples Collected from Fresh Water Ecosystems in Schirmacher Oasis, Antarctica

Physicochemical properties	Glacier melt water stream	Snow drifts melt water stream	Priyadarshini lake	Pond
pH	**7.4**	**7.4**	**7.2**	**7.6**
Chloride (mgl^{-1})	**1.2**	**1.1**	**1.8**	**1.3**
Sulphate (μgl^{-1})	2.9	3.0	2.6	2.6
Nitrate (μgl^{-1})	13.0	12.0	16.3	17.9
Ammonium (μgl^{-1})	6.9	4.9	5.2	5.4
Orthophosphate (μgl^{-1})	1.5	1.1	1.8	2.0

inorganic phosphate (P_i). The pH of all the water bodies was slightly alkaline ranging from pH 7.3 to 7.8. In Priyadarshini lakes, which represent a nearly stable fresh water ecosystem depicted higher Cl^-, NO_{3-} and PO_{4-}^{3} contents, however, SO_{4-}^{2} and NH_4^+ were present at moderately low levels compared with stream water. Pond water, a highly unstable habitat, showed a higher amount of NO_3^-, which it might have received through seepage or leaching and run-off. The Cl- and NH_4^+ were present in low amounts; however, this water body represented quite a high level of PO_{4-}^{3} (Table-8.3). The high PO_{4-}^{3} content in pond water might have supported the good growth of cyanobacteria both in terms of species richness and biomass.

8.9 LIMNOLOGICAL ASSESSMENT OF PRIYADARSHINI WATER BODY

Eight water bodies: S1a (70° 45' 53.039"S; 11° 43' 936"E), S3 (70° 45.322'S; 11° 44.210'E), S4(70° 45.553'S; 11° 44.269'E), S9(70° 46.118'S; 11° 46.507'E), S19(70° 45.683'S; 11° 43.896'E), S37(70° 45.551'S; 11° 38.887'E), S43(70° 46.687'S; 11° 33.970'E) and S53(70° 44.424'S; 11° 27.780E) have only been covered during the austral summer (2006–7) to understand spatial variations and diurnal variations in physicochemical and biological characteristics (phytoplankton and zooplankton).

8.9.1 Spatial Variation During Austral Summer 2006–7

8.9.1.1 Physicochemical Characteristics

Data procured on freshwater bodies of Schirmacher Oasis during the austral summer of 2006–7. The atmospheric temperature during the present investigation in Schirmacher Oasis varied from –0.6 °C to –4.4 °C. Among Schirmacher water bodies surface water temperature fluctuated from –0.8 °C (S19 water body) to 3.3 °C (S3). In Schirmacher water bodies, pH ranged from 7.37 (S19 water body) to 7.9 (S37 water body). Specific conductivity in Schirmacher oasis water bodies varied from 10.07 µS/cm (S19 water body) to 25.59 µS/cm (S53 water body) while, TDS in Schirmacher water bodies ranged from 4.62 ppm (S19 water body) to 11.57 ppm (S53 water body).

Except for Priyadarshini (1.4 mg/l) and S22 water body (0.6 mg/l), water bodies (S3, S4, S9, S19, S37, and S53) recorded a uniform free Carbon dioxide value of 0.4 mg/l. Likewise, a uniform total alkalinity value of 4 mg/l was recorded in Schirmacher Oasis water bodies except for Priyadarshini water body in which a value of 8mg/l. Chloride concentration varied from 3.99 mg/l (S 43 water body) to 8.99 mg/l (S1a).

Total hardness in East Schirmacher Oasis water bodies ranged from 18 mg/l (S3, S4, S19, S37, S53, S43 water bodies) to 20 mg/l (S1a, S9 water bodies). Except for Priyadarshini (12.6 mg/l) and S9 (18.9 mg/l) water bodies, all Schirmacher Oasis water bodies recorded a uniform Calcium hardness value of 16.8 mg/l. Nitrate nitrogen in Schirmacher water bodies ranged from 0.2 mg/l (S1a water body) to 0.6 mg/l (S43 water body).

Phosphate phosphorus in Schirmacher water bodies ranged from 0.01 mg/l (S43 water body) to 0.09 mg/l (S19 water body). In Schirmacher water bodies total Iron varied from 0.01 mg/l (S19, S37, S53 water bodies) to 0.03 mg/l (S3, S4, S43 water bodies). Sulphate in Schirmacher water bodies during the present investigation was found to be below detection level while Silica in Schirmacher water bodies varied from 0.01 mg/l (S53 water body) to 0.248 mg/l (S3 and S4 water bodies).

8.9.1.2 Biological Characteristics

8.9.1.2.1 Phytoplankton

Qualitatively phytoplankton community of Schirmacher Oasis water bodies was mainly represented by Chlorophyceae, Bacillariophyceae and Cyanophyceae only. Cyanophyceae dominated the group with a maximum contribution of 6 sps. (38%) followed by Chlorophyceae and Bacillariophyceae, contributing 5 sps., (31%) each (Figure 8.3).

Among water bodies of Schirmacher Oasis, S53 water bodies recorded a maximum phytoplankton population density of 448 units/l followed by S9, S4, and S19 water bodies which documented population density values of 124 units/l, 50 units/l and 42 units/l respectively. The phytoplankton population density in the rest of the water bodies ranged from (6–24 units/l) (Figure 8.4).

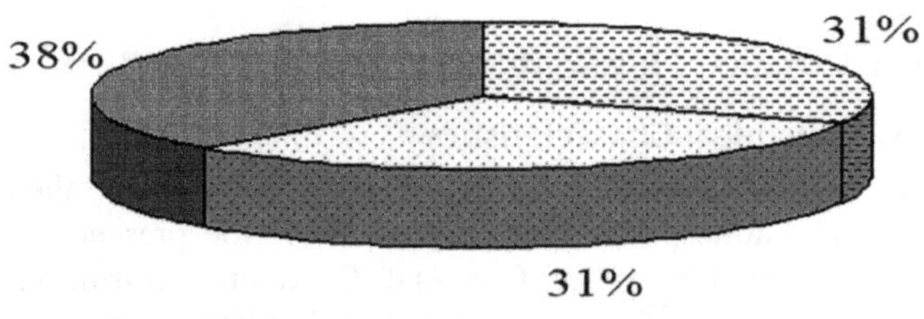

FIGURE 8.3 Phytoplankton species contribution in water bodies of Schirmacher Oasis.

As depicted in Figure 8.4, the order of dominance of various phytoplankton classes in water bodies is:

S9 water body:

- Cyanophyceae > Chlorophyceae > Bacillariophyceae

S4 water body:

- Chlorophyceae >Bacillariophyceae > Cyanophyceae

S3 water body:

- Cyanophyceae represented cent% dominance

Among Chlorophycean representatives in Schirmacher Oasis, *Cosmarium* sp. recorded relatively the high population density (60 units/l) in S9 water body. The population density of other Chlorophycean representatives remained very low with their population density values ranging from below detection level to 2 units/l only.

Among Bacillariophyceae, *Tabillaria* sp. recorded the maximum population density of 10 units/l in the S4 water body followed by *Nitzschia* sp. which documented a population density of 4 units/l in S4 water body and 2 units/l in S9 water body.

Nostoc sp. (below detection level to *420* units/l) followed by *Gleotrichia* sp. (below detection level to 44 units/*l*) and *Oscillatoria* sp. (below detection level to 16 units/l), recorded relatively maximum population density among Cyanophycean representatives in Schirmacher Oasis.

8.9.1.2.2 Zooplankton

During the present investigation zooplankton population in Schirmacher Oasis water bodies was represented by *Centropyxis* sp. only. The species, representing the class

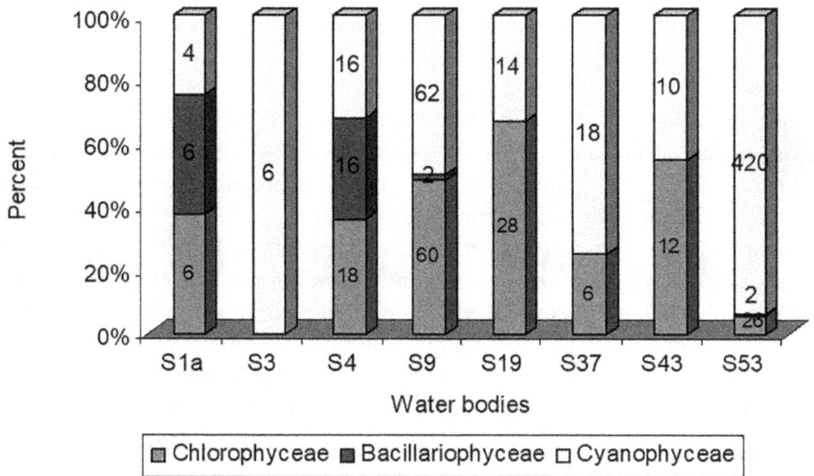

FIGURE 8.4 Phytoplankton population density (units/l) in freshwater bodies of Schirmacher Oasis during austral summer 2006–7.

protozoa, recorded its presence in an S9 water body in Schirmacher Oasis with a population density of 2 units/l during the present investigation.

8.9.2 Diurnal Variation During Austral Summer 2006–7

8.9.2.1 Physicochemical Characteristics

The diurnal study made on Priyadarshini water body during the austral summer of 2006–7 reveals that atmospheric temperature near the water surface recorded a diurnal fluctuation ranging between –2.8 °C to 0 °C with the minimum value recorded at 0.00 hrs and the maximum value obtained at 8.00 hrs. Surface water temperature in Priyadarshini water body also documented a diurnal variation ranging from –0.3 °C (at 0.00 hrs) to 0 °C (12.00 hrs).

The wind velocity along with the sunshine hours are responsible for bringing diurnal variation in both atmospheric and water temperature.

Secchi transparency recorded more or less a marked diurnally variation ranging from 58.7 inch (0.00 hrs) to 89.2 inch (12.00 hrs.) which is obvious due to changing the position of the sun on the horizon. Little change with respect to the diurnal fluctuation pattern of Priyadarshini surface water pH was recorded. The minimum pH value of 7.26 units was recorded at 16.00 hrs and the maximum value of 7.48 units was observed at 12.00 hrs.

Specific conductivity and TDS of surface water in the Priyadarshini water body recorded a little diurnal variation, ranging from 13.34 µS/cm (20hrs) to 16.66 µS/cm (12.00 hrs) and from 8.95 ppm (20.00–0.00 hrs) to 9.06 ppm (12.00 hrs) respectively. The shift in water temperature might be responsible for the enhancement in values towards noon time.

Diurnal variation with respect to dissolved oxygen in surface water of Priyadarshini water body was more or less significant ranging from 7.2 mg/l to 9 mg/l. Minimum and maximum values were recorded at 0.00 hrs and 12.00 hrs, respectively.

A near-uniform concentration value of 1.2 mg/l for carbon dioxide in surface water of Priyadarshini has been recorded for most of the time during the diurnal study except at 12.00 hrs when it rose to 1.4 mg/l.

A uniform value of 6 mg/l for total alkalinity and 18 mg/l for total hardness was observed in the surface waters of Priyadarshini water body throughout the study period except, at 12.00 hrs and 16.00 hrs when a common value of 8 mg/l for total alkalinity and 20 mg/l for total hardness was observed.

Calcium hardness diurnally varied from 12.6 mg/l to 16.8 mg/l with its minimum values recorded at 12.00 hrs and maximum values observed at 16 hrs and 0.00 hrs. Total iron in Priyadarshini surface water recorded its diurnal variation pattern ranging from 0.02 mg/l (12.00 hrs) to 0.03 mg/l (16.00 hrs, 0.00 hrs).

Priyadarshini surface water did not record any diurnal variation in its chloride content, rather, a value of 8.99 mg/l appeared throughout the 24-hour study.

Among biologically important nutrients phosphate phosphorus varied, diurnally, within a narrow range of 0.01–0.05 mg/l while, nitrate-nitrogen fluctuated from 0.3–0.5 mg/l. The minimum concentration of phosphate phosphorus was recorded at 16 hrs and 20 hrs while the maximum was observed at 4.00 hrs and 12.00 hrs. On the other hand, Nitrate nitrogen documented its minimum concentration at 16.00 hrs while its maximum concentration (0.5 mg/l), however, remained uniform throughout the study period.

Silica concentration of Priyadarshini water body during the present diurnal study varied from 0.02 mg/l to 0.029 mg/l with the minimum value recorded at 8.00 hrs, 12.00 hrs, and 16.00 hrs and maximum values observed at 0.00 hrs.

8.9.2.2 Biological Characteristics

8.9.2.2.1 Phytoplankton

Diurnal variation in population density of phytoplankton in Priyadarshini water body during the present investigation reveals its minimum (6 units/l) and maximum (18 units/l) peaks at 4.00 hrs and 16.00 hrs, respectively (Figure 8.5). Diurnally Chlorophyceae recorded minimum (2 units/l) and maximum (6units/l) population densities at 4.00hrs and 16.00 hrs, respectively (Figure 8.5). At 0.00 hrs, the class, however, remained below its detection level. Bacillariophyceae, on the other hand, documented a minimum population density of 2 units/l at 8.00 hrs and a maximum density of 12 units/l at 16.00 hrs. Minimum, 2 units/l of Cyanophyceae during the diurnal study was recorded at 8 hrs and its maximum (4 units/l) was found between 20.00–0.00 hrs. The class remained below detection level at 12.00 hrs, 16.00 hrs and 4.00 hrs (Figure 8.5).

During the present diurnal study, *Cosmarium* sp. among Chlorophyceae recorded its presence thrice; at 12.00 hrs (4 units/l), at 16.00 hrs (6 units/l) and 4.00 hrs (2 units/l). The other two species, like *Elkatothrix* sp. and *Scenedesmus* sp. representing Chlorophyceae, together appeared twice at 20.00 hrs and 8.00 hrs with a population density of 2 units/l.

Cymbella sp. among Bacillariophyceae appeared four times; at 16.00 hrs with 4 units/l and at 0.00 hrs, 4.00 hrs, and 8.00 hrs with a population density of 2 units/l each.

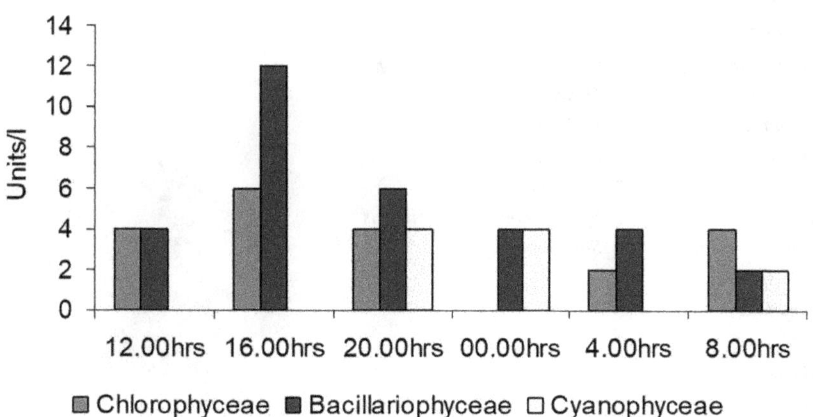

FIGURE 8.5 Diurnal variations in phytoplankton community in Priyadarshini water body during austral summer 2006–7.

Nitzschia sp. recorded its presence four times during the diurnal cycle with a population density of 2 units/l recorded at 12.00 hrs, 0.00 hrs, and 4.00hrs and 4 units/l recorded at 16.00 hrs.

A single Cyanophycean species, *Gloeotrichia* sp. appeared thrice during the diurnal cycle with a population density of 4 units/l recorded at 20.00 hrs and 0.00 hrs and 2 units/l observed at 8.00 hrs.

8.9.2.2.2 Zooplankton

Centropyxis sp., belonging to protozoa, recorded its presence at 20.00 hrs only with a population density of 2 units/l. Since only surface water samples were collected for the diurnal study as such, species present below surface waters throughout have not appeared in the samples collected for the diurnal study.

The overall phytoplankton taxa identified from the fresh water bodies of Schirmacher Oasis are provided in Table 8.4. A total of 57 species of phytoplankton

TABLE 8.4
Phytoplankton Taxa of Fresh Water Bodies of Schirmacher Oasis

Bacillariophyceae	Cyanophyceae	Chlorophyceae	Euglenophyceae
Achnanthes minutissima	*Anabaena sp.*	*Chlorella sp.*	*Euglenomorpha sp.*
Amphora microcephala	*Aphanocapsa sp.*	*Cosmarium sp.*	*Euglena sp.*
Cyclotella sp.	*Aphanothece nidulans*	*Elkatothrix sp.*	
Cymbella sp.	*A. microscopica*	*Gloeotrichia sp.*	
Diatoma sp.	*Chroococcus sp.*	*Mougeotia sp.*	
Fragillaria sp.	*Gloeocapsa sp.*	*Selenastrum sp.*	
Gomphonema sp.	*Lyngbya sp.*	*Westella sp.*	
G.parvulum	*L. versicolor*	*Tetradesmus sp.*	
G. lucas rankala	*Nostoc sp.*		
G. intricatum	*Oscillatoria sp.*		
G. sphaerophorum	*O. tenuis*		
G. lanceolatum	*O. rubescence*		
G. olivaceum	*O. subbrevis*		
Mastogloia sp.	*Phormidium sp.*		
Melosira sp.	*P. tenue*		
Navicula sp.	*Synechocystis sp.*		
N. halophila	*Synechococcus aeruginosus*		
N. subrhyncocephala	*Spirulina sp.*		
Nitzschia sp.			
N. cummutata			
N. palea			
Pinnularia sp.			
Synedra sp.			
S. ulna			
S. rumpens			
Penium sp.			
Opephora sp.			

were recorded from the water bodies present in Schirmacher Oasis. Bacillariophyceae contributed 28 species while Cyanophyceae contributed 18 species followed by Chlorophyceae (nine species). Euglenophyceae was represented only by two species (Table 8.4).

Ariyadej *et al.* (2004) have a detailed study on phytoplankton diversity and its relationships to the physicochemical environment in the Banglang Reservoir, Yala Province.

According to them, the phytoplankton in a reservoir is an important biological indicator of water quality. While phytoplankton is an important primary producer and the basis of the food chain in open water, some species on the other hand can be harmful to humans and other vertebrates by releasing toxic substances (hepatotoxins or neurotoxins etc.) into the water. The proliferation of harmful organisms, particularly species should be monitored.

The variation of phytoplankton succession is strongly linked to meteorological and water stratification mixing processes, patterns in temperate ecosystems differ considerably from those of tropical waters (Wetzel, 2001). The dynamics of phytoplankton are a function of many of the environmental processes that affect species diversity. For example, the onset of the spring bloom in dimitic lakes is controlled by the relief of light limitation at a time when nutrient concentrations are high and growing abundance is low (Roelke and Buyukates, 2002).

Similarly, according to Gannon and Stemberger (1978), zooplanktons have potential value as assessors of tropic conditions. They respond quickly to environmental change and may be effective indicators of subtle alterations in water quality. Since most species are widely distributed in diverse environments, those with the greatest value are ones limited to extremes of trophic lake types (i.e., oligotrophy, eutrophy, and dystrophy). In the wide range of ill-defined intermediate lake types, quantitative data on zooplankton community composition offers more potential than qualitative information on the presence or absence of certain species. The ratio of calanoid copepods to other major groups of zooplankton appears to be of value in identifying relative differences in trophic conditions. Multivariate analyses based on the distribution and abundance of rotifer and crustacean species have proved useful in delineating major water masses of different trophic conditions in large lake systems. But caution must be exercised in establishing one-to-one causal relationships between zooplankton composition and trophic conditions since other factors, especially toxic pollutants and size-selective predation, may exert considerable influence on changes in the community composition.

The zooplankton because of its central position between the autotrophs (algae, phytoplankton) and other heterotrophs (fish and other carnivores), forms an important link in the food web of the aquatic ecosystem (Gulati, 1983). Its strategic position both in terms of feeding and energy flow in the ecosystem as well as its sensitivity to anthropogenic and natural changes makes zooplankton quite suitable for biological monitoring of water quality (Gulati, 1983).

According to Gulati (1983), important abiotic factors causing changes in zooplankton community composition are nutrient loading, water temperature, dissolved oxygen and salinity etc.

Bacterial biomass and heterotrophic potential of water, ice and sediment microbial populations of freshwater lakes of Schirmacher Oasis have been studied by

Matondkar and Gomes (1983), Matondkar, (1986) and Ramaiah (1995). The epifluorescence counts of bacteria were observed to be lower by a factor when compared with ultra-oligotrophic Antarctic lakes. The study was very useful in recognizing the relative abundance and productivity of the limnetic microbial population.

The presence of lower invertebrates, like nematodes, turbellarians, mites, tardigrades, rotifers and kinorhynchs in the Schirmacher Oasis area has been reported for the first time by Ingole and Parulekar (1990). This gave the lead to many more investigations that followed in subsequent expeditions. The luxuriant growth of benthic vegetation (moss and algae) in and around the water body increases the amount of organic matter and is very important for the survival of microorganisms. According to Ingole and Dhargalkar (1998), the scale and magnitude of the probable impact of human activities over a decade (1983–1994) on the Priyadarshini Lake at Schirmacher Oasis, East Antarctica, was assessed through an ecological study conducted over an annual cycle during January 1993 to January 1994.

Chlorophyll-a and microbenthos showed a bimodal pattern of fluctuations with one maximum in austral summer and another in spring (late winter). Fluctuations in the hydrobiological parameters were closely related to meteorological conditions over the area. The biological productivity of the lake was dependent on the availability of ice-free water and an increase in atmospheric temperature (Ingole and Dhargalkar, 1998). An in-depth analysis and comparison of certain lake environmental aspects over a decadal period indicate that the lake environment is in a healthy condition. However, intensive human activities in the catchment area may deteriorate the ecosystem and hence regular monitoring and stringent regulations are needed to maintain its quality.

The lake water temperature varied from 1.0–7.9 °C; dissolved nutrients fluctuated from lake to lake showing the very low levels of nitrate and phosphate (Ingole *et al.*, 1987). The solubility of oxygen in Antarctic waters is high and is generally higher in the lakes than in seawater. The chlorophyll was low varying from 0.07–0.65 μg/l and surface primary productivity from 17–3.95 mg. C m^{-3} hr^{-1}. The particulate organic carbon in different lakes showed considerable variation (Ingole *et al.*, 1987; Shirodkar *et al.*, 1992). This signifies that the organic inputs to these lake ecosystems are from different sources. Glacial melt ice-free water is the key environmental variable and main source of dissolved nutrients in the oasis lake ecosystem (Ingole and Parulekar, 1993).

Despite the extreme climatic conditions that persist in the icy Antarctic continent, microorganisms have been detected in all the distinctive habitats (such as lakes, ponds, streams, rocks, and Antarctic soils) which differ from one another with respect to the range of temperature, availability of water and richness of the nutrients. Since these factors influence the survival and growth of bacteria present in the cold deserts of Antarctica are bound to be different from those in the coastal fringe, which is relatively warmer.

Soil and water samples collected at Schirmacher Oasis, Antarctica have been analyzed for the presence of bacteria and yeast and have also been studied with respect to biomass over a period of time (Matondkar, 1986; Ramaiah, 1995; Loka Bharati *et al.*, 1999). The results indicated the presence of Gram –ve and Gram +ve bacteria. All these microorganisms were psychrophilic and have fastidious nutrients requirements.

The bacteria belong to the genera Pseudomonas, Flavobacterium, Sphingobacterium, Micrococcus, Planococcus, and Arthrobacter. However, the above studies have been confined to only certain restricted regions of Schirmacher Oasis and they need to be extended to appreciate the extent of bacterial and yeast biodiversity. The total direct anaerobic viable counts (TDAnVC) were one order more than the total direct aerobic viable counts (TDAeVC). Laboratory experiments with one of the lake isolates indicated that there was a tendency to express higher viability of 61% at redox potential (e^H) ranging from –281 to –335 mv. It is suggested that the disposition to express increased viability under reducing conditions is a strategy to counteract stress due to the supersaturation of oxygen in the cold lacustrine environment. A comparison of results revealed a lower endergonic reaction of alkaline phosphatase activity in the Antarctic isolate. These results provide a basis for future comparative studies on the energy of alkaline phosphatase activity in cyanobacteria.

The ecological and underwater diving survey of the Priyadarshini (ZUB Lake) conducted in 1986–87 by Ingole and Parulekar, (1990, 1993) was not only critical to knowing the bathymetry. Water carrying capacity of the lake and portability of the lake water for human beings but also to understand the abundance, diversity, survival strategies and community dynamics of planktonic and benthic flora and fauna and their relationship with the oasis environment. It was concluded that the microfaunal composition and abundance in the supra littoral zone of lakes were related to the sediment texture. The dispersal of microinvertebrates depends on the glacial melt water and occasional breaching of lakes. More recent studies recognized that the majority of the biomass standing crop accumulation in Oasis ecosystems is attributable to the benthos (flora and fauna) and benthic vegetation plays an important role in maintaining high faunal diversity (Ingole and Parulekar, 1993; Ingole and Dhargalkar, 1998). In addition, these fragile ecosystems are unique in several respects: most of them have a perennial 2–6 m ice cover, variable temperature profile and abundant microflora (Matondkar, 1986; Ramaiah, 1995; Loka Bharati *et al.*, 1999) and fauna (Ingole and Parulekar, 1987, 1990, 1993, Ingole and Dhargalkar, 1998) but they lack grazers.

Moss communities of the Schirmacher Oasis were investigated earlier by Wafer and Untawale (1983). The terrestrial mosses, though quite widespread in the Schirmacher Oasis colonizing a range of habitats from soils just outside the snow field, the sheltered situation in the exposed areas, patterned grounds, biogenic remains and around the nests of south polar Skua (*Catharacta macromicii)* etc., but they are most abundant along the banks of water bodies and the melt water streams.

Investigations relating to the ecology of moss-felt communities confined to the Subantartica maritime zones. Only ten species, representing six genera of nematodes, have so far been recorded from the continental Antarctic zone. Five genera of nematodes viz., *Tylechorhynchus, Drylaimellus, Aporcelaimellus, Dorylamoides,* and *Paramylochulus* were recorded for the first time from the Schirmacher Oasis by Ingole and Parulekar (1987). The population of nematodes, including juveniles, was maximum in January and minimum in February suggesting that the nematodes at Schirmacher Oasis complete their life cycle within a short period of ice-free conditions and is a survival strategy in Antarctic microinvertebrates. The physio-chemical factors of soil viz., temperature, relative humidity, nitrate, pH, and organic carbon were also studied. The peak population was associated with the higher levels of temperature, nitrate, organic carbon and relative humidity and a significant correlation exist between soil nematode population and soil factors. Although some nesting sites of the snow petrel and Antarctic skua are reported at Schirmacher Oasis, this migratory bird population is seen only during the fair season. Also, few Adele penguins do visit the area occasionally.

The variation in the chemical composition of various water bodies in Schirmacher Oasis signifies the role of respective catchment areas. The inflow from glaciers or accumulated snow has been found to be responsible for governing the metabolic activity of the Antarctic water bodies. Besides this, the hydrological detention governing the nutrient accumulation in water bodies has been found to be the main parameter responsible for the chemical and biological status of these water bodies. Besides anthropogenic activity, nutrient input is on account of the frost action accompanied by chemical and biological weathering, of rocks. A second source is an aerial deposition in the form of precipitation and marine-derived aerosols, besides the deposition from gases emanating from aero-planes, vehicles and garbage burning etc. The third source of nutrients (especially nitrogen and Phosphorous) is from the feces of birds and seals, which come to land to breed and molt.

Though temperature and light intensity are the chief factors in the Antarctic yet the enhancement in nutrient levels has been found to give a shift in the primary production towards the higher side in the Antarctic water bodies. Each year water bodies remain frozen for around 8–10 months each year while some of the water bodies remain frozen for the whole of the year. The water bodies present in Schirmacher Oasis in general are small and are mostly shallow. During austral summer the water bodies in Schirmacher Oasis swell with snow melt water which gets considerably reduced with the lowering of atmosphere temperature as the flow of melt water from glaciers gets completely stopped.

Among zooplankton, only two classes, Rotifera and Protozoa were recorded in the present investigation. Towards the west end of Schirmacher Oasis Cyanophyceae was followed by Chlorophyceae, Euglenophyceae and Bacillariophyceae (Figure 8.6a).

Moving from the west towards the East the dominance of Chlorophyceae was recorded more over Cyanophyceae. Bacillariophyceae and Euglenophyceae followed Cyanophyceae in order of dominance at this geographical location (Figure 8.6b).

The water bodies on the Eastern side of the Schirmacher Oasis recorded a higher density of Cyanophyceae followed by Bacillariophyceae and in-turn followed by Chlorophyceae (Figures 8.6c and d).

The abundance of Cyanophyceae in the water bodies towards the Eastern side of Schirmacher Oasis may be on account of enhancement in the anthropogenic activities as, mostly filamentous blue-green forms (Oscillatoria sp. in general, was followed by Nostoc, Phormidium, and Anabaena sp.) were often recorded in relatively high numbers in the samples collected from various water bodies. Species of Oscillatoria, Nostoc, Phormidium, Lyngbya, Gloeocapsa, Aphanocapsa, Aphanothece, and Chroococcus were also reported by Gupta and Kashyap (1998).

Besides the dominance of Cyanophyceae, Bacillariophyceae was mainly represented in order of density dominance by Nitzschia sp., Gomphonema sp., and Navicula sp. The abundance of these species signifies the organic enrichment of the said water bodies. However, the dominant presence of Cosmarium and Gloeotrichia sp. among Chlorophyceae reveals the Oligotrophic nature of the water bodies.

The work conducted during three consecutive austral summer months reveals no significant variation in their physicochemical and biological characteristics. However, small digression in the plankton abundance is on account of variation in the start of austral summer/melting of ice covers on the water bodies. Short summer plankton pulse follows prolonged austral winter resting stages. Wanganeo and Wanganeo (1991) while working on High mountain lakes of Kashmir Himalayas reported a low plankton population in these water bodies on account of low water temperature and low nutrient status.

Sommer (1989) revealed no evidence of consistent regular seasonal changes in abundance among Rotorua cycles. His findings were in contrast with those summarized in the PEG (Plankton Ecology Group) model.

The present investigation reveals a well-established column plankton community in comparison to earlier findings. The presence of filamentous cyanobacteria among the plankton population signifies the increasing anthropogenic pressure. No filamentous blue-green algae were recorded in studies undertaken before 1980 (Spaulding *et al.* 1994). These changes in the plankton over the past 15 years were suggested by the scientists to be caused by a greater inflow of melt water- flushing species from the

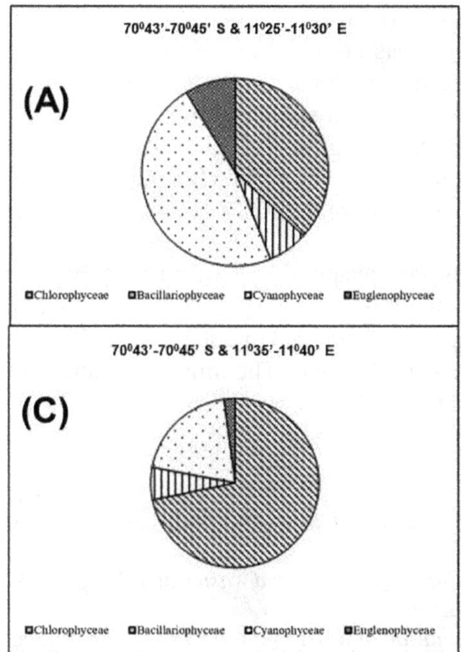

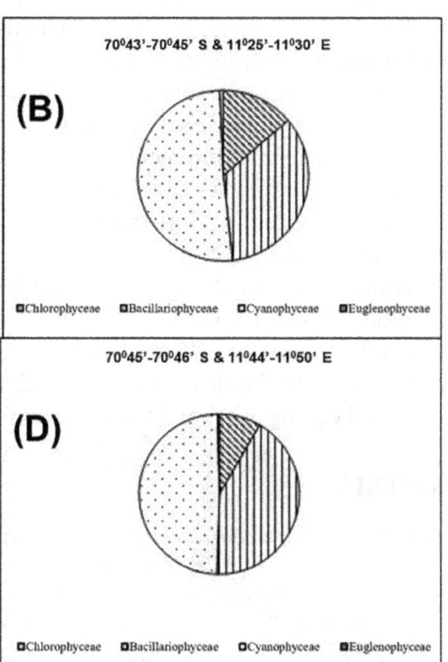

FIGURE 8.6 Variations in phytoplankton classes in water bodies placed at different geographical coordinates in Schirmacher Oasis. (A) 70^{0}43'-70^{0}45' S & 11^{0}25'-11^{0}30' E (B) 70^{0}44'-70^{0}46' S & 11^{0}40'-11^{0}45' E. (C) 70^{0}43'-70^{0}45' S & 11^{0}35'-11^{0}40' En (D) 70^{0}45'-70^{0}46' S & 11^{0}44'-11^{0}50' E.

littoral zone into the main body of the lake, thereby facilitating the establishment of "new" planktonic species.

The periodical variations in the abundance of various species among Cyanophyceae and Chlorophyceae reveal that the water bodies in Schirmacher Oasis have entered into a transition phase, oscillating between Oligotrophic and near mesotrophic conditions. Present work signifies the sensitive nature of fresh water bodies of Schirmacher oasis towards the anthropogenic pressure.

Atmospheric and water temperatures have been found to profoundly govern the metabolic conditions of Antarctic water bodies. As a result, the water bodies of this polar continent experience a short growth period for the biota they sustain.

A comparison of data collected during three austral summers 2005, 2006, and 2007, revealed an interesting trend in various physicochemical parameters of Antarctic water bodies, for example, water temperature showed a decreasing trend (similar to the trend followed by atmospheric temperature) with minimum values (–0.6 °C) observed in 2007 and maximum (5 °C) observed in 2005. Secchi transparency was found to vary from 1.57 m to 1.75 m with minimum values recorded in 2005 and maximum recorded during 2007. However, the water bodies falling on their way towards the convoy route (close to Maitri) experience very low Secchi transparency towards the distributed shore areas wherefrom a lot of eroded material finds its way into the system. Similarly, the pH of water samples varied from 6 to 7.46. The waters of freshwater bodies within Schirmacher Oasis are as such, near neutral.

On the other hand, specific conductivity varied from 13.86 μS/cm to 20 μS/cm and a steep rise from 2.04 mg/l to 8.9 mg/l was observed in the case of Chloride content. The minimum value was observed during 2005 and the maximum during 2007. Similarly, the dissolved oxygen content of fresh water bodies showed fluctuations during its measurement and varied from 6.8 mg/l to 11.2 mg/l with minimum values observed during 2005 and maximum during 2006. Total alkalinity and Total hardness of water fluctuated from 4 mg/l (2005) to 8 mg/l (2007) and from 2–14 mg/l (2005) to 50 mg/l (2006) respectively.

Among biologically important nutrients Phosphate phosphorous and Total Iron recorded minimum values of 0.004 mg/l and 0.03 mg/l respectively, in 2007 and maximum values of 0.75 mg/l and 0.11 mg/l were observed in 2005. NO_3-N in these Antarctic water bodies varied from 0.2 mg/l to 0.5 mg/l. The minimum value was observed during 2005 and the maximum during 2007.

REFERENCES

APHA (1985): *Standard methods for the examination of water and waste water.* APHA, AWWA and WPCF, New York.

APHA (1989): *Standard methods for the examination of water and wastewater.* 17th edition, American Public Health Association, Washington, D.C.

APHA (1992): *Standard methods for the examination of water and wastewater.* 18th edition, American Public Health Association (APHA), American Water Works Association (AWWA) and Water Pollution Control Federation (WPCF), Washington, D.C.

APHA, AWWA and WEF (2005): *Standards methods for the examination of water and waste water.* 21st edition, American Public Health Association (APHA), American Water Works Association (AWWA), and Water Environment Federation (WEF), Washington, DC.

Ariyadej, C., Tansakul, R., Tansakul, P. and Angsupanich, (2004): Phytoplankton diversity and its relationships to the physicochemical environment in the Banglang Reservoir, Yala Province, Songklanakarin. *Journal of Science and Technology* 26(5): 595–607.

Boyd, C.E. (1979): *Water quality in warm water fish ponds.* Craftmaster Auburn, Alabama, USA, Printers Inc.

Chaturvedi, S.K. and Khare, N. (2006): *Antarctic lake survey for paleoclimatic study: Achievements and Future plan (Abstract) National Workshop on Emerging areas of Scientific investigations during XXVI Indian Antarctic Expedition*, June 29–30, 2006, Goa.

Downing, J.A. and Rigler, F.H. (1984): *A manual on methods for the assessment of secondary Productivity of fresh waters.* I.B.P. Handbook No. 17, Blackwell Sci. Pub., London.

Edmondson, W.T. (1959): *Freshwater biology.* 2nd edition, John Wiley and Sons, New York.

Gannon, J.E. and Stemberger, R.S. (1978): Zooplankton (especially crustaceans and rotifers) as indicators of water quality. *Transactions of American Microscopy Society* 97(1): 16–35.

Goldman, C.R. and Horne, A.J. (1983): *Limnology.* McGraw-Hill Book Co., New York, 464 p.

Golterman, H.L. (ed.) (1969): *Methods of chemical analysis of freshwaters.* IBP Handbook No. 8, Blackwell, Oxford.

Golterman, I.L. and Clymo, R.S. (1979): *Methods for physical and chemical analysis of fresh waters.* I.B.P. (handbook No.8) Blackwell Scientific pub, Oxford.

Gulati R.D. (1983): Zooplankton and its grazing as indicators of trophic status in Dutch lakes. *Environmental Monitoring Assessment* 3: 343–354.

Gupta, R.K. and Kashyap, A.K. (1998): Algal colonization of Schirmacher Oasis, Antarctica. *Fourteenth IAE Scientific Reports DoD. Tech. Pub.* 12: 187–192.

Hemasundaram, A., Dhanalakshmi, K., Prasad, B. and Naidu, N.V.S. (2003): Assessment of water quality with regard to surfactants in pilgrim town—a case study of Tirupati. *Journal of Ultra Scientist of Physical Sciences* 15(2): 189–194. http://dx.doi.org/10.2307/3225681

Ingole, B.S. and Dhargalkar, V.K. (1998): Ecobiological assessment of a freshwater lake at Schirmacher Oasis, East Antarctica, with reference to human activities. *Current Science* 74: 529–534.

Ingole, B.S. and Parulekar, A.H. (1987): Micro-fauna of Schirmacher Oasis, Antarctica: 1. Water moss communities. *DoD, Tech. Pub.* 4: 139–148.

Ingole, B.S. and Parulekar, A.H. (1990): Limnology of Priyadarshini Lake, Schirmacher Oasis, Antarctica. *Polar Record* 26: 13–17.

Ingole, B.S. and Parulekar, A.H. (1993): Limnology of fresh water lakes at schirmacher Oasis, East Antarctica. *Proc. Indian National Science Academy B* 59(6): 589–600.

Ingole, B.S., Verlecar, X.N. and Parulekar, A.H. (1987): Microfauna of Priyadarshini Lake, Schirmacher Oasis, Antarctica. *DoD, Tech. Pub.* 9: 301–311.

Khare, N., Chaturvedi, S.K. Srivastava, R., Saraswat, R. and Wanganeo, A. (2008): Some Morphometric characteristics of the Priyadarshini Lake of Schirmacher Oasis, central Dronning Maud Land, Antarctica with special reference to its bathymetry. *Indian Journal of Marine Sciences, New Delhi* 37 (4): 435–438.

Lind, O.T. (1979):*A handbook of limnological methods.* C.V. Mosby, St. Louis, MO, p. 199.

Loka Bharti, P.A., Nair. S, Desouza, M.J.B.D. and Chandramohan, D. (1999): Truce with oxygen- Anaerobiosis outcompete aerobiosis in the Antarctic lacustrine bacteria. *Current Science* 76 (12): 1585–1587.

Matondkar, S.G.P. (1986): *Microbiological studies in Schirmacher Oasis Antarctica: Effect of temperature on bacterial population.* Report of Third Indian Expedition to Antarctica, pp. 133–147.

Matondkar, S.G.P. and Gomes, H.R. (1983): *Biological studies on the ice shelf and in a fresh water lake at Princess Astrid Coast, droning Maud Land, Antarctica.* Scientific report of first Indian Expedition to Antarctica, pp. 186–190.

Michael, P. (1984): *Ecological methods for field and laboratory investigations.* Tata-McGraw Hill Pub. Com. Ltd., New Delhi, p. 404.

Pandey, K.D, Kashyap, A.K and Gupta, R.K. (1995): Nutrient status, algal and cyanobacterial flora of six fresh water streams of Schirmacher Oasis. *Antarctica* 299: 83–91.

Ramaiah, N. (1995): Summer abundance and activities of bacteria in the fresh water lake of Schirmacher Oasis, Antarctica. *Polar Biology V.* 15(18): 547–553.

Roelke, D. and Buyukates, Y. (2002): Dynamics phytoplankton succession coupled to species diversity as a system-level tool for the study of *Microcystis* population dynamics in eutrophic lakes. *Limnology and Oceanography* 47(4): 1109–1118.

Sengupta, R. and Qasin, S.Z. (1983): *Chemical studies on the ice shelf in a freshwater lake and in a polynya at Princess Astrid Coast, Dronning Maud Land, Antarctica.* Scientific Report- First Indian Expedition to Antarctica, Technical Publication No. 1, DoD, New Delhi, pp. 62–68.

Shirodkar,P.U, Algarswamy, R., Goes, J.F and Fondekav, S.P. (1992): *Major elements, nutrients, and plankton biomass in the ice edge and an offshore region of the Indian Ocean sector of the Southern Ocean. Polar record.* Vol. 28, Cambridge University Press, Cambridge, pp. 127–136.

Sommer, U. (Ed.) (1989): *Plankton ecology. Succession in plankton communities.* Berlin, Springer-Verlag, p. 369.

Spaulding, S.A., Mcknight, D.M., Smith R.L. and Dufford, R. (1994): Phytoplankton population dynamics in perennially ice-covered Lake. Fryxell, Antarctica. *The Journal of Pl ankton Research* 16:527–541.

Swain, A.K. and Chaturvedi, A. (2017): *Bathymetric profiling of lakes of Schirmacher Oasis, East Antarctica, Twenty Seventh Indian Antarctic Expedition 2007–2009.* Ministry of Earth Sciences, Technical Publication No. 25, pp. 45–72.

Trivedi, R.K. and Goel, P.K. (1984): *Chemical and biological methods for water pollution studies.* Environmental Publishers, Karad, India, p. 304.

Trivedi, R.K. and Goel, P.K. (1986): *Chemical and biological methods for water pollution studies.* Environmental publication, Karad, Maharashtra.

Wafer, S. and Untwale, A.G. (1983): *Flora of Dakshin Gangotri in Antarctica.* Scientific Report of first Indian Expedition to Antarctica (DOD Tech. Publ.; 1) 182–185.

Wanganeo, A. and Wanganeo, R. (1991): Algal population in valley lakes of Kashmir Himalaya. *Archiv für Hydrobiologie (Stuttgart).* 121(2): 219–233.

Wanganeo, A. and Wanganeo, R. (2006): Variation in Zooplankton population in two morphologically dissimilar rural lakes of Kashmir Himalayas. *Proceedings of the National Academy of Sciences of India.* 76(B): III.

Wanganeo, A., Wanganeo, R., Khare, R. and Kumar, P. (2018): Variations in physicochemical characteristics of water bodies placed at different geographical coordinates in Antarctica. *Polar Science* 18: 48–56.

Wetzel, R.G. (2001): *Limnology.* 3rd edition. Academic Press, California.

Wetzel, R.G. and Likens, G.E. (1991): *Limnological analysis.* 2nd edition, Springer-Verlag, New York, pp. 391, ISBN 0-387-97331-1. https://doi.org/10.1002/rrr.3450070410

Wood, R.B. and G.E. Gibson (1974): Eutrophication and Loch Heagen. *Water Research* 7: 163–287.

9 Glacial Morpho-Sedimentology and Processes of Landscape Evolution in Gangotri Glacier Area, Garhwal Himalaya, India

Anoop Kumar Singh,[1] *Dhirendra Kumar,*[1,2] *Chetan Anand Dubey,*[1] *Pawan Kumar Gautam,*[1] *Balkrishan Vishawakarma,*[1] *and Dhruv Sen Singh*[1]

[1] Department of Geology, University of Lucknow, Lucknow

[2] Department of Geology, Central University of South Bihar, Gaya, Bihar

CONTENTS

DOI: 10.1201/9781003284413-9

9.1 INTRODUCTION

Glaciers occupy about 10% of the Earth's land surface (~16 million Km^2) and hold roughly 70% of the freshwater. About 96% of glacier ice lies in the Polar regions; Arctic and Antarctica. The mountains of the Himalayas are the largest ice mass outside the polar regions, commonly termed the third pole. The Himalayan glaciers are the source of innumerable rivers of the Indo-Gangetic plain and provide fresh water to millions of people.

The Gangotri Glacier is one of the most prominent glacier of the Himalayas. The study of Gangotri Glacier dates back to 1891 when the first documentation of the snout (Gangotri Glacier) in the form of a rough sketch was made by Carl Greisbach of the Geological Survey of India. In recent years, however, renewed interest in the region, especially with the aid of remote-sensing technologies (Naithani et al., 2001; Tangri et al., 2004; Bahuguna et al., 2007; Ambinakudge, 2010; Bhambri et al., 2012; Rawat et al., 2015; Bhattacharya et al., 2016; Singh et al., 2017, 2022) and newly developing numerical dating techniques (Sharma and Owen, 1996; Barnard et al., 2004; Juyal et al., 2004, 2009; Mehta et al., 2014; Singh et al., 2019, 2022; Dubey et al., 2022), have provided new insights into the nature of latest Pleistocene and Holocene glacial oscillations. The recent studies reflect that the Gangotri Glacier has been retreating not at alarming rate (~10–20 m/yr) during the past few years as compared to the last century (Singh et al., 2017, 2022; Kumar et al., 2021a,b).

Gangotri Glacier is one of the best studied glaciers and is analyzed for the glacial history (Sharma and Owen, 1996; Singh et al., 2022), fluctuation of snout by Geological Survey of India, tree rings study (Singh et al., 2020), retreat and processes of sedimentation (Singh and Mishra, 2001; Singh, 2018; Singh et al., 2017, 2022), morphometry and morphotectonics (Naithani et al., 2001; Bali et al., 2003), pollen based paleoclimatic study (Kar et al., 2002), role of tributary glaciers on landscape modification (Singh and Mishra, 2002a,b), morpho-sedimentary processes (Singh, 2004), hydrological characteristics (Singh et al., 2006), frontal recession (Bhambri et al., 2012), snow melt ephemeral streams (Singh, 2013a), ice thickness (Gantayat et al., 2014), pattern of retreat and related morphological zones (Singh et al., 2017, 2022), monsoon variability and major climatic events (Singh et al., 2019, 2022; Dubey et al., 2022), study of temporal response and associated mass movement event (Singh et al., 2020), rapid retreat (Singh et al., 2020), recessional pattern of Thelu and Swetvarn Glaciers (Kumar et al., 2021b) etc. Other parts of the Himalayas have been studied for mass balance and ELA (Benn and Lehmkuhl, 2000), timing and style of late Quaternary glaciation in the eastern Hindu Kush (Owen et al., 2002), causes of the Kedarnath tragedy (Singh, 2013b), chronology of Quaternary glaciation (Ali and Juyal, 2013), surface processes and causes of Kedarnath (Singh, 2014), climate change (Singh, 2015), elevational changes of ablation region (Zhang et al., 2015), climate change and its societal impact (Singh and Agnihotri, 2016; Singh and Singh, 2005; Singh et al., 2015) and Spatio-temporal fluctuations over Chorabari Glacier (Kumar et al., 2021a) etc. Despite so many analyses of the Gangotri Glacier region, no attention has been paid to explaining the morpho-sedimentology of the Glacial landforms. The present chapter is focused on the glacial landforms and the process of landform modifications. This study helps to reconstruct the landscape evolution of the Gangotri Glacier region.

9.2 STUDY AREA

Gangotri Glacier is located between 30^0 44'- 30^0 56' N latitude and 79^0 40'- 79^0 15 E longitude in Uttarkashi district of Garhwal Himalaya, India (Figure 9.1). It originates from the northwestern slopes of Chaukhamba peaks at an elevation of about 7100 msl. The snout known as Gaumukh is situated at 4000 msl. This glacier comprises three main tributaries, namely Raktvarn, Chaturangi (including Kalandini bamak) and Kirti and more than 18 other tributary glaciers (Singh and Mishra, 2001, 2002a, 2002b; Singh et al., 2017, 2019, 2022; Dubey et al., 2022). The upper surface of the glacier is full of crevasses which are 5–10 m in length and 1–3 m in width. The glacier surfaces are often overlain by supra-glacial moraines.

9.3 GEOLOGY OF THE AREA

Geologically the area comprises the Central Crystalline zone fringing the Tethyan sedimentary sequences, part of which can be seen at the top of Sudarshan peak and the ridge southwest of Gaumukh. Thick veins of coarse-grained biotite granite

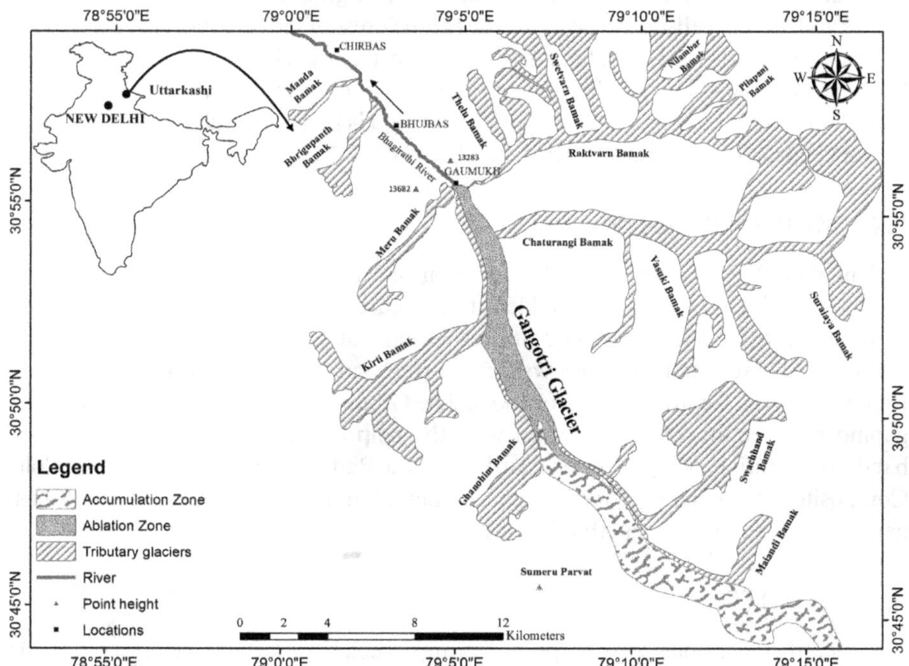

FIGURE 9.1 Location map showing Gangotri Glacier its snout, accumulation and ablation zone, tributary glaciers, Bhagirathi River, Bhujbas, Chirbas and few point heights.

TABLE 9.1
Lithostratigraphy of Parts of Bhagirathi and Jadhganga Valley, Uttarkashi District, Uttarakhand (Srivastava, 2012)

Age	Group	Formation	Lithology	Environment
Late Cambrian–Mid-Miocene or only Mid-Miocene			Tourmaline rich leucocratic granite	Intrusive in the granotoid rocks
Early Ordovician	HIMANTA	Badrinath granitoid	Granite-granodiorite	Intrusive in the Batal Formation
Early–Mid-Cambrian		Kunzam La	Light gray to greenish slates, siltstones and arenites	Subtidal–intertidal
Neoproterozoic		Batal	Gray pyritous slates, subarkose, carbonaceous slates, gritty and pebbly lenses	Peritidal

are intruded across the foliation of banded gneiss south of Jangla. The litho units exposed in the vicinity of Gaumukh are bands of quartz-biotite schist, quartzite, and biotite granulite (Table 9.1). In addition, the southeast part of the high hill ranges comprises tourmaline granite intrusive into the composite rock of migmatite schist association. On the left bank of Kirti Glacier, the migmatite contains coarse, banded gneiss and shows well-developed mullions at the crest of megascopic east-west trending isoclinal folds with greater amplitude than the wavelength (Srivastava, 2012). Most parts of the Gangotri Glacier are situated in a high seismic zone (zone IV) and show numerous neo-tectonic activities. The region is located in the north of the MCT (Main Central Thrust: passing south of Bhatwari).

9.4 METHODOLOGY

The methodology includes the identification of various geomorphic features/landforms in the field and marking the identified features on the satellite data. The orthorectified LISS IV image having 5.8m resolution was used to mark the geomorphic features. The selected imagery was cloud-free and purchased from the National Remote Sensing Centre (NRSC) of India. The Ground Control Points (GCP) of these geomorphic features were collected with the help of Garmin handheld GPS. The band combination of 3, 2, 1 were utilized as a Red, Green, Blue for False Color Composite (FCC) and preferred onscreen manual digitization process to mark these geomorphic features on satellite imagery.

9.5 RESULTS

We identified various geomorphic features such as; Lateral Moraines (LM), Terminal/Recessional Moraines (RM), Outwash Plain Deposits (OWP) and Kame Terraces during the geological fieldwork. The lithological characteristics and their relationship with depositional conditions are mentioned below.

9.5.1 Lateral Moraines (LM)

LM are the most common geomorphic feature which explains the dynamics of the glacier. At least three stages (Figure 9.2a) of LM can be identified on the left bank and two stages on the right bank. The first stage of the lateral morainic ridge is on the top of the Gangotri Glacier valley wall about 75–80 m above the present Gangotri Glacier surface. The second stage of the lateral morainic ridge is about 40–50 m above the glacier surface, while the third stage of the lateral morainic ridge is about 20–25 m above the present glacier surface. The lateral moraine is unconsolidated, unstratified, immature, devoid of any sedimentary structures, matrix-supported boulders in which the percentage of matrix varies from 30–40% and boulders 60–70%. The grains are angular to subangular, non-spherical, primary, poorly sorted and unoriented.

9.5.2 Terminal/Recessional Moraines (RM)

RM are ridges of unconsolidated debris (Figure 9.2b) deposited at the end of the glacier which reflect the shape of the glacier terminus and mark the maximum advance of the glacier. At least three stages of recessional moraines are present between Bhujbas and snout. A series of recessional morainic ridges lying between 900 m to 1400 m downstream of the present snout cut across the valley and indicate the various stages of the recession of the glacier in the past (Singh et al., 2017, 2022) (Figure 9.2b). These recessional moraines are equivalent to the LIA 200–300 BP (Sharma and Owen, 1996).

The RM consists of unconsolidated, unstratified, devoid of any sedimentary structures and matrix-supported boulders. The grains are sub-angular to sub-rounded, non-spherical, primary, immature and poorly sorted without any orientation of boulders.

9.5.3 Outwash Plain Deposits (OWP)

OWP also known as sandur deposits (Figure 9.3a) are formed beyond the snout of a glacier by meltwater streams. Glaciers contain a huge amount of debris and erode

FIGURE 9.2 (a) Showing temporal and spatial positions of three main stages of lateral moraines and so the shrinking of the Gangotri Glacier. (b) Positions of different stages of recessional moraines indicating retreat of the Gangotri Glacier.

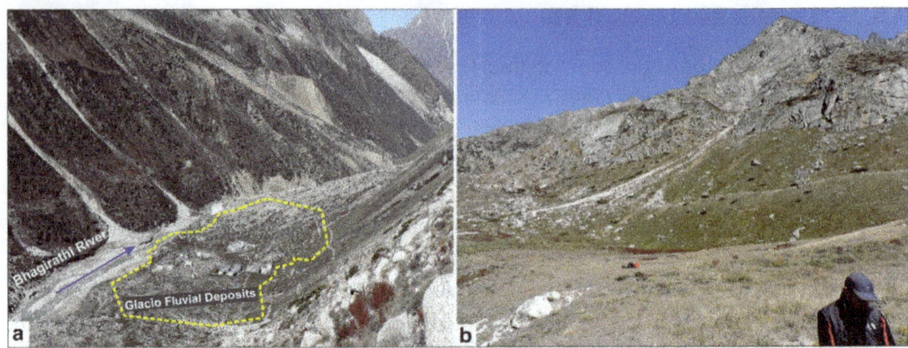

FIGURE 9.3 (a) Bhagirathi River and glacio-fluvial deposits. (b) Kame Terraces.

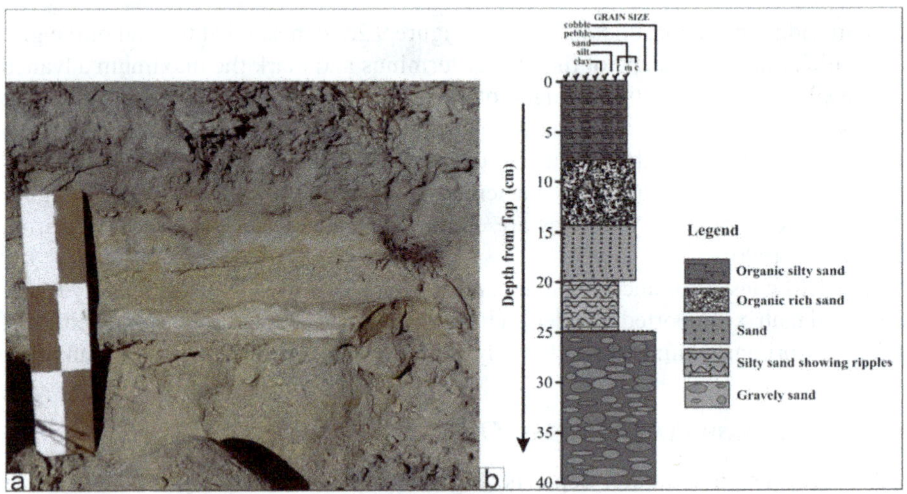

FIGURE 9.4 (a) A trench in the outwash plain. (b) Lithology of the trench.

the underlying rock when they move down. Near the snout of a glacier, meltwater can carry the debris away from the glacier in the downstream direction and deposit it on a flat plain. Such deposits are sorted or graded into sizes by the water runoff of the melting glacier, the coarsest being found near the ice, becoming progressively finer as the distance from the glacier increases. Outwash plains are glacio-fluvial deposits.

A 40 cm deep trench was made at the right bank of the Bhagirathi River on the flat terrain deposits of the Bhujbas area (Figure 9.4a). It is located at latitude N 30^{0}56'59.2" and longitude E 79^{0}03'3.6" at an elevation of 3796 m AMSL. The basal unit is 15 cm thick gravely sand which is underlain by 5 cm silty (Figure 9.4b) sand rippled unit. It is followed by the deposition of a 6 cm thick sandy unit. Above this unit 6 cm, thick organic-rich sand was deposited with sharp contact. At the top 8 cm, thick organic silty sand was deposited.

The OWP deposits are consolidated to semi consolidated, stratified, coarse to fine-grained sand and silt with boulders and exhibit primary sedimentary structures. The presence of ripples explains the fluvial activity and the rounded gravels describe the high degree of transportation from the place of weathering. The flat terrain of the Bhujbas area is outwash plain deposits evolved by glacio-fluvial processes under fluctuating energy conditions.

9.5.4 Kame Terraces

On the right bank of the Bhagirathi river a flat surface known as kame terraces. It is composed of light color coarser sediment deposited during melting/summer season and dark color fine sediments during non-melt/winter season. A 120 cm deep trench (Figure 9.5a) was made at the right bank of the Bhagirathi River on the kame terraces. It is located at latitude N30°57′00" and longitude E79°3′28" at an elevation of 3945 m. The basal unit is a 15 cm thick sandy unit which is followed by 5 cm gravely sand (Figure 9.5b) with erosional contact. The gravely sand unit is underlain by 15 cm yellow color clayey sand which shows an erosional contact with the previous unit. It is followed by the deposition of 25 cm brown color clayey silt with gradational contact. Above this 15 cm, yellow color clayey silt was deposited with the gradational contact. This unit is overlain by a 25-cm sandy unit that was deposited. This unit is overlain by the deposition of 15 cm organic-rich gray color clay with sharp

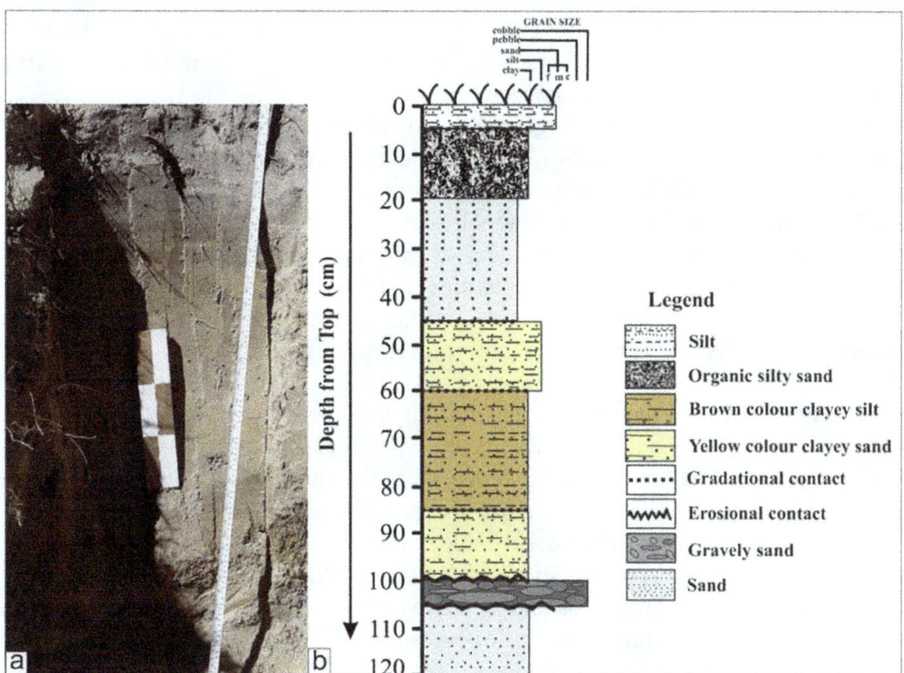

FIGURE 9.5 (a) Trench in kame terraces. (b) Lithology of the trench.

contact. At the top 5 cm, thick silt was deposited. These units reflect low to moderate energy of the depositional environments for a long period with fluctuating energy conditions. Therefore, sediments were deposited under the dominance of low energy environment under stagnant ponding conditions.

9.5.5 LLOF AND GLOF

It has been observed that some catastrophic events modify the landforms. These catastrophic events such as the Bhujbas LLOF of June 2000 and the Gaumukh debris flow event (LLOF) of October 2017 are identified as major processes which modify the landforms and landscape in the Gangotri Glacier region. Heavy rain in a glaciated terrain fills the glacial lake and may induce landslides. The landslides may block the river and form ephemeral lakes. The incessant rain causes the bursting of such ephemeral landslide lakes and glacial lakes, which create the LLOF and GLOF and redistributes the huge amount of sediments and affects the pre-existing landforms in the region.

The temperature also plays an important role in the modification process of geomorphic features in Gangotri Glacier. The Gangotri Glacier region is completely covered with snow during the winter period. The satellite data also supports that the snowfall during the winter season is about ten times higher than the permanent snow in the Himalayas (Tangri, 2002). It has been observed that the building up of snow cover starts from October to December, develops during January to March and starts melting due to an increase in temperature from April onwards. The melting of snow generates ephemeral streams (Singh, 2013a) and contributes to the discharge of the Bhagirathi river. These streams erode the finer sediments from the pre-existing landforms and accelerate the mass movements.

Heavy rains transported a large number of sediments from the left bank tributary glaciers (Singh and Mishra, 2002a) and deposited an enormous amount of material into the valley on June 6, 2000, near Bhujbas. The debris consisting of boulders, pebbles, cobbles, sand, and silt dammed the Bhagirathi River to form a short-lived extensive lake. The bursting of this lake caused Lake Outburst Flood (LLOF) in which the entire area, including DST huts and tents located at Bhujbas, were affected (Figure 9.6). The tributary glaciers continuously modify all these landforms. Thus, tributary glaciers are important in giving final shape to the Gangotri Glacier landforms.

Satellite images (Figures 9.6a and b) and field evidence (Figures 9.7c and d) indicate that the lower part of the glacier experienced intensive mass movement. This mass movement transported and deposited the enormous debris material from the upper reaches to the lower area of the glacier and breached the continuity of the left morainic wall of Bhagirathi valley (Figure 9.7c). This deposited material altered the initial course of the Bhagirathi river and shifted from a minimum 2 m to a maximum 175 m (Figures 9.6a and b). Nearly about 1559 m length of the river and around 360714.11 m^2 area of the valley were affected (from Gaumukh downward) by the deposited material. This deposited material created favorable conditions for the development of a dammed lake adjacent to the snout of the Gangotri Glacier (Gaumukh) (Figure 9.7d). The detailed analysis of the IMD data with Google Earth

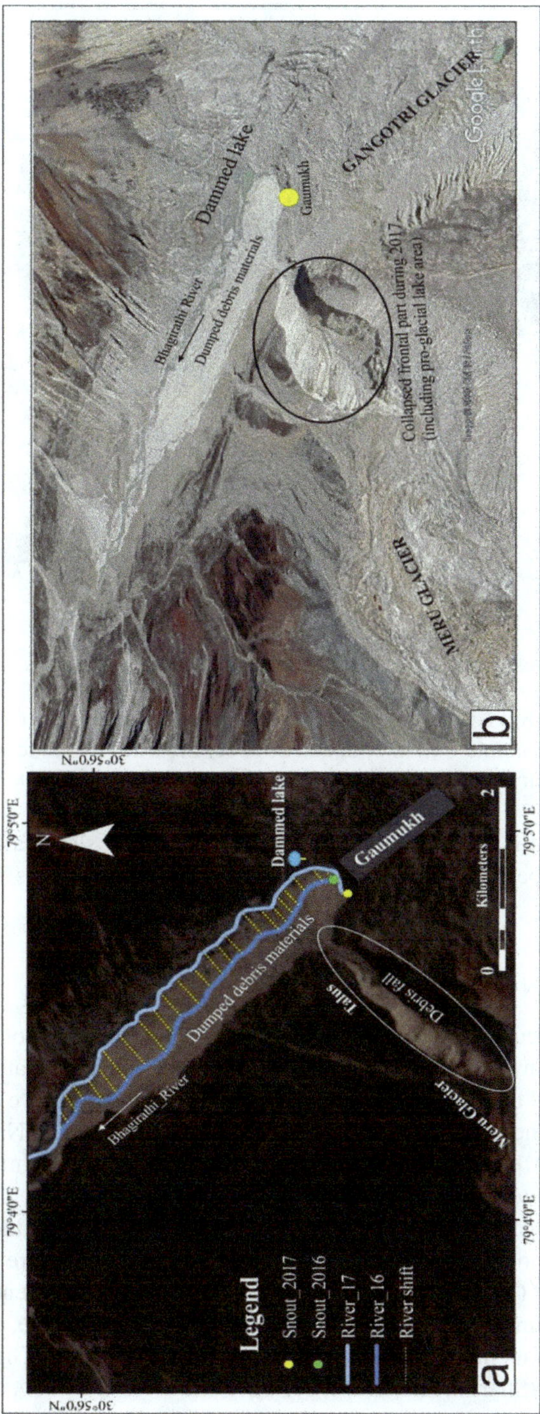

FIGURE 9.6 (a) Event of intense mass movement during 2017 at Bhagirathi valley. (b) Google Earth image (07/10/2017) showing collapsed frontal part of the Meru Glacier and dumped debris material within the Bhagirathi valley.

FIGURE 9.7 (a) and (b) Field photographs showing the snout position of Gangotri Glacier and continuity of left morainic wall in 2015 and 2016. (c) Breached morainic wall and debris flow during 2017. (d) Dammed lake adjacent to the snout of Gangotri Glacier.

images suggested that intense precipitation, large debris flow, slope failure, and collapse of the frontal part of the glacier, were the responsible factors for this mass movement.

9.6 DISCUSSION

The present study documents various stages of lateral moraines on both sides of the valley. The previous investigations documented three main stages of lateral moraines on the left valley wall at an elevation of 5100, 4900, 4700 m which are about 75–80, 40–50, 20–25 m higher than the present valley floor respectively (Naithani et al., 2001; Singh et al., 2017). In addition, a series of recessional morainic ridges also lying between 900 m to 1400 m downstream of the present snout cut across the valley and indicate the various stages of the recession of the glacier in the past (Singh et al., 2017) well supports to our study. These recessional moraines are equivalent to the LIA 200–300 BP (Sharma and Owen, 1996). The maximum extent of Gangotri Glacier occurred around 63 ka BP known as Bhagirathi Glacial Stage, then Shivling Glacial advanced during mid-Holocene (< 5 ka BP), and the Bhujbas Glacial advance about 200 to 300 BP (Sharma and Owen, 1996).

The OWP lithology reflects the stratified deposits of sand and silt sediments with gravels (Figure 4b). The deposition of coarse-grained sediments with boulders under high energy fluctuating conditions is attributed to glacio-fluvial activity. Similar lithological parameters have been identified from other regions for outwash plain deposits (Singh and Awasthi, 2011a,b; Singh and Ravindra, 2011a,b; Singh et al., 1999, 2013, 2015, 2017, 2018, 2022). The presence of clay units in the litholog of the kame terrace (Figure 9.7b) from bottom to top explains the stagnant low energy ponding conditions. Similar lithological parameters have been described in Himalaya in the Gangotri Glacier region (Singh et al., 2019, 2022; Dubey et al., 2022) and other regions (Bali et al., 2017a, b) and also in the Ganga Plain region (Trivedi et al., 2011; Singh et al., 2015; Saxena and Singh, 2017). Therefore, the deposition of fine-grained suspended sediments under low energy stagnant conditions is attributed to a lacustrine environment.

In the Gangotri Glacier area, LLOF deposits were reported in 2002 (Singh and Mishra, 2002a) which resulted due to heavy rain on 6th June 2000 in Bhujbas and distributed huge amounts of sediments. In addition, the Gaumukh region faced the problem of landslides and LLOF in October 2017. Similar events of modification have been well reported in the Kedarnath region. On June 16 and 17, 2013, heavy incessant rains overfilled the Gandhi Sarovar and induced landslides which blocked the Mandakini River system in the Kedarnath area (Singh, 2013b, 2014) and formed the ephemeral lakes. The bursting of Gandhi Sarovar and the temporary lakes caused LLOF and GLOF in the area. These catastrophic processes differentially eroded the sediments from mountain walls, moraines, outwash plains, valley walls and river banks and deposited 1–1.5 m thick, poorly sorted sediments on the outwash plain and river valley. The differential erosion, redistribution, and deposition of the huge amount of poorly sorted sediments in the outwash plain, river valley and the entire Kedarnath region modified the pre-existing glacial landforms and landscape.

The meteorological data of the Uttarkashi region from 1960–2000 (Figure 9.8a) reflects that the retreat rate between 1976 and 1990 was directly influenced by the catastrophic variations in temperature and precipitation in the Himalayan region (Bhutiyani et al., 2009; Singh et al., 2017; Singhvi and Krishnan, 2014; Dyurgerov and Meier, 2000). The rise in temperature ~2 °C and a reduction in winter precipitation in the Himalayas during 1974–2006 was responsible for the enhanced melting of the glaciers (Dimri and Mohanty, 2007; Bhutiyani et al., 2009; Singhvi and Krishnan, 2014; Kumar et al., 2017).

Figure 9.8a and b depict that the monsoon (ISM) precipitation dominated in the study area rather than the westerlies precipitation. Scherler et al. (2011) reported that ~65% of monsoon-influenced Himalayan glaciers are retreating. The temperature-precipitation graph of the Uttarkashi area reflects the anti-correlation with each other. In addition, the increasing trend of temperature supports the high rate of melting of the Himalayan glaciers. Therefore, intense precipitation, large debris flow and slope failure were the responsible factors for these catastrophic events (Dobhal et al., 2013; Singh, 2014).

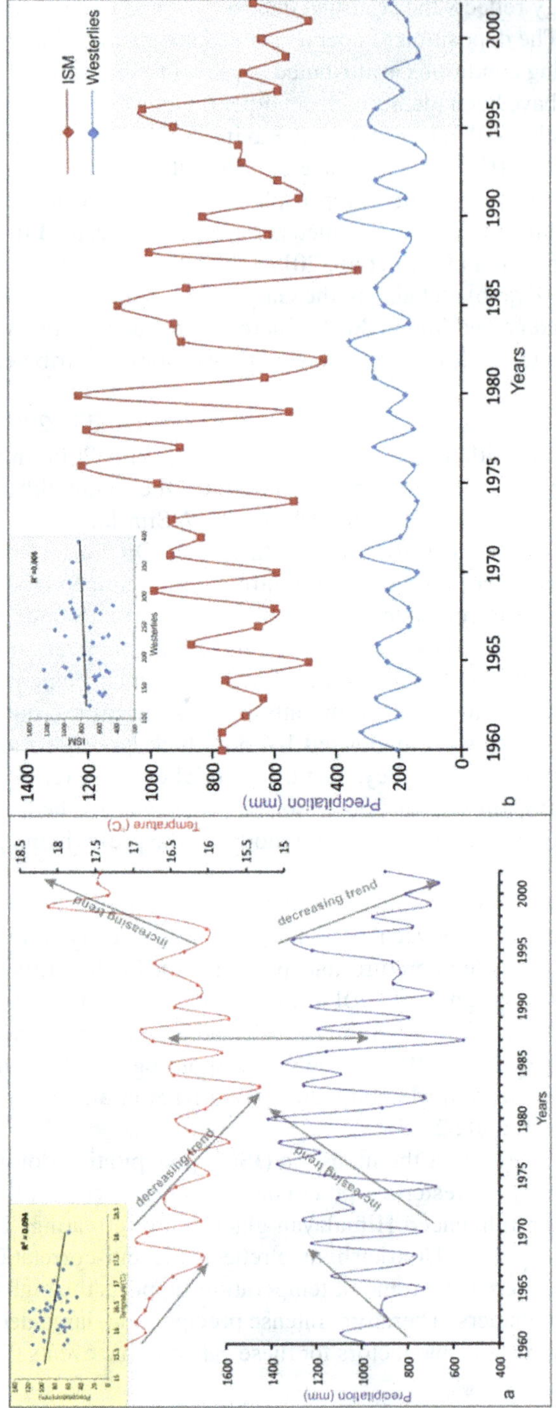

FIGURE 9.8 (a) Precipitation and temperature inverse relationship during 1960–2002 of Uttarkashi district. (b) ISM and westerlies decrease rainfall patterns in recent years.

9.7 CONCLUSIONS

The present study recognizes the various glacial geomorphic features such as Lateral Moraines (LM), Terminal/Recessional Moraines (RM), Outwash Plain Deposits (OWP), Kame Terraces in the study area. The Glacial (Gl) sedimentary environment is dominantly active in the area during ice ages and evolves the landforms. The field observations and lithological analysis reflect that these morphological features are evolved by the Glacial and glacio-fluvial and lacustrine processes. The glacial landforms become complex due to the modification of landforms by catastrophic events such as LLOF and GLOF. Therefore, geomorphic features/landforms in a glaciated terrain are modified by non-glacial factors also. These LLOF and GLOF processes are helpful to understand the evolutionary trend of the glaciated terrain.

ACKNOWLEDGMENTS

The Department of Science and Technology (SERB), Government of India is highly acknowledged for financial assistance (SB/DGH-68/2013). We are thankful to the Head, Centre of Advanced Study in Geology, University of Lucknow, Lucknow for providing the working facilities. Anoop Kumar Singh is thankful to SERB N-PDF (PDF/2020/000251) and Dhirendra Kumar to UGC for financial assistance (PDFSS-2015–17-UTT-10685) for financial support. Pawan Kumar Gautam gratefully acknowledged to the UGC RGNF Fellowship SRF (RGNF-2017–18-SC-UTT-34263). We are also thankful to Dr N. Khare, Scientist G, Ministry of Earth Sciences Prithvi Bhawan, New Delhi for allowing us to submit this chapter to this prestigious Book.

REFERENCES

Ali, S.N., Juyal, N., 2013. Chronology of late Quaternary glaciations in Indian Himalaya: a critical review. *J. Geol. Soc. India.* 82, 628–638.

Ambinakudge, S., 2010. A study of the Gangotri glaciers retreats in the Himalayas using Landsat satellite images. *Int. J. Geoinf.* 6 (3), 7–12.

Bahuguna, I.M., Kulkarni, A.V., Nayak, S., Rathore, B.P., Negi, H.S., Mathur, P., 2007. Himalayan glacier retreat using IRS 1C PAN stereo data. *Int. J. Rem. Sens.* 28 (2), 437–442.

Bali, R., Awasthi, D.D., Tiwari, N.K., 2003. Neotectonic control on the geomorphic evolution of the Gangotri glacier Valley, Garhwal Himalaya. *Gond. Res.* 6, 829–838.

Bali, R., Chauhan, M.S., Mishra, A.K., Ali, S.N., Tomar, A., Khan, I., Singh, D.S., Srivastava, P., 2017a. Vegetation and climate change in the temperate-subalpine belt of Himachal Pradesh since 6300 cal. yrs. BP inferred from pollen evidence of Triloknath palaeolake. *Quat. Int.* 444, 11–23.

Bali, R., Khan, I., Sangode, S.J., Mishra, A.K., Ali, S.N., Singh, S.K., Tripathi, J.K., Singh, D.S., Srivastava, P., 2017b. Mid- to late Holocene climate response from the Triloknath palaeolake, Lahaul Himalaya based on multiproxy data. *Geomorphology* 284, 206–219.

Barnard, P.L., Owen, L.A., Finkel, R.C., 2004. Style and timing of glacial and paraglacial sedimentation in a monsoon-influenced high Himalayan environment, the upper Bhagirathi Valley, Garhwal Himalaya. *Sediment. Geol.* 165, 199–221.

Benn, D.I., Lehmkuhl, F., 2000. Mass balance and equilibrium-line altitudes of glaciers in high-mountain environments. *Quatern. Int.* 65–66, 15–29.

Bhambri, R., Bolch, T., Chaujar, R.K., 2012. Frontal recession of Gangotri Glacier, Garhwal Himalayas, from 1965 to 2006 measured through high-resolution remote sensing data. *Curr. Sci.* 102, 489–494.

Bhattacharya, A., Bolch, T., Mukherjee, K., Pieczonka, T., Krop´ačEk, J., Buchroithner, M.F., 2016. Overall recession and mass budget of Gangotri Glacier, Garhwal Himalayas, from 1965 to 2015 using remote sensing data. *J. Glaciol.* 62 (236), 1115–1133.

Bhutiyani, M.R., Kale, V.S., Pawar, N.J., 2009. Climate change and the precipitation variations in the north-western Himalaya: 1866–2006. *Int. J. Climatol.* 30, 535–548.

Dimri, A.P., Mohanty, U.C., 2007. Location-specific prediction of maximum and minimum temperature over the Western Himalayas. *Meteorol. Appl.* 14, 79–93.

Dobhal, D.P., Gupta, A.K., Mehta, M., Khandelwal, D.D., 2013. Kedarnath disaster: facts and plausible causes. *Curr. Sci.* 105(2), 171–174.

Dubey, C.A., Singh, D.S., Singh, A.K., Sangode, S.J., Kumar, D., Kumar P., 2022. Sedimentation pattern of kame terraces and its implication to climatic events in the Gangotri glacier region since 25 Ka BP, Garhwal Himalaya, India. *J. Asian Earth Sci.* 229, 1–9. https://doi.org/10.1016/j.jseaes.2022.105160

Dyurgerov, M.B., Meier, M.F., 2000. Twentieth-century climate change: evidence from small glaciers. *Proc. Natl. Acad. Sci.* 97 (4), 1406–1411.

Gantayat, P., Kulkarni, A. V., Srinivasan, J., 2014. Estimation of ice thickness using surface velocities and slope: a case study at Gangotri Glacier, India. *J. Glaciol.* 60, 277–282.

Juyal, N., Pant, R.K., Basavaiah, N., Bhushan, R., Jain, M., Saini, N.K., Yadava, M.G., Saini, N.K., Singhvi, A.K., 2009. Reconstruction of last Glacial to early Holocene monsoon variability from relict lake sediments of the higher Central Himalaya, Uttrakhand, India. *J. Asian Earth Sci.* 34, 437–449.

Juyal, N., Pant, R.K., Basavaiah, N., Yadava, M.G., Saini, N.K., Singhvi, A.K., 2004. Climate and seismicity in the higher Central Himalaya during 20–10 ka: evidence from the Garbayang basin, Uttaranchal, India. Palaeogeogr. *Palaeoclimatol. Palaeoecol.* 213, 315–330

Kar, R., Ranhotra, P.S., Bhattacharayya, A., Sekar, B., 2002. Vegetation vis-a-vis climate and glacial fluctuations of the Gangotri glacier since last 2000 years. *Curr. Sci.* 82, 347–351.

Kumar, D., Singh, A.K., Singh, D.S., 2021b. Spatio-temporal fluctuations over Chorabari glacier, Garhwal Himalaya, India between 1976 and 2017. *Quatern. Int.* 575–576, 178–189.

Kumar, D., Singh, A.K., Taloor, A.K., Singh, D.S., 2021a. Recessional pattern of Thelu and Swetvarn glaciers between 1968 and 2019 Bhagirathi basin, Garhwal Himalaya, India. *Quatern. Int.* 575–576, 227–235.

Kumar, V., Mehta, M., Mishra, A., Trivedi, A., 2017. Temporal fluctuations and frontal area change of Bangni and Dunagiri glaciers from 1962 to 2013, Dhauliganga Basin, central Himalaya, India. *Geomorphology* 284, 88–98.

Mehta, M., Dobhal, D.P., Pratap, B., Majeed, Z., Gupta, A. K., Srivastava, P., 2014. Late Quaternary glacial advances in the Tons River Valley, Garhwal Himalaya, India and regional synchronicity. *Holocene* 24(10), 1336–1350.

Naithani, A.K., Nainwal, H.C., Sati, K.K., Prasad, C., 2001. Geomorphological evidence of retreat of the Gangotri glacier and its characteristics. *Curr. Sci.* 80, 87–94.

Owen, L.A., Kamp, U., Spencer, J.Q., Haserodt, K., 2002. Timing and style of Late Quaternary glaciation in the eastern Hindu Kush, Chitral, northern Pakistan: a review and revision of the glacial chronology based on new optically stimulated luminescence dating. *Quatern. Int.* 97–98, 41–45.

Rawat, M. S., Uniyal D.P., Dobhal, R., Joshi, V., Rawat, B.S., Bartwal, A., Singh D., Aswal, A., 2015. Study of landslide hazard zonation in Mandakini Valley, Rudraprayag district, Uttarakhand using remote sensing and GIS. *Curr. Sci.* 109, 158–170.

Saxena, A., Singh, D.S., 2017. Multiproxy records of vegetation and monsoon variability from the lacustrine sediments of eastern Ganga Plain since 1350 A.D. *Quat. Int.* 444 (A), 24–34.

Scherler, D., Bookhagen, B., Strecker, M.R., 2011. Spatially variable response of Himalayan glaciers to climate change affected by debris cover. *Nat. Geosci.* 4 (3), 156–159.

Sharma, M.C., Owen, L.A., 1996. Quaternary glacial history of NW Garhwal, central Himalayas. *Quat. Sci. Rev.* 15(4), 335–365.

Singh, A.K., Kumar, D., Kumar, V., Singh, D.S., 2020. Study of temporal response (1976–2019) and associated mass movement event (during 2017) of Meru glacier, Bhagirathi valley, Garhwal Himalaya, India. *Quatern. Int.* 565, 12–21.

Singh, D.S., 2004. Late quaternary morpho-sedimentary processes in the Gangotri Glacier area garhwal Himalaya, India. *Geological Survey of India* (Special Publication) 80, 97–103.

Singh, D.S., 2013a. Snowmelt ephemeral streams in the Gangotri glacier area, Garhwal Himalaya, India. In: Kotlia, B.S. (Ed.), *Holocene: Perspectives, Environmental Dynamics and Impact Events.* Nova Science Publishers, pp. 157–164.

Singh, D.S., 2013b. Causes of Kedarnath tragedy and human responsibilities. *J. Geol. Soc. India.* 82 (3), 303–304.

Singh, D.S., 2014. Surface Processes during flash floods in the glaciated terrain of Kedarnath, Garhwal Himalaya and their role in the modification of landforms. *Curr. Sci.* 106(4), 594–597.

Singh, D.S., 2015. Climate change: past present and future. *J. Geol. Soc. India.* 85, 634–635.

Singh, D.S., 2018. Concept of rivers: an introduction for scientific and socio-economic aspects. In: Singh, D.S. (Eds.), *The Indian Rivers: Scientific and Socio-Economic Aspects.* Springer publication, pp. 1–24.

Singh, D.S., Agnihotri, R., 2016. Climate change in the Indian perspective and its societal impacts. *Curr. Sci.* 110 (6), 964.

Singh, D.S., Awasthi, A., 2011a. The implication of drainage basin parameters of Chhoti Gandak river, Ganga Plain, India. *J. Geol. Soc. India.* 78 (2), 370–378.

Singh, D.S., Awasthi, A., 2011b. Natural hazards in the Ghaghara river area, Ganga Plain, India. *Nat. Hazards.* 57, 213–225.

Singh, D.S., Dubey, C.A., Kumar, D., Kumar, P., Ravindra, R., 2018. Climate events between 47.5 and 1 ka BP in glaciated terrain of the Ny-Alesund region, Arctic, using geomorphology and sedimentology of diversified morphological zones. *Polar Sci.* 18, 123–134.

Singh, D.S., Dubey, C.A., Kumar, D., Vishawakarma, B., Singh, A.K., Tripathi, A., Gautam, P.K., Bali, R., Agarwal, K.K., Sharma, R., 2019. Monsoon variability and major climatic events between 25 and 0.05 ka BP using sedimentary parameters in the Gangotri Glacier region, Garhwal Himalaya, India. *Quatern. Int.* 507, 148–155.

Singh, D.S., Gupta, A.K., Sangode, S.J., Clemens, S.C., Srivastava, P., Prajapati, S.K., Prakasam, M., 2015. Multiproxy record of monsoon variability from the Ganga Plain during 400–1200 A.D. *Quat. Int.* 371, 157–163.

Singh, D.S., Mishra, A., 2001. Gangotri glacier characteristics, retreat and processes of sedimentation in the Bhagirathi valley. *Geological Survey of India* 65 (3), 17–20.

Singh, D.S., Mishra, A., 2002a. Role of tributary glaciers on landscape modification in the Gangotri glacier area, Garhwal Himalaya, India. *Curr. Sci.* 82 (5), 101–105.

Singh, D.S., Mishra, A., 2002b. Gangotri Glacier system: an analysis using GIS technique. In: Pant, C.C., Sharma, A.K. (Eds.), *Aspects of Geology and Environment of the Himalaya. Gyanoday Prakashan*, pp. 349–358.

Singh, D.S., Prajapati, S.K., Kumar, D., Awasthi, A., Bhardwaj, V., 2013. Sedimentology and channel pattern of the Chhoti Gandak River, Ganga Plain, India. *Gondwana Geol. Mag.* 28(2), 171–180.

Singh, D.S., Ravindra, R., 2011a. Geomorphology of the MidreLoven glacier, Ny-Alesund, Svalbard, Arctic. In: Singh, D.S., Chhabra, N.L. (Eds.), *Geological Processes and Climate Change*. Macmillan Publishers India Ltd., pp. 269–281.

Singh, D.S., Ravindra, R., 2011b. Control of glacial and fluvial environments in the Ny-Alesund region. *Arctic. Mausam*. 62(4), 641–646.

Singh, D.S., Singh, I.B., 2005. Facies architecture of the Gandak Megafan, Ganga Plain, India. *Palaeontol. Soc. India Spec. Pub.* 12, 125–140.

Singh, D.S., Singh, A.K., Dubey, C.A., Kumar, D., Sangode, S.J., Trivedi, A., Agnihotri, R., Singh, J., 2022. Multi-proxy analysis in the Gangotri Glacier region, Garhwal Himalaya: Glacial stratigraphy and the overview of snout retreat, geomorphic evolution, and palaeoclimate signatures. *J. Palaeontol. Soc. India*. 67(1), 158–182.

Singh, D.S., Tangri, A.K., Kumar, D., Dubey, C.A., Bali, R., 2017. Pattern of retreat and related morphological zones of Gangotri glacier, Garhwal Himalaya, India. *Quat. Int.* 444, 172–181.

Singh, I.B., Srivastava, P., Sharma, S., Sharma, M., Singh, D.S., Rajgopalan, G., Shukla, U. K., 1999. Upland interfluve (Doab) deposition: alternative model to muddy overbank deposits. *Facies* 40, 197–210.

Singh, J., Yadav, R.R., Negi, P.S., Rastogi, T., 2020. Sub-alpine trees testify late 20th-century rapid retreat of Gangotri glacier, Central Himalaya. *Quatern. Int.* 565, 31–40.

Singh, P., Haritashya, U.K., Kumar, N., Singh, Y., 2006. Hydrological characteristics of Gangotri glacier, Central Himalayas, India. *J. Hydro.* 327 (1–2), 55–67.

Singhvi, A.K., Krishnan, R., 2014. *Past and the present climate of India. Landscapes and landforms of India*. Springer, Netherlands, pp. 15–23.

Srivastava, D., 2012. *Status Report on Gangotri glacier. Himalayan Glaciology Technical Report.* Science and Engineering Research Board Department of Science & Technology, pp. 1–116.

Tangri, A.K., 2002. Shrinking glaciers of Uttaranchal: cause of concern and hope for the future. In: Pant, C.C., Sharma, A.K. (Eds.), *Aspect of geology and environment of the Himalaya*. Gyanodaya Prakashan, Nainital, pp. 335–348.

Tangri, A.K., Chandra, R., Yadav, S.K.S., 2004. Temporal monitoring of the snout, Equilibrium line and Ablation zone of Gangotri Glacier through remote sensing and GIS techniques—an attempt at deciphering the climatic variability. *Geological Survey of India, Special Publication* 80, 145–153.

Trivedi, A., Singh, D.S., Chauhan, M.S., Arya, A., Bhardwaj, V., Awasthi, A., 2011. Vegetation and climate change around Ropan Chhappra Tal in Deoria District, The Central Ganga Plain During the Last 1350 Years. *Palaeontol. Soc. India*. 56 (1), 39–43.

Zhang, G., Pan, B., Cao, B., Wang, J., Cui, H., Cao, X., 2015. Elevation changes measured during 1966–2010 on the monsoonal temperate glaciers' ablation region, Gongga Mountains, China. *Quatern. Int.* 371, 49–57.

Glossary

Above Mean Sea Level (AMSL): The term *above mean sea level* (AMSL) refers to the elevation (on the ground) or altitude (in the air) of any object, relative to the average sea level datum. AMSL is used extensively in radio (both in broadcasting and other telecommunications uses) by engineers to determine the coverage area a station will be able to reach. It is also used in aviation, where all heights are recorded and reported to AMSL (though also see flight level), and in the atmospheric sciences.

Annular Solar Eclipse: An annular solar eclipse occurs when it is a new moon, and the moon is near or at a lunar node. During this eclipse, the moon is usually far from the Earth and is near a point called the moon's apogee. At the same time, the Earth, the sun, and the new moon line up in a nearly straight line.

Antarctic Amplification: Polar amplification is the phenomenon that which any change in the net radiation balance (for example greenhouse intensification) tends to produce a larger change in temperature near the poles than in the planetary average. This is commonly referred to as the ratio of polar warming to tropical warming. On a planet with an atmosphere that can restrict the emission of longwave radiation to space (a greenhouse effect), surface temperatures will be warmer than a simple planetary equilibrium temperature calculation would predict. Where the atmosphere or an extensive ocean can transport heat poleward, the poles will be warmer and equatorial regions cooler than their local net radiation balances would predict. The poles will experience the most cooling when the global-mean temperature is lower relative to a reference climate; alternatively, the poles will experience the greatest warming when the global-mean temperature is higher.

Antarctic Bottom Water: Antarctic bottom water (AABW) is a type of water mass in the Southern Ocean surrounding Antarctica with temperatures ranging from −0.8 to 2 °C (35 °F) and salinities from 34.6 to 34.7 psu.

Antarctic Circumpolar Current (ACC): The Antarctic Circumpolar Current, or ACC, is the strongest ocean current on our planet. It extends from the sea surface to the bottom of the ocean and encircles Antarctica. It is vital for Earth's health because it keeps Antarctica cool and frozen. It is also changing as the world's climate warms.

Antarctic Cold Reversal (ACR): The Antarctic Cold Reversal (ACR) was an important episode of cooling in the climate history of the Earth during the deglaciation at the close of the last ice age. It illustrates the complexity of the climate changes during the transition from the Pleistocene to the Holocene Epochs.

Antarctic Intermediate Water (AAIW): Antarctic Intermediate Water (AAIW) is a cold, relatively low salinity water mass found mostly at intermediate depths in the Southern Ocean. The AAIW is formed at the ocean surface in the Antarctic Convergence zone or more commonly called the Antarctic

Polar Front Zone. This convergence zone is normally located between 50 °S and 60 °S, hence this is where almost all of the AAIW is formed.

Asian Monsoon: The Asian monsoon is one of the most important components of the global climate system. It dominates large densely-populated areas extending from the Indian sub-continent eastwards to Southeast and East Asia. Its evolution and variability can exert significant influences on the vegetation, populations, economies, and even cultures that inhabit Asian monsoon regions.

Australasian Mediterranean Water: The Australasian Mediterranean Sea is a Mediterranean sea located in the area between Southeast Asia and Australasia. It connects the Indian and Pacific oceans. It has a maximum depth of 7,440 m and a surface area of 9.08 mil. km^2.

Bathymetric Map: A bathymetric map is a type of isarithmic map that depicts the submerged topography and physiographic features of ocean and sea bottoms. Their primary purpose is to provide detailed depth contours of ocean topography as well as provide the size, shape, and distribution of underwater features.

Central Dronning Maud Land (cDML): Queen Maud Land (Norwegian: Dronning Maud Land) is a roughly 2.7-million-square-kilometer (1.0-million-square-mile) region of Antarctica claimed by Norway as a dependent territory. It borders the claimed British Antarctic Territory at 20° west and the Australian Antarctic Territory at 45° east. In addition, a small unclaimed area from 1939 was annexed on June 12, 2015.

Chloro-alkaline Indices: The chloro-alkaline indices are calculated for the groundwater samples in the study area and it has been found that CAI 1 values range between −0.02 and 0.89 with mean values of 0.35, while CAI 2 values lie between −0.13 and 0.95 with mean values of 0.40.

Climate System: The climate system is a highly complex global system consisting of 5 major components: the atmosphere, the oceans, the cryosphere (snow and ice), the land surface, the biosphere, and the interactions between them.

Electron Density: Electron density is a representation of the probability of finding an electron in a specific location around an atom or molecule. In general, the electron is more likely to be found in regions with high electron density.

ENSO: El Niño—Southern Oscillation (ENSO) is an irregularly periodic variation in winds and sea surface temperatures over the tropical eastern Pacific Ocean, affecting the climate of much of the tropics and subtropics. The warming phase of the sea temperature is known as El Niño and the cooling phase as La Niña.

Eocene-Oligocene Transition (EOT): The Eocene-Oligocene transition (EOT) was a climate shift from a largely ice-free greenhouse world to an icehouse climate, involving the first major glaciation of Antarctica and global cooling occurring ~34 million years ago (Ma) and lasting ~790 kyr.

Epiglacial Lakes: Epiglacial lakes are characterized as lakes in contact with a glacier and with inflow from glacial meltwater. They are an important habitat of life in the polar world.

Equilibrium Fractionation Value: The equilibrium fractionation factors (al-v) for the water liquid-vapor phase transition are 1.0098 and 1.084 at 20 °C and 1.0117 and 1.111 at 0 °C for 18O and 2H, respectively.

Fluvial System: *Fluvial* is a term used in geography and Earth science to refer to the processes associated with rivers and streams and the deposits and landforms created by them. Water evaporates from water bodies such as rivers, lakes and seas, and plants and trees. The water vapor rises, cools, and condenses to form clouds.

Fragile Ecosystem: A fragile ecosystem is an ecosystem or community which lacks resilience or which is so heavily impacted by an 'unnatural' (human?) event that it changes in unexpected and undesirable ways leading to the conditions that are often termed as a natural catastrophe.

Glacial Erosion: Glacial erosion includes processes that occur directly in association with the movement of glacial ice over its bed, such as abrasion, quarrying, and physical and chemical erosion by subglacial meltwater, as well as from the fluvial and mass wasting processes that are enhanced or modified by glaciation.

Glacier Dynamics: Dynamics of glaciers are the resulting ice flow processes caused by the interplay of the physical mechanism causing a glacier to move (gravity) and forces resisting this movement (e.g., friction). Alternatively, it can be defined as a glacier's response to climate change or other external or internal forces.

Glacier Lake Outburst Flooding (GLOF): Glacial lake outburst floods (GLOFs) occur from an unstable natural dam formed from a glacial retreat. Glaciers are large bodies of ice moving slowly. So, when a glacier retreats, it leaves behind a large impression on the ground, filling it with water and a lake is formed.

Glaciomorphic Landforms: Geomorphology is the study of landforms, their processes, form, and sediments at the surface of the Earth (and sometimes on other planets). The study includes looking at landscapes to work out how the Earth's surface processes, such as air, water, and ice, can mold the landscape.

Global Circulation Models: Airflow for no rotation and no water on a planet. Global Circulations explain how air and storm systems travel over the Earth's surface. The global circulation would be simple (and the weather boring) if the Earth did not rotate, the rotation was not tilted relative to the sun and had no water.

Gradual Recovery: A gradual recovery has begun and is expected to continue. Activity is now higher than a year ago and the gradual recovery is expected to continue. That is essential to the gradual recovery to which the forecast points.

Greenhouse Gas (GHG): A greenhouse gas (GHG or GhG) is a gas that absorbs and emits radiant energy within the thermal infrared range, causing the greenhouse effect. The primary greenhouse gases in Earth's atmosphere are water vapor (H2O), carbon dioxide (CO2), methane (CH4), nitrous oxide

(N_2O), and ozone (O_3). Without greenhouse gases, the average temperature of Earth's surface would be about −18 °C (0 °F), rather than the present average of 15 °C (59 °F). The atmospheres of Venus, Mars and Titan also contain greenhouse gases.

Greenland Stadials: The climate over the last glacial period was characterized by a series of abrupt, millennial-scale climate oscillations known as Dansgaard-Oeschger (DO) events, marked by rapid warming followed by slow cooling, referred to as Greenland Interstadial (GIS) and Greenland Stadial (GS).

Ground Penetrating Radar: Ground-penetrating radar (GPR) is a geophysical method that uses radar pulses to image the subsurface. It is a non-intrusive method of surveying the sub-surface to investigate underground utilities such as concrete, asphalt, metals, pipes, cables, or masonry.

Humboldt Current: Peru Current, also called Humboldt Current, cold-water current of the southeast Pacific Ocean, with a width of about 900 km (550 mi).

Hydrological Cycle: The hydrologic cycle begins with the evaporation of water from the surface of the ocean. As moist air is lifted, it cools and water vapor condenses to form clouds. Moisture is transported around the globe until it returns to the surface as precipitation.

Hydrothermal Alteration: Hydrothermal alteration is defined as any change in the mineralogic composition of a rock due to the action (by either physical or chemical means) of hydrothermal fluids in an open system.

Ice Cores: An ice core is a core sample that is typically removed from an ice sheet or a high mountain glacier. Since the ice forms from the incremental buildup of annual layers of snow, lower layers are older than the upper, and an ice core contains ice formed over a range of years. Cores are drilled with hand augers (for shallow holes) or powered drills; they can reach depths of over two miles (3.2 km), and contain ice up to 800,000 years old.

Incoherent Scatter Radar (ISR): This technique is a powerful ground-based tool used to measure various properties of the ionized part of the upper atmosphere called the ionosphere.

Ionospheric Layer: There are three main regions of the ionosphere, called the D layer, the E layer, and the F layer. These regions do not have sharp boundaries, and the altitudes at which they occur vary during a day and from season to season.

Isotopic Ratio: Isotopic ratio refers to the ratio of the atomic abundances of two isotopes of the same element, e.g., $^{18}O/^{16}O$ or $^{143}Nd/^{144}Nd$. An advantage of using ratios rather than absolute abundances of a particular nuclide is better precision.

Katabatic Wind: katabatic wind, also called downslope wind, or gravity wind, is a wind that blows down a slope because of gravity. It occurs at night when the highlands radiate heat and are cooled.

Lake Water Line: a line that marks the level of the surface of the water on something, such as (1) the point on the hull of a ship or boat to which the water rises or (2) a line marked on the outside of a ship that corresponds with the water's surface when the ship is afloat on an even keel under specified conditions of loading.

Landslide Lake Outburst Flooding (LLOF): Landslide Lake Outburst Floods (LLOFs) are common in the Himalayan river basins. These are caused by the breaching of lakes created by landslides. The active and paleo-landslide mapping along the Satluj and Spiti Rivers indicate that these rivers were blocked and breached at many places during the Quaternary period.

Limnological: Limnology is the study of inland aquatic ecosystems. The study of limnology includes aspects of the biological, chemical, physical, and geological characteristics and functions of inland waters. This includes the study of lakes, reservoirs, ponds, rivers, springs, streams, wetlands, and groundwater.

Marine Isotope Stage (MIS): Marine isotope stages (MIS), marine oxygen-isotope stages, or oxygen isotope stages (OIS), are alternating warm and cool periods in the Earth's paleoclimate, deduced from oxygen isotope data reflecting changes in temperature derived from data from deep-sea core samples. Working backwards from the present, which is MIS 1 in the scale, stages with even numbers have high levels of oxygen-18 and represent cold glacial periods, while the odd-numbered stages are lows in the oxygen-18 figures, representing warm interglacial intervals. The data are derived from pollen and foraminifera (plankton) remains in drilled marine sediment cores, sapropels, and other data that reflect historic climate; these are called proxies.

Meridional Ocean Circulation: The Atlantic meridional overturning circulation (AMOC) is the zonally integrated component of surface and deep currents in the Atlantic Ocean. It is characterized by a northward flow of warm, salty water in the upper layers of the Atlantic, and a southward flow of colder, deep waters that are part of the thermohaline circulation. These "limbs" are linked by regions of overturning in the Nordic and Labrador Seas and the Southern Ocean. The AMOC is an important component of the Earth's climate system and is a result of both atmospheric and thermohaline drivers.

Metamorphic Rocks: Metamorphic rocks started as some other type of rock, but have been substantially changed from their original igneous, sedimentary, or earlier metamorphic form. Metamorphic rocks form when rocks are subjected to high heat, high pressure, hot mineral-rich fluids or, more commonly, some combination of these factors.

Microbes and Microfossils: Microfossils are the microscopic remains of organisms. The organisms may be prokaryotic cells of the Bacteria or Archaea domains, unicellular eukaryotes (protists), whole multicellular eukaryotes, or parts of multicellular microscopic or macroscopic eukaryotes.

Microorganisms: An organism that can be seen only through a microscope. Microorganisms include bacteria, protozoa, algae, and fungi. Although viruses are not considered living organisms, they are sometimes classified as microorganisms.

Mid-Pleistocene Transition (MPT): The Mid-Pleistocene Transition (MPT), also known as the Mid-Pleistocene Revolution (MPR), is a fundamental change in the behavior of glacial cycles during the Quaternary glaciations. The transition happened approximately 1.25–0.7 million years ago, in the Pleistocene epoch. Before the MPT, the glacial cycles were dominated by a 41,000-year periodicity with low-amplitude, thin ice sheets and a linear relationship to the Milankovitch forcing from axial tilt. After the MPT there have been strongly asymmetric cycles with long-duration cooling of the climate and build-up of thick ice sheets, followed by a fast change from extreme glacial conditions to a warm interglacial. The cycle lengths have varied, with an average length of approximately 100,000 years.

Middle Miocene: The Middle Miocene is a sub-epoch of the Miocene Epoch made up of two stages: the Langhian and Serravallian stages. The Middle Miocene is preceded by the Early Miocene. The sub-epoch lasted from 15.97 ± 0.05 Ma to 11.608 ± 0.005 Ma (million years ago). During this period, a sharp drop in global temperatures took place. This event is known as the Middle Miocene Climate Transition.

Moisture Flux Convergence (MFC): Moisture flux convergence (MFC) is a term in the conservation of water vapor equation and was first calculated in the 1950s and 1960s as a vertically integrated quantity to predict rainfall associated with synoptic-scale systems.

Partial Solar Eclipse: During a partial solar eclipse, the moon, the sun, and Earth don't align in a perfectly straight line, and the Moon casts only the outer part of its shadow, the penumbra, on Earth. From our perspective, this looks like the Moon has taken a bite out of the Sun.

Polar Frontal Zone: The polar front is the boundary between the Ferrell and the polar cells, located at 60 degrees latitude in each hemisphere. There is a sharp rise in temperature between the two air masses in this zone.

Sedimentary Records: Sedimentary records collected through coring or geologic outcrops represent the most commonly studied paleoclimate archive. Because the oldest sedimentary rocks are about 3.9 billion years old, sediment records provide a means to study past climates throughout most of Earth's history.

Silicate Rocks: Silicate minerals are rock-forming minerals made up of silicate groups. They are the largest and most important class of minerals and make up approximately 90% of the Earth's crust. In mineralogy, silica (silicon dioxide) SiO 2 is usually considered a silicate mineral.

Snow Water Equivalent (SWE): Snow Water Equivalent (SWE) is a common snowpack measurement used by hydrologists and water managers to gauge the amount of liquid water contained within the snowpack. It is equal to the amount of water contained within the snowpack when it melts.

Southern High Latitudes: The high latitude area in the Southern Hemisphere is located between the Antarctic Circle, at 66 degrees 33 minutes south latitude, and the South Pole, at 90 degrees south latitude. Antarctica is located at the South Pole.

Spectroscopy System: Spectroscopy is a branch of science concerned with the spectra of electromagnetic radiation as a function of its wavelength or frequency measured by spectrographic equipment, and other techniques, to obtain information concerning the structure and properties of matter.

Sub-Antarctic Front: The northern front of the Antarctic Circumpolar Current that separates the polar frontal zone in the south from the Subantarctic zone in the north. It is characterized by sea surface temperatures near 7°–9°C and a salinity minimum of 33.8–34.0 psu produced by high rainfall.

Sun's Photosphere: The photosphere is the visible "surface" of the Sun. The Sun is a giant ball of plasma (electrified gas), so it doesn't have a distinct, solid surface like Earth.

Supracrustal Rocks: Supracrustal rocks are rocks that were deposited on the existing basement rocks of the crust, hence the name. They may be further metamorphosed from both sedimentary and volcanic rocks.

Thermohaline Circulation (THC): Thermohaline circulation (THC) is a part of the large-scale ocean circulation that is driven by global density gradients created by surface heat and freshwater fluxes.

Total Solar Eclipse: Total solar eclipses happen when the New Moon comes between the Sun and Earth and casts the darkest part of its shadow, the umbra, on Earth. During a total eclipse of the Sun, the Moon covers the entire disk of the Sun.

Younger Dryas: The Younger Dryas is one of the most well-known examples of abrupt change. About 14,500 years ago, Earth's climate began to shift from a cold glacial world to a warmer interglacial state. Partway through this transition, temperatures in the Northern Hemisphere suddenly returned to near-glacial conditions.

Index

www.ingramcontent.com/pod-product-compliance
Lightning Source LLC
Chambersburg PA
CBHW060350220326
41598CB00023B/2865
* 9 7 8 1 0 3 2 2 5 6 5 7 3 *